Catalogue of Primates in the British Museum (Natural History) and elsewhere in the British Isles

PART III: FAMILY CERCOPITHECIDAE, SUBFAMILY COLOBINAE

Catalogue of Primates in the British Museum (Natural History) and elsewhere in the British Isles

PART III: FAMILY CERCOPITHECIDAE, SUBFAMILY COLOBINAE

Prudence Hero Napier

British Museum (Natural History)
London 1985

Publication Number 894
ISBN 0 565 00894 3

British Library Cataloguing in Publication Data
British Museum (Natural History)
Catalogue of primates in the British Museum (Natural History) and elsewhere in the British Isles.
Pt. 3 : Family cercopithecidae, subfamily colobinae
1. Primates—Catalogs and collections 2. Type specimens (Natural History)—Catalogs and collections
I. Title II. Napier, Prudence Hero
599.8'074'02134 QL737.P9

ISBN 0 565 00894 3

Printed at the University Press, Oxford

CONTENTS

Acknowledgments

I would like to thank Dr G B Corbet, who initiated the project of cataloguing the primate collections, for his continued encouragement and support. I am also very grateful to the members of the Electronic Data Processing Section for their efficiency in handling the data, and particularly to Dr D B Williams for his advice and help throughout and for contributing the section on the production of the catalogue. My thanks are also due to Dr P Andrews for his chapter on the origin of the Cercopithecoidea which puts the fossil material into perspective, thus adding greatly to the completeness of the volume; and to Miss P D Jenkins for her invaluable help in cataloguing the African Colobinae.

I would also like to express my grateful thanks to the Trustees of the Boise Fund of Oxford for financial support; to Professor J R Napier for allowing me to include his unpublished data on limb proportions; to Dr C P Groves for reading the text and making valuable constructive comments; and to Mr D Brandon-Jones who placed his extensive knowledge of the Asian colobines and their associated literature unreservedly at my disposal.

I am very grateful to all the officers of museums in the British Isles whose co-operation has helped to make this catalogue more complete and comprehensive.

I would also like to thank the following for their generous help on various aspects of the project:

Mr I R Bishop, O.B.E.
Dr F L Brett
Mr G Bridson of the Linnean Society of London
Mr B H Collins
Dr E Delson
Dr J Fooden
Dr A Fox Maule
Dr C Grigson
Dr C P Groves
Dr C O Hansen
Mr J Hewish of the Science Reference Library
Mr J Edwards Hill
Dr D B Hull
Mr G C J MacPherson
Mr R V Melville
Professor A E W Miles
Miss T I Molleson
Mr K Nicklin, Curator of the Powell-Cotton Museum
Dr T R Olson
Miss V Rapu
Dr N K B Robson
Mr V Weitzel

P H Napier

British Museum (Natural History)
London
November, 1981

INTRODUCTION

Scope

This catalogue lists all the specimens of the subfamily Colobinae of the Old World Monkeys (family: Cercopithecidae) in the collections of the British Museum (Natural History), and is complete up to the end of 1980. The collections consist mainly of study skins and skulls but skeletons, wet specimens and casts are included, as well as mounted specimens exhibited in the public galleries in London or in the Zoological Museum, Tring, Hertfordshire (formerly the Rothschild Museum), and specimens in the Comparative Anatomy Collection which is used for teaching. Fossil specimens in the collections of the Sub-department of Anthropology, Department of Palaeontology, are also included, among them some that could only be identified as Cercopithecoidea (see p. 96). Also recorded is a selection of important specimens available for study at other museums in the British Isles (see Appendix I, p. 97).

All the available data on these specimens, derived from labels and from the Museum Register, together with the co-ordinates of the great majority of localities, have been stored in a computer-based retrieval system. The published catalogue gives only a selection of the available data which also include information relating to the collection of the specimen, the nature and source of acquisition, any earlier identifications with identifiers, the habitat and any field measurements (see Appendix II, p. 98).

Major collections of colobines were acquired by the British Museum (Natural History) through the Bombay Natural History Society's Mammal Survey of India and Sri Lanka; through Major P H G Powell-Cotton who, together with F G Merfield, made extensive collections in Central Africa for the Powell-Cotton Museum, Birchington, Kent; through H C Robinson of the Federated Malay States Museum; through A J Haddow and his colleagues at the Yellow Fever Research Institute at Entebbe, Uganda; and through Sir Frank Colyer of the Odontological Museum of the Royal College of Surgeons. The primate skulls in the Odontological Museum were catalogued in 1975-76 by Dr Caroline Grigson, and the data entered into a computer-based retrieval system which contains detailed information on dentition, as well as correlating skulls with skins held in other museums (e.g. the British Museum (Natural History) and the National Museum of Victoria, Melbourne, Australia).

Information on specimens at the Powell-Cotton Museum was compiled from the cardfile at the British Museum (Natural History) and those of the Royal College of Surgeons (Odontological Museum) from Dr Grigson's catalogue, a copy of which is also available here.

Classification

The classification employed in the catalogue is based primarily on G G Simpson's *The Principles of Classification and a Classification of Mammals* (1945) with modifications following the systematic lists of Thorington & Groves (1970) for living species, and Szalay & Delson (1979) for the fossils. Subspecies follow Napier & Napier (1967) for the Asian colobines and Rahm (1970) for the African colobines. Any departures from the foregoing are documented and discussed in the appropriate places. Lists of the principal synonyms are given under genus, species and subspecies names, and have been compiled from the check-lists of Allen (1939), Chasen (1940), Ellerman & Morrison-Scott (1951), and Laurie & Hill (1954).

As far as possible, specimens have been identified to subspecies level where the presence of both skin and skull made this possible, except in the case of specimens from other museums where the original identification has been accepted.

Six genera (and four subgenera) of living colobines are recognised and the description of each includes the following information: number of recognised species, type species, list of synonyms and a résumé of any taxonomic and nomenclatural problems. Then follows a short summary of morphology, range, ecology, social behaviour and reproduction, together with a key to the species. Listed under each species are its synonyms with their type localities, its range and dimensions, and a key to subspecies, followed by the formal lists of specimens. At the end of the volume can be found appendices which include field measurements of species and a table of characters of colobine genera, subgenera and species groups.

A new taxonomic arrangement of the subfamily Colobinae, in which 6 genera and 8 subgenera are recognised, has recently been published by Brandon-Jones (1984).

In order to put the fossil specimens into their true perspective, a short account of the origin of the Cercopithecoidea by Peter Andrews of the Subdepartment of Anthropology, Department of Palaeontology, is included on p. 2.

Explanation of headings and abbreviations

The following notes refer to the information given under each heading of the catalogue. All items that have had to be inferred are enclosed in square brackets, e.g. the country, if the original documentation is limited to a village, town or region.

Locality Spelling of geographical place names follows the conventions of *The Times Atlas*. Many of the older specimens have no locality data, in which case the country was, where possible, inferred from the known range of the species or subspecies concerned. Co-ordinates were taken from *The Times Atlas Gazetteer* or, if not available there, from the *Official Standard Names Gazetteer* of the United States Board on Geographic Names. The remainder were estimated from 1:1 000 000 maps.

Date of collection This signifies collection in the wild. If it is preceded by the word 'died', this indicates that the animal died in captivity.

Sex and age Three sex categories are used: M, F and I (indeterminate). The sex given on the label was confirmed, where possible, by examining the dentition, genitalia and nipples. Square brackets denote inferences made with regard to sex.

Three categories, infant, juvenile and adult, are used to indicate the age of specimens, based on the development of the dentition, as follows:

Infant: Any stage of dental development up to complete deciduous dentition.

Adult: Complete permanent dentition.

Juvenile: Any stage of dental development between these two.

An animal with complete permanent dentition erupted, but with canines not yet fully grown, is classed as adult.

Nature of specimen The following categories have been recorded in the catalogue: skin, skull, skeleton and 'other'. The categories 'skull' and 'skeleton' are employed when a major portion of the skull or postcranial skeleton is present. 'Other' includes smaller skeletal specimens such as bacula, hyoids, fragments of fossil material, and individual bones and teeth. Tooth nomenclature follow that of Simons (1972, p. 20) except that the words 'upper' and 'lower' are utilised rather than superscript and subscript numerals. Also included under 'other' are casts, photographs, radiographs and wet specimens. The latter are preserved in a one per cent solution of phenoxytol or in spirit.

Accession number The first part of the number records the year in which the specimen was registered. The letters ZD preceding the number indicate registration in the main collections of the Mammal Section in the Department of Zoology; ZE and ZG indicate registration in the Osteology Section (see below) and in the Comparative Anatomy Collection respectively. Other departments in the Museum and other museums in the British Isles are indicated by suitable abbreviations, a list of which appears in Appendix I, p. 97. For further information on the registration of specimens, see below.

Citations, Remarks, etc Type specimens and others that have been cited or figured in the literature are recorded here. Full bibliographical references will be found at the end of the volume on page 102. Where the scientific name on the label of the specimen differs from the taxon under which it is catalogued, it is included in this column, in brackets.

Registration of specimens

The registration of specimens in the Zoology Department of the British Museum was initiated in 1837 by J E Gray. In April 1837 he began the consecutive numeration of accessions by the four-part system familiar to students of the collections — a system which shows unequivocally the exact date on which each specimen was incorporated in the collections. Thus the first specimen registered under this system was given the number 1837.4.28.1. (28 April, 1837, No. 1). The accession number is commonly abbreviated by omitting the first two digits of the year, e.g. 37.4.28.1.

Concurrently, osteological material, which had previously formed the bulk of the mammal collections, was catalogued by assigning a number to each recognised species. The first skull or skeleton of the species was given the letter 'a', the second 'b', and so on. Gerrard's *Catalogue of the Bones of Mammalia in the Collections of the British Museum*, published in 1862, is based on these records. As will be seen from the present catalogue, many skulls and skeletons still bear these old

Osteology Catalogue numbers and no other. The dual task of cataloguing the increasing numbers of specimens that were coming into the possession of the Museum as well as recording their accession in the Register proved to be impossible. The Osteology Catalogue seems gradually to have lapsed after the publication of Gray's *Catalogue of Monkeys, Lemurs and Fruit-eating Bats* in 1871 (see Sherborn, 1934). From that time to the present day, over one hundred years, no definitive catalogue of the specimens has been available.

Gray's registration system has withstood the test of time and it was not until one hundred years later in 1939, when it became apparent that duplication of numbers might lead to confusion, that a change was made. A two-part registration system was introduced, the year being followed by a single number (e.g. 1939.1, 1939.2, and so on). This two-part system has not been modified by the introduction of electronic data processing of all accessions in the Mammal Section which was begun in October, 1970.

In 1909, the Osteology Section, handling all skeletal material, including large collections of marine mammals, was separated from the Mammal Section, and skeletons were stored and registered separately. The four-part system of registration was, in this case, continued beyond 1939 until the two sections were again amalgamated in 1969.

Production of the catalogue

This published catalogue represents an abstract from a file which is compatible with the Museum's computer-held catalogue of mammal accessions. Raw data on input forms modified from those used for Part I was converted to computer-readable form either on a paper tape typewriter, or a computer terminal attached to the Museum's computer. The transcribed data was then submitted to the Sedgwick Museum's 'CGDS' program package for further processing. CGDS was used to edit the raw machine-readable data to incorporate corrections discovered during proof-reading, and the data was then converted to a CGDS 'standard file' on magnetic tape.

The data relating to one specimen constitutes one 'record' in the standard file. Each record is broken down into 'fields' which largely correspond to the divisions and subdivisions of the headings in the tabulated parts of this volume. As an aid to further checking, CGDS was used to produce an index to records, ordered by field type and then alphabetically on field contents. In this type of index, misplaced and mis-spelt fields are more prominent than in the normal raw-input data, and errors which had slipped through proof-reading could be corrected. Once these corrections had been made, the file was processed by another program in the CGDS package to produce an output magnetic tape suitable for the printers. This program, in simplified terms, performed the following actions. It read a record from the input tape. It examined the record field by field to determine if the field was a type required for printing. If the field was not needed, the program passed to the next field. If it was needed, depending on the field type, the program wrote appropriate leading special items to the output tape, followed by the data in the field, followed by appropriate trailing special items, and the program then passed on to the next field. The special items included comments for the printer's computer typesetting program to tell it which column on the printed page was to contain the data which followed or which type style to use for the data, and any characters (such as headings or punctuation) which should precede or follow the data in the final publication.

The magnetic tape resulting from this program was sent to the printer where it was processed with a program which produced yet another tape (punched paper or magnetic) which was used to drive the phototypesetter, and the bromide print produced by the phototypesetter was used for offset printing of the catalogue.

Superfamily: CERCOPITHECOIDEA Gray, 1821
Family: Cercopithecidae Gray, 1821

Taxonomy and Morphology
Schultz (1970) has justly stressed the basic homogeneity of the Cercopithecidae, a fact which is perhaps most acutely appreciated when attempting to provide a key to the two major groups which comprise it.

The dental formula of $\frac{2\ 1\ 2\ 3}{2\ 1\ 2\ 3} \times 2 = 32$, and the characteristic bilophodont molars, are found throughout the family. Schultz draws particular attention to the uniformity of the postcranial skeleton in which the reduction or absence of the tail in some of the cercopithecines (and *Simias*) and the reduction or absence of the thumb in the colobines are the most striking variations. His comments strongly reinforce Simpson's decision (1945) to separate the leaf-eaters from the omnivorous Old World Monkeys at subfamily rather than family level. Hill on the other hand gave them full family status (1966) as did Maier (1970*a*).

In the past this very homogeneity gave rise to numerous differing classifications, based at first on the most obvious but most misleading character — tail length. For instance Linnaeus (1758) included in his sub-group *Cercopitheci* (= tailed apes) a hotch-potch of primates from Old and New Worlds, equipped with tails of varying length and type, from the lion-tailed *Macaca silenus,* the long-tailed Diana monkey, *Cercopithecus diana,* the prehensile-tailed spider monkey, *Ateles paniscus,* and even the feather-tailed *Tarsius syrichta.* A little later Buffon (1766) noted the differences distinguishing New from Old World Monkeys, their widely spaced nostrils and their lack of ischial callosities and cheek pouches, but many years were to pass before the distinction between the omnivorous and leaf-eating groups of Old World Monkeys was clearly recognised. This was undoubtedly due to the fact that there is no external character by which to distinguish them.

Martin (1841) was probably the first to note that the langurs of southeast Asia lacked cheek pouches but possessed a complex stomach and molars with high cusps. He provisionally included them with the Colobus monkeys from Africa in his second subfamily of Old World Primates, the apes comprising the first. His third subfamily, with cheek pouches and a simple stomach, included the guenons, macaques and baboons. Since that time some minor characters of the skull have been added, but few are absolutely clear-cut.

The following list of characters separating the Colobinae from the Cercopithecinae has been adapted from Verheyen (1962), Hill (1966), Vogel (1968) and Delson (1975).

Colobinae	**Cercopithecinae**
No cheek pouches	Cheek pouches
Complex sacculated stomach	Simple stomach
Lacrimal fossa partly formed by maxilla	Lacrimal fossa wholly formed by enlarged lacrimal bone
Interorbital width/muzzle length index 16.3 — 33 (except *Nasalis* and *Simias*: 13.3 — 19.6)*	Interorbital width/muzzle length index 4.2 — 16.8*
Nasal bones short and broad (except *Nasalis* and *Simias*)	Nasal bones long and narrow
Median mental foramen generally absent (except *Procolobus verus*)	Median mental foramen generally present
Molar teeth with high pointed cusps separated by deep notches	Molar teeth with low rounded cusps (except *Theropithecus*)

*The interorbital width/muzzle length index (Verheyen, 1962) is obtained as follows: $\frac{\text{anterior interorbital width} \times 100}{\text{glabella — prosthion}}$

Preliminary studies (Hiiemae, personal communication) indicate that there may well be a difference between the two subfamilies in the thickness of enamel on the lingual aspect of the lower incisors, as suggested by Delson (1975), colobines having thicker enamel than cercopithecines. The problem is at present under investigation.

Fossil Cercopithecidae
The allocation of extinct catarrhines to the two subfamilies of the Cercopithecidae has altered somewhat since Simpson's (1945) classification as a result of further study and the description of several new genera. Simpson included *Parapapio* and *Dinopithecus* in the Cercopithecinae, to which Simons (1972) added *Gorgopithecus* and *Procynocephalus.* All the foregoing, together with *Paradolichopithecus,* are assigned by Szalay & Delson (1979) to the tribe Papionini. They have been included in Part II of the catalogue (Cercopithecinae, 1981).

Part III includes the living Colobinae, together with the following extinct genera: *Mesopithecus* and *Dolichopithecus,* following both Simpson (1945) and Simons (1972); *Libypithecus,* now usually treated as a colobine (Jolly, 1967; Delson, 1977); and *Cercopithecoides* and *Paracolobus,* now referred to the Colobinae by Maier (1970*a*) and Delson (1977). Two Early Miocene monkeys, *Prohylobates* Fourtau, 1918, and *Victoriapithecus* von Koenigswald, 1969, are also included as Cercopithecidae *incertae sedis,* following Szalay & Delson (1979).

The taxonomic position of *Parapithecus* and *Apidium* of the Late Oligocene is at present the subject of some difference of opinion. Simpson (1945) considered *Parapithecus* to represent a separate family in the Hominoidea and *Apidium* a subfamily *incerta sedis* within the Cercopithecidae. Simons (1972) united them in the subfamily Parapithecinae of the Cercopithecidae but later (1974) relegated them to the family Parapithecidae in the Cercopithecoidea. From studies of molar occlusion Kay (1977) has suggested that a *Parapithecus*-related stock may have given rise to the Cercopithecidae. Szalay & Delson (1979) distinguish both *Parapithecus* and *Apidium* from all other catarrhines, placing them in the superfamily Parapithecoidea.

Subfamily: COLOBINAE Jerdon, 1867 (1825)*

The Colobinae are more homogeneous than the Cercopithecinae, both in size and morphology. The largest leaf monkeys, *Pygathrix* and *Nasalis,* are comparable with the largest cercopithecines in size, though they are probably lighter in weight. However, the smallest colobine, the olive colobus (*Procolobus verus*), is considerably larger than the smallest cercopithecine (*Miopithecus talapoin*), and about three times its weight. External differences are minimal, the most noticeable being the shortened tail in *Simias,* the Mentawai Snub-nosed Leaf Monkey, and the vestigial thumb of the African colobines. As pointed out by Jouffroy & Lessertisseur (1977), even this latter feature is part of a general trend in the subfamily, with *Nasalis,* the subgenus *Rhinopithecus* and the *Presbytis entellus* group showing some slight thumb reduction compared with *Cercopithecus;* subgenus *Pygathrix* and the *P. cristata, P. vetulus* and *P. melalophos* groups show a more substantial reduction of the thumb while retaining both phalanges. In *Colobus* and *Procolobus* the thumb is either absent or at most represented by a small tubercle which sometimes bears a vestigial nail. Only a single phalanx is present in the skeleton, and even this is occasionally absent.

Ischial callosities are variable in size and shape, and are typically contiguous in males, separate in females. There are exceptions to this in both African and Asian groups; male *Pygathrix, Colobus* (*Piliocolobus*) and *Procolobus* have the callosities separated by a strip of furred skin. The presence of a sexual swelling in females however is confined to the African *Colobus* (*Piliocolobus*) and *Procolobus.*

The stomach is characterised by its large size and complex structure – specialisations for the digestion of a bulk diet of leaves. In olive and red colobus, the stomach is divided into four parts, but in the black-and-white colobus and in *Presbytis* the division of the saccus gastricus into two sections is less pronounced (Kuhn, 1964). In *Nasalis,* the saccus gastricus has a conical hook-like extension dorsally (Berenberg-Gossler, 1911, cited in Hill, 1958).

The chromosome diploid number is 44 in every species so far examined, with the exception of *Nasalis* with 48 (Chiarelli, 1973).

Szalay & Delson (1979) separate Asian and African colobines, giving them subtribal status.

* For information on the family-group names of leaf monkeys, see Delson (1976) and Brandon-Jones (1978).

The Origin of the Cercopithecoidea

By Peter Andrews

The living species of the Cercopithecoidea can be divided into three main groups. The Colobinae are characterized by generally primitive characters of the skull (Delson, 1975) and by highly specialized characters in the stomach and dentition especially, for instance, their squarish molars with very high cusps. The Cercopithecinae as a whole have simple stomachs but have a feeding specialization in their cheek pouches and specializations of the skull such as their relatively narrow orbits. Within the Cercopithecinae and sharing these characters are two main groups of species, the Papionini, which appear to have retained the primitive condition of the dentition almost exactly, although *Theropithecus* is a possibly separate group with its reduced incisors and canine, and the Cercopithecini, which have smaller, more elongated teeth and reduced third molars. Studies of the ancestral cercopithecoid morphotype (Delson and Andrews, 1975) suggest that it combined colobine-like characters of the skull and mandible with papionin-like characters of the dentition (except for the incisors).

Turning now to the fossil record, the earliest supposed fossil cercopithecoids have been reported from Oligocene and Lower Miocene deposits of Africa. Simons (1974) has recently described a species of *Parapithecus* which is known from Oligocene deposits in the Egyptian Fayum. He claims that it shows evidence of relationship to the ancestry of the Cercopithecoidea based on its great similarity in cheek tooth morphology to living monkeys. This view has been contested (Delson, 1975) and it can be shown that the *Parapithecus* material shows little resemblance to the predicted ancestral cercopithecoid morphotype (Delson and Andrews, 1975). The M_3 is reduced, which is similar to the derived condition in living cercopithecines, but as it predates the divergence of colobines and cercopithecines it must therefore be considered a convergence; and the elongated P_4, with reduced metaconids, and the P_2 honing action against the upper canine are derived characters not present in any cercopithecoid. It is probable that *Parapithecus* represents an early catarrhine lineage that was convergent on monkeys but had no close relationship with them. Other primates from the Fayum, such as *Oligopithecus* and *Propliopithecus,* are sufficiently primitive in morphology to be consistent with the ancestral cercopithecoid morphotype, although, as they show no particular affinity with the Cercopithecoidea, it is necessary to look towards the Miocene to find the first cercopithecoids.

Definite cercopithecoid monkeys are known from two early Miocene localities in Africa. These are at Napak, in Uganda (Pilbeam and Walker, 1968) and Wadi Moghara, in Egypt (Simons, 1969). The material is fragmentary and shows no clear affinities with any particular cercopithecoid group, but it is consistent with the predicted ancestral morphotype (Delson, 1975). The earliest record of division within the Cercopithecoidea comes from the middle Miocene deposits at Maboko Island in Kenya, where two species of *Victoriapithecus* are known (Von Koenigswald, 1969). The more common of these shows some evidence of colobine relationships in its squarish molars and deep mandibles, while the less common second species retains the primitive cercopithecine-like dentition. It would appear, therefore, that the Cercopithecoidea had evolved by the beginning of the Miocene period, that is 22-20 million years ago, and that it underwent an early divergence into the Colobinae and Cercopithecinae by the beginning of the middle Miocene when the Maboko Island deposits were laid down, that is 16-14 million years ago.

By the beginning of the late Miocene, about 11 million years ago, the Colobinae had become completely distinct. One of the earliest species was *Mesopithecus,* a large colobine which was partly terrestrial and which is very common in several late Miocene localities. There is a good collection of this species from Pikermi in the collections of the British Museum (Natural History). Later European colobines such as *Dolichopithecus* were even more terrestrially adapted and must have resembled baboons in general appearance rather than the present day colobines. It would appear that these partly terrestrial fossil colobines must be more closely related to the present day Asian colobines, while the living African species, with their specialized thumb reduction indicating a total commitment to arboreal life, have had a largely unknown history within Africa. It seems probable that the divergence of the two groups must have occurred before 11 million years ago, when the first European semi-terrestrial *Mesopithecus* is known (Delson and Andrews, 1975), but no African colobines are known until the late Pliocene to Pleistocene, about 4 million years ago, when *Paracolobus* and *Cercopithecoides* of East and South Africa appeared. These genera appear to be rather distinct from the later Pleistocene *Libypithecus* and the living African colobines, and the exact origin of the latter is still unknown.

Most of the colobine species known from the fossil record are rather large in size, and where postcranial remains are present there appear to be well established adaptations for ground living. Moreover, the associated faunas at many of the colobine localities appear to indicate open, non-forest, conditions, varying from woodland to savanna, and this is consistent with the terrestrial adaptations of the fossil colobines and supports the evidence for their being ground living. This evidence contrasts with the present distribution of living colobine species, most of which are forest living and some of which appear to be well adapted for an arboreal forest existence, particularly the African colobines. It has recently been suggested, however, from analyses of the living species of cercopithecoid monkey, that the divergence and initial radiation of the old world monkeys was in woodland savanna habitats (Andrews, 1981) and that the monkey species have limb proportions indicative of terrestrial locomotion, even when today they are fully arboreal in habit (Aiello, 1981). These results are much more in accord with the fossil evidence, and considered together they suggest that much of the early evolution of the cercopithecoid monkeys was as ground living species in open woodland and savanna habitats. This trend persisted even after the divergence of the colobines and cercopithecines, and only more recently have they become more arboreal again.

African Colobinae

COLOBUS Illiger, 1811

Colobus Monkeys
(2 subgenera: *Colobus* Illiger, 1811 and *Piliocolobus* Rochebrune, 1886-87)

Taxonomic note The inter-relationships of the African colobines are discussed by Groves (1970) who separates *Procolobus* (the olive colobus) at generic level. The taxonomy of the group is complicated by a number of anatomical characters, some of which are tabulated below, in which *Piliocolobus* (the red colobus) resembles *Procolobus* more closely than *Colobus* (the black-and-white colobus).

Colobus	Piliocolobus	Procolobus
No sexual swelling in ♀♀	Sexual swelling in ♀♀	Sexual swelling in ♀♀
No perineal organ in ♂♂	Perineal organ in ♂♂	Perineal organ in ♂♂
Ischial callosities contiguous in ♂♂	Ischial callosities separate in ♂♂	Ischial callosities separate in ♂♂
Stomach 3-chambered	Stomach 4-chambered	Stomach 4-chambered
Larynx large	Larynx small	Larynx small
Sub-hyoid sac present	Sub-hyoid sac absent	Subhyoid sac absent

(Data from Pocock, 1936; Hill, 1952; Hill & Booth, 1957; Kuhn, 1967, 1972)

Many zoologists (e.g. Pocock, 1936; Kuhn, 1967) have therefore united *Piliocolobus* and *Procolobus* in the genus *Procolobus* (the earlier name), while keeping *Colobus* separate. Others (e.g. Verheyen, 1962) have considered that subgeneric status for all three taxa within the genus *Colobus* best expresses their inter-relationships. The catalogue follows the systematic list of Thorington & Groves (1970), with two subgenera in the genus *Colobus*.

Morphology Large leaf-eating monkeys with typical colobine characters: molar teeth with high cusps and long sharp crests, a complex stomach modified for the digestion of mature leaves, and no cheek pouches such as are present in the Cercopithecinae. The thumb is absent or reduced to a small tubercle. *Colobus* differs from *Procolobus* in its larger size, and in many details of the visceral anatomy and male genitalia (Hill, 1952). The following key separating the skulls of the two genera here recognised is freely translated from Verheyen (1962: 168):

A mediosagittal mental foramen; 3rd mandibular molar showing 6 distinct cusps (sometimes 5); on the lingual aspect of the incisors a distinct cingulum with projecting cusplet; length of the superior dental arcade less than 35mm and dental length less than 24.5mm .... *Procolobus*

No mediosagittal mental foramen; 3rd mandibular molar with 5 distinct cusps (exceptionally 6); on the lingual aspect of the incisors a slight cingulum, without cusplet; length of the superior dental arcade greater than 35.5mm and dental length greater than 24.5mm .... *Colobus*

Range Widely distributed in the central African forest block with extensions into forested areas of W. Africa, S. Sudan, Ethiopia, Kenya and Tanganyika.

Subgenus: COLOBUS Illiger, 1811

Black-and-white Colobus, Guereza, Black Colobus
(4 species: type species *Simia polycomos* Schreber, *c.* 1800 = *Cebus polykomos* Zimmermann, 1780).

Principal synonyms *Guereza* Gray, 1871 (type species: *Guereza rueppelli*); *Stachycolobus* Rochebrune, 1886-87 (type species: *Colobus satanas*); *Pterycolobus* Rochebrune, 1886-87 (type species: *Colobus vellerosus*).

Taxonomic note Four species are recognised, following Thorington & Groves (1970). However, from a recent study of male loud calls, skull characters, coat pattern and geographic distribution, Oates & Trocco (1983) suggest that *C.p. vellerosus* should be given specific status.

Morphology Large black monkeys with striking white ornamentation except in the case of *C. satanas* which is completely black. The infant is white at birth, even in *C. satanas*, and gradually acquires adult colouration during the first three to six months of life (Mearns & Pidgeon, 1978). Characters in which *Colobus* (*Colobus*) differs from *Colobus* (*Piliocolobus*) are cited in the taxonomic note above. The following key separating the skulls of the subgenera, based mainly on material from Zaire, is freely translated from Verheyen (1962:168):

Pterygoid fossa deep, narrow, perforated at base; base of the lamina medialis separated by a groove from the side of the basi-sphenoid; no foramen in the posterior part of the lamina lateralis; choanae high and narrow; apex pyramidis pointed, with crest extending to the lamina medialis; orbits always angulated with very thick superciliary ridges and distinct foramina supraorbitalia or frontal incisurae; pronounced suborbital depressions; ♂ skull almost always bearing mediosagittal crest .... *Piliocolobus*

Pterygoid fossa wide, shallow, not perforated at base; base of the lamina medialis continuous with the lateral part of the basi-sphenoid; always one or several foramina in the posterior part of the base of the lamina lateralis; choanae low and wide; apex pyramidis, obtuse and without crest, does not extend to the lamina medialis of the pterygoid process; oval orbits with narrow superciliary ridges; never any foramina supraorbitalia (sometimes a small frontal incision); no suborbital depression; ♂ skull never bears mediosagittal crest .... *Colobus*

Dandelot & Groves (in preparation) find, however, that some of these characters do not apply when material from outside Zaire is examined.

The limb proportions are typically colobine; the intermembral index is 79 (n = 40), considerably lower than that of the subgenus *Piliocolobus* 86 (n = 21).

Chromosome diploid no. 44 (*C. polykomos*; Chiarelli, 1973).

Range Africa, south of the Sahara, from Guinea and Sierra Leone to Ethiopia, and southwards through the Zaire basin to NE. Angola (Machado, 1969), and S. Tanganyika, possibly entering NW. Zambia (Dandelot, 1971).

Ecology The genus *Colobus* has been studied mainly in East Africa, in rain forest and montane forest in Uganda, and in mosaic forest in Tanganyika (Clutton-Brock, 1975). Both subgenera are leaf-eaters, but *Colobus* (*Colobus*) have a more monotonous diet, utilising the products of only a few tree species, whereas *Colobus* (*Piliocolobus*) are more eclectic (Struhsaker & Oates, 1975). *Colobus* (*Colobus*) are capable of surviving on mature leaves alone, which enables them to occupy regions with pronounced dry seasons such as the Lado region, southern Sudan, where Heller (1913) observed them near the banks of the Nile. They ran along the ground between small scattered clumps of acacias, hence Heller's choice of subspecific name, *terrestris*, now synonymised with *occidentalis*. *Colobus* (*Colobus*) are found at higher altitudes than *Colobus* (*Piliocolobus*), up to 9300ft (*C.g.matschiei* on Mt Elgon, BMNH 1931.11. 1. 1).

Social behaviour *Colobus* (*Colobus*) live in cohesive groups of between 3-15 individuals (Groves, 1973). There is usually only one adult male and three to four adult females (Struhsaker & Oates, 1975). Home ranges are small (15-16 hectares), and core areas are defended. Spectacular jumping displays by the adult male are accompanied by a loud 'snorting and roaring' vocalisation, and help to maintain distance between groups.

Reproduction Because their specialised dietary requirements were not understood, *Colobus* (*Colobus*) have until recently rarely bred in zoos, so their reproductive physiology is still largely unknown. Mating has been described by Hollihn (1973); the female always initiates copulation which takes place only at oestrus. Unlike the red colobus, there is no sexual swelling. Gestation is estimated at 165 days (147-178: n = 5) in *C.p.polykomos* at Jersey Zoo (Mallinson, 1973). The infant is white at birth with slightly woolly crinkly fur. The change to adult colouration lasts about 112 days on average (Mearns & Pidgeon, 1978). The transfer of newborn infants to other females in the group has been observed in the wild (*C. guereza:* Struhsaker & Oates, 1975). In captivity, infant transfer to other females was observed in *C. guereza* and *C. polykomos* by Horwich & Manski (1975) and the transfer of infants to adult males has been noted in *C. guereza* (Anon., 1964). Longevity is estimated at 27-30 years in San Diego Zoo (Hill, 1975).

Key to the species of the subgenus Colobus

1a Coat completely black; (tail untufted); .... *C. satanas*
1b Coat black with conspicuous white markings........ 2
2a Long white mantle fringing flanks is continuous across midline of rump, but well separated from short white fur encircling face; long flowing tail; (crown hairs short, upright, forming bonnet; tail tuft white)........ *C. guereza*
2b White mantle absent, tail untufted 3
3a Long white epaulettes are continuous with white hairs of cheeks and throat; tail varies from almost completely white to almost completely black........ *C. angolensis*
3b Long white or greyish hairs encircle face; tail completely white; epaulettes greyish, sometimes reduced or absent (where absent, replaced by white thigh patches) .... *C. polykomos*

Dandelot (1971: 30, fig. 1) notes that *C. polykomos* and *C. guereza* have the longest noses and *C. satanas* the shortest, with *C. angolensis* intermediate. This is correlated with the degree of prognathism of the skulls.

COLOBUS (COLOBUS) POLYKOMOS (Zimmermann, 1780)

King Colobus, Ursine Colobus, Full-bottom Monkey
(3 subspecies)

Taxonomic note Both Rahm (1970) and Dandelot (1971) recognise three subspecies. Rahm (1970) has suggested that *C.p.dollmani*, from between Rs Sassandra and Bandama, may be a hybrid form between *C.p.polykomos* and *C.p.vellerosus*. There is undoubtedly much variation in BM(NH) material, notably Booth's specimens from Bouaflé Forest Reserve to the west of R. Bandama, catalogued as *C.p.vellerosus*?

Range Forested areas of West Africa from Guinea to W. Nigeria.

Dimensions Adult specimens in the museum collections have the following field measurements (see Appendix II for comparative table):

h & b	10 ♂♂	633	(584-680)	: 8 ♀♀	600	(570-648)
tail	10 ♂♂	874	(718-1000)	: 8 ♀♀	886	(806-959)
wt	5 ♂♂	8420	(6600-10 000)	: 6 ♀♀	7133	(6100-8500)

Subspecies All have glossy black fur on body and limbs, with long black hairs on shoulders and flanks, a completely white or dirty white tail, a narrow ring of white hairs surrounding the callosities and a small white pubic patch The black face is encircled by white, grizzled or yellowish white fur which extends to the chest.

Key to the subspecies of *Colobus polykomos*

C.p.polykomos (Guinea to R. Sassandra, Ivory Coast)	Grizzled white or yellowish white browband, cheek whiskers and throat: grizzled area continues down sides of neck to shoulders, forming long epaulettes.
C.p.dollmani (between R. Sassandra and R. Bandama, Ivory Coast)	Pure white browband, cheek whiskers and throat. Epaulettes vary in size, from a few white hairs among the long black hairs of the shoulders to a grizzled area almost as extensive as in *C.p.polykomos*.
C.p.vellerosus (Ivory Coast, from W. bank of R. Bandama, to Ghana, Togo, Dahomey, W. Nigeria)	Pure white browband, cheek whiskers and throat. No white epaulettes but large grizzled white patches on back of thighs, continuous with white pubic patch.

Colobus polykomos polykomos (Zimmermann, 1780)

Principal synonyms *Simia(Cercopithecus)regalis* Kerr, 1792:74; *Simia tetradactyla* Link, 1795:62; *Simia polycomos* Schreber, *c.* 1800, pl.10D; *Simia comosa* Shaw, 1800:59; *Colobus ursinus* Bennett, 1832: 122.

Type localities Of *polykomos*, Sierra Leone; of *tetradactyla*, none cited; of *polycomos*, as for *polykomos* — Sierra Leone; of *comosa*, Sierra Leone; of *ursinus*, said to be from Algoa Bay, South Africa.

Colobus polykomos dollmani Schwarz, 1927.

Type locality Bandama, Bandama R., Ivory Coast.

Colobus polykomos vellerosus (I. Geoffroy, 1830).

Principal synonyms *Semnopithecus bicolor* Wesmael, 1835:237; *Colobus leucomeros* Ogilby, 1838:69.

Type localities Of *vellerosus*, unknown; of *bicolor*, 'Côtes d'Afrique'; of *leucomeros*, Africa, corrected by I. Geoffroy (1831:37) to India or neighbouring archipelago. The specimen is said to have come from the R. Gambia, but this is outside the known range of the subspecies.

Locality	Date of Collection	Sex	Age	Nature of Specimen	Accession no.	References, Remarks, etc.
Colobus (Colobus) polykomos polykomos (Zimmermann 1780)						
[GUINEA]						
French Guinea	—	[M]	[Ad]	Skin	ZD.1939.3473	(*Colobus ursinus*)
IVORY COAST						
Guiglo, between Rs. Nzo and Sassandra	19 Jul 1953	M	[Ad]	Skin	ZD.1956.345	
" " "	19 Jul 1953	F	Ad	Skin, Skull	ZD.1956.346	Listed Grubb,P. and Booth,A.H. (unpubl.): 4
LIBERIA						
50 miles inland of Monrovia	—	M	Ad	Skin, Skull	ZD.1904.6.2.1	Cited Schwarz,E. 1929: 588
Farmington R.	4 Jul 1888	[F]	Ad	Skin, Skull	ZD.1939.1155	(*Colobus ursinus*)
Upper St.Paul's R.	—	I	[Ad]	Skin	ZD.1919.7.11.2	(*Colobus ursinus*), Cited Schwarz,E. 1929: 588
SIERRA LEONE						
Near Gegbwema, 40 miles from Liberian Border, Eastern Province, [c.7°39′N 11°03′W]	11 May 1965	F	Ad	Skull, Skeleton	ZD.1979.39	Correspondence Walters,L.S. (unpubl.), Cited Walters,L.S. (unpubl.)a
Mamansu, Kunike Fulawusu Chiefdom	18 Oct 1946	M	Ad	Skin, Skull	ZD.1947.584	
Potoru, Barri Chiefdom, [7°31′N 11°29′W]	30 Aug 1950	F	Ad	Skin, Skull	ZD.1952.24	(*Colobus polykomos ursinus*)
Zimmi, Makpele Chiefdom, [Pujehu], S.W.Sierra Leone, [7°19′N 11°18′W]	18 Feb 1946	F	Juv	Skin, Skull	ZD.1947.585	
—	—	[M]	[Juv]	Skin	ZD.1855.12.24.2	(*Colobus ursinus*)
—	—	[M?]	[Ad]	Skin	ZD.1928.4.5.1	" " " "
NO DATA						
—	—	[M]	[Ad]	Skin	ZD.1855.12.24.406	(*Colobus ursinus*), Cited Schwarz,E. 1929: 588
—	—	F	Ad	Skull, Skeleton	ZD.1864.6.23.5	(*Colobus ursinus*), Listed Gray,J.E. (unpubl.): 1429
—	Died 28 Apr 1906	F	[Juv]	Skin	ZD.1939.576	(*Colobus ursinus*)
—	Died 2 Oct 1907	[M]	Juv	Skin, Skull	ZD.1939.1156	" " " "
—	Died May 1979	M	Inf	Skin	ZD.1979.1691	Background Walters,L.S. (unpubl.)a
—	Died 22 Oct 1977	M		[Juv], Wet cadaver,lacking head and skin.	ZD.1979.2056	Correspondence Walters,L.S. (unpubl.), Cited Walters,L.S. (unpubl.)a
—	Died 24 Feb 1976	[M]	[Ad]	Wet,cadaver in spirit	ZD.1981.193	Cited Walters,L.S. (unpubl.)

Locality	Date of Collection	Sex	Age	Nature of Specimen	Accession no.	References, Remarks, etc.
Colobus (Colobus) polykomos polykomos? (Zimmermann 1780)						
NO DATA						
—	—	[I]	[Ad]	Skin	ZD.1980.801	Correspondence H.M.Customs and Excise (unpubl.)
Colobus (Colobus) polykomos dollmani Schwarz 1927						
IVORY COAST						
Bandama, R.Bandama	31 Dec 1922	M	Ad	Skin, Skull	ZD.1923.2.3.4	Holotype *Colobus polykomos dollmani* Schwarz 1927, Cited Schwarz,E. 1927: 155, Cited Schwarz,E. 1929: 589, Listed Lowe,W.P. (unpubl.)b
Bandama, R.Bandama, 600ft	28 Dec 1922	F	Ad	Skin, Skull	ZD.1923.2.3.5	Paratype *Colobus polykomos dollmani* Schwarz 1927, Cited Schwarz,E. 1927: 155, Cited Schwarz,E. 1929: 589, Listed Lowe,W.P. (unpubl.)b
25km W. of Daloa	13 Jul 1953	F	Ad	Skin, Skull	ZD.1956.344	Listed Grubb,P. and Booth,A.H. (unpubl.): 4
40km W. of Daloa	1 Apr 1954	M	Ad	Skin, Skull	ZD.1956.342	Listed Grubb,P. and Booth,A.H. (unpubl.): 5
,, ,, ,,	1 Apr 1954	F	Ad	Skin, Skull	ZD.1956.343	Listed Grubb,P. and Booth,A.H. (unpubl.): 4
18 miles E. of R.Sassandra on Daloa-Man road	1 Apr 1954	M	[Ad]	Skin	ZD.1955.379	Listed Brandon-Jones,D. (unpubl.)
18 miles E. of R.Sassandra on Daloa-Man road	1 Apr 1954	M	[Juv]	Skin	ZD.1955.380	,, ,, ,, ,,
Colobus (Colobus) polykomos vellerosus (Geoffroy,I. 1830)						
GHANA						
Ashanti	12 Mar 1953	M	Inf	Skin, Skull	ZD.1956.357	Listed Grubb,P. and Booth,A.H. (unpubl.): 14
15 miles E. of Bibiani	5 Jul 1952	M	Ad	Skin, Skull	ZD.1956.358	Listed Grubb,P. and Booth,A.H. (unpubl.): 5
Enchi, [5°53′N 2°48′W]	17 Jul 1952	F	Ad	Skin, Skull	ZD.1956.355	Listed Grubb,P. and Booth,A.H. (unpubl.): 4
Gambia [=Kenyasi Mohu], 30 miles N. of Goaso, [7°04′N 2°39′W]	2 Sep 1953	M	Ad	Skin, Skull	ZD.1956.349	Listed Grubb,P. and Booth,A.H. (unpubl.): 5
Gambia [=Kenyasi Mohu], 30 miles N. of Goaso, Ashanti, [7°04′N 2°39′W]	27 Dec 1952	M	Juv	Skin, Skull	ZD.1956.348	Listed Grubb,P. and Booth,A.H. (unpubl.): 13
,, ,, ,,	30 Aug 1953	F	Juv	Skin, Skull	ZD.1956.350	,, ,, ,, ,,
On Sorri R., between Yapei and Damongo, Gonja	5 Mar 1954	F	Juv	Skin, Skull	ZD.1956.353	Listed Grubb,P. and Booth,A.H. (unpubl.): 4
R.Sorri, between Damongo and Yapei, Gonja	5 Mar 1954	F	Juv	Skin, Skull	ZD.1956.351	,, ,, ,, ,,
,, ,, ,,	5 Mar 1954	F	Ad	Skin, Skull	ZD.1956.352	,, ,, ,, ,,
,, ,, ,,	5 Mar 1954	F	Juv	Skin, Skull	ZD.1956.354	,, ,, ,, ,,
Sukusuku, W.Ghana, [6°34′N 3°14′W]	2 May 1969	M	[Ad]	Skin	ZD.1970.2625	
Tano Lodge, 10 miles N. of Sefwi Wiawso, [c.6°22′N 2°29′W]	11 Sep 1953	M	Ad	Skin, Skull	ZD.1956.359	Listed Grubb,P. and Booth,A.H. (unpubl.): 4
Tano Lodge, Sefwi Wiawso, [c.6°22′N 2°29′W]	—	M	Ad	Skull	ZD.1971.2363	Listed Grubb,P. and Booth,A.H. (unpubl.): 4-5
,, ,, ,,	—	F	Ad	Skull	ZD.1971.2364	,, ,, ,, ,,
,, ,, ,,	—	F	Ad	Skull	ZD.1971.2365	,, ,, ,, ,,
,, ,, ,,	—	F	Ad	Skull	ZD.1971.2366	,, ,, ,, ,,
,, ,, ,,	—	M	Ad	Skull	ZD.1971.2367	,, ,, ,, ,,
,, ,, ,,	—	F	Ad	Skull	ZD.1971.2368	,, ,, ,, ,,
[GHANA]						
Bibiani, Gold Coast, [6°28′N 2°20′W], 700ft	1 Jun 1908	M	[Inf]	Skin, Skull	ZD.1908.8.6.1	
Goaso, Ashanti, [6°49′N 2°27′W]	19 Dec 1933	M	Ad	Skin, Skull	ZD.1935.1.30.11	Listed Lowe,W.P. (unpubl.)c
,, ,, ,,	22 Dec [1933]	M	Juv	Skin, Skull	ZD.1935.1.30.13	,, ,, ,, ,,
,, ,, ,,	23 Dec 1933	M	Ad	Skin, Skull	ZD.1935.1.30.14	,, ,, ,, ,,
,, ,, ,,	18 Dec 1934	F	Ad	Skin, Skull	ZD.1935.10.22.5	,, ,, ,, ,,
,, ,, ,,	18 Dec 1934	F	[Inf]	Skin	ZD.1935.10.22.6	,, ,, ,, ,,
,, ,, ,,	—	M	Ad	Skull	ZD.1935.10.22.7	,, ,, ,, ,,
Goaso, Ashanti, Gold Coast, [6°49′N 2°27′W]	20 Dec 1933	M	[Inf]	Skin	ZD.1935.1.30.12	,, ,, ,, ,,
,, ,, ,,	28 Dec 1933	M	Juv	Skin, Skull	ZD.1935.1.30.15	,, ,, ,, ,,
,, ,, ,,	29 Dec 1933	M	Ad	Skin, Skull	ZD.1935.1.30.16	,, ,, ,, ,,
,, ,, ,,	29 Dec 1933	M	Ad	Skin, Skull	ZD.1935.1.30.17	,, ,, ,, ,,
,, ,, ,,	19 Dec 1933	F	Ad	Skin, Skull	ZD.1935.1.30.18	,, ,, ,, ,,
,, ,, ,,	30 Dec 1933	F	Ad	Skin, Skull	ZD.1935.1.30.19	,, ,, ,, ,,
,, ,, ,,	3 Jan 1934	F	Inf	Skin, Skull	ZD.1935.1.30.20	,, ,, ,, ,,
,, ,, ,,	—	M	[Ad]	Skull	ZD.1935.1.30.21	,, ,, ,, ,,
Gold Coast	Died 30 Nov 1904	[M]	Juv	Skin, Skull	ZD.1939.1159	(*Colobus vellerosus*)
,, ,, ,,	1925-6	[M]	[Inf]	Skin, Skeleton in skin	ZD.1975.585	
Kpetchu, L. bank of Oti R., 20 miles N.E. of Kete Krachi, Togoland, [7°56′N 0°06′E]	17 Aug 1953	M	Inf	Skin, Skull	ZD.1956.360	Listed Grubb,P. and Booth,A.H. (unpubl.): 4
,, ,, ,,	17 Aug 1953	F	Ad	Skin, Skull	ZD.1956.361	,, ,, ,, ,,
Near Kumasi, Gold Coast	Early 1948	I	[Inf]	Skin	ZD.1952.1255	
Mangoasi, E.Province, Gold Coast, 500ft	23 Apr 1948	F	[Juv]	Skin	ZD.1952.1254	

Locality	Date of Collection	Sex	Age	Nature of Specimen	Accession no.	References, Remarks, etc.
Colobus (Colobus) polykomos vellerosus *continued*						
Odumassi, Ashanti[?]	—	M	Ad	Skin, Skull	ZD.1939.1157	
Sekondi, Ashanti, [4°56′N 1°42′W]	—	I	[Ad]	Skin	ZD.1939.1158	(*Colobus vellerosus*)
IVORY COAST						
—	—	F	Ad	Skull	ZD.1971.2357	Listed Grubb,P. and Booth,A.H. (unpubl.): 4
—	—	F	Ad	Skull	ZD.1971.2358	,, ,, ,, ,,
—	—	M	Ad	Skull	ZD.1971.2359	,, ,, ,, ,,
—	—	F	Juv	Skull	ZD.1971.2360	,, ,, ,, ,,
NIGERIA						
R.Okpara, Ilaro Division, Abeokuta Province	15 Jun 1937	M	Juv	Skin, Skull	ZD.1938.11.22.2	
NO DATA						
—	—	[F]	Juv	Skin, Skull, Skeleton	ZD.1392a	(*Colobus bicolor*), Listed Gray,J.E. (unpubl.): 1392
—	—	M	Ad	Skull, Skeleton	ZD.1392b	,, ,, ,, ,,
—	—	I	[Ad]	Skin	ZD.1841.1755	(*Colobus ursinus?*), Listed Anon. (unpubl.)b: 82,Cited Schwarz,E. 1929: 589
—	—	I	[Ad]	Skin	ZD.1841.1756	(*Colobus bicolor*), Listed Anon. (unpubl.)b: 82,Cited Schwarz,E. 1929: 589
—	—	F	Ad	Skin	ZD.1850.5.7.2	Associated with next specimen (1850.8.21.15)
—	—	F	Ad	Skull	ZD.1850.8.21.15	(*Colobus bicolor*), Listed Gray,J.E. (unpubl.): 944
—	—	I	[Inf]	Skin, Skull in skin	ZD.1855.11.26.11	(*Colobus bicolor*)
—	11 Mar 1837	I	[Ad]	Skin	ZD.1855.12.24.405	Holotype *Colobus leucomeros* Ogilby 1838, (*Colobus bicolor*), Cited Ogilby,W. 1838: 69 Cited Schwarz,E. 1929: 589
—	—	M	Ad	Skin, Skull	ZD.1859.5.5.2	(*Colobus ursinus*), Figured Elliot,D.G. 1913: Pl.18
—	—	[F]	[Juv]	Skin, Skull in skin	ZD.1862.3.19.1	
—	—	M	Juv	Skull	ZD.1971.2361	
—	—	M	Juv	Skull	ZD.1971.2362	
—	Died 23 Oct 1912	[M]	[Juv]	Skin, Skull, Skeleton,	ZD.1980.768	
—	—			Skull, Skeleton	CMZ E7309A	
—	—	I	Juv	Skull	RCS(OM) A68.1	Cited Colyer,J.F. 1943: 183-188
—	—	[F]	Juv	Skull	RCS(OM) A68.11	
—	—			Skull, Skeleton	RSM 1861/52/1a	
—	—			Skull, Skeleton	RSM 1861/52/18	
—	—			3 Skeletons with skulls	RSM	
Colobus (Colobus) polykomos vellerosus? (Geoffroy,I. 1830)						
GHANA						
Bia tributaries, N.Forest Reserve, 100ft	Nov 1970	M	Ad	Skull	ZD.1976.1793	
IVORY COAST						
W. edge of Bouaflé forest reserve, near Daloa	15 Jul 1953	M	Ad	Skin, Skull	ZD.1956.356	Listed Grubb,P. and Booth,A.H. (unpubl.): 5
[IVORY COAST]						
W. edge of Bouaflé forest reserve, nem Daloa	31 Mar 1954	F	[Ad]	Skin	ZD.1956.347	(*Colobus polykomos vellerosus x dollmani*)
Colobus (Colobus) polykomos? vellerosus? (Geoffroy,I. 1830)						
GHANA						
E.Region	—	M	Ad	Skull	ZD.1981.58	
Tano Lodge, Sefwi Wiawso, [c.6°22′N 2°29′W]	—	F	Ad	Skull	ZD.1971.2369	Listed Grubb,P. and Booth,A.H. (unpubl.): 4-5
,, ,, ,,	—	F	Ad	Skull	ZD.1971.2370	,, ,, ,, ,,
,, ,, ,,	—	M	Ad	Skull	ZD.1971.2371	,, ,, ,, ,,
,, ,, ,,	—	F	Ad	Skull	ZD.1971.2372	,, ,, ,, ,,
,, ,, ,,	—	F	Ad	Skull	ZD.1971.2373	,, ,, ,, ,,
,, ,, ,,	—	I	Juv	Skull	ZD.1971.2374	Listed Grubb,P. and Booth,A.H. (unpubl.): 4
Colobus (Colobus) polykomos (Zimmermann 1780)						
GHANA						
Tano Lodge, Sefwi Wiawso	—	I	Juv	Skull	ZD.1971.2403	(*Procolobus verus?*), Listed Grubb,P. and Booth,A.H. (unpubl.): 5
Colobus (Colobus) polykomos? (Zimmermann 1780)						
[IVORY COAST]						
—	—	I	Juv	Skull, Skeleton	ZD.1981.110	(*Cercopithecus sp.*)

COLOBUS (COLOBUS) ANGOLENSIS Sclater, 1860

Angolan Colobus
(8 subspecies)

Taxonomic note Eight subspecies were recognised by Rahm (1970): *angolensis, adolfifriederici, cordieri, cottoni, palliatus, prigoginei, ruwenzorii* and *sharpei*. Dandelot (1971) recognised six, synonymising *adolfifriederici* with *ruwenzorii* and *sharpei* with *palliatus*. Hull's (1979) craniometric studies led him tentatively to accept seven subspecies, uniting *adolfifriederici* with *ruwenzorii*, but keeping *prigoginei* and *cordieri* separate pending further study. In view of this uncertainty, the eight subspecies recognised by Rahm have been catalogued separately.

Range NE. Angola, Zaire (except the northwest and southeast regions), Rwanda, Uganda, Kenya coast, north as far as Shimba Hills, some coastal and montane forests including the Mahali Mts, Tanganyika (Rodgers, 1981; Nishida *et al.*, 1981), possibly NW. Zambia (Dandelot, 1971).

Dimensions Adult specimens in the museum collection have the following field measurements (see p. 98 for comparative table):

h & b	7 ♂♂	590	(553-660)	: 7 ♀♀	529	(483-593)
tail	7 ♂♂	829	(762-920)	: 7 ♀♀	706	(629-763)
wt	2 ♂♂	9679	(9072, 10 285)	: 5 ♀♀	7403	(6804-8895)

Subspecies The stiff black brow hairs are separated from the black crown by a white brow-band which narrows medially (disappearing completely in some races) but broadens laterally, meeting the downy white hairs of the cheeks and throat. These in turn are confluent with the white shoulder hairs, or 'epaulettes'. A white or grizzled patch or stripe extends from the callosities to the pubis in most races. The untufted tail has a black basal and white (or grizzled) terminal portion. The tail may be as much as 95% white as in *prigoginei*, or less than 5% white as in *ruwenzorii*.

Key to the subspecies of *Colobus angolensis*

Subspecies	White brow band	White epaulettes	White pubic patch	Terminal portion of tail	Other characters
angolensis (NE. Angola, Zaire on L. bank of R. Zaire)	Absent in majority of specimens	Long, profuse	White, small	White (33-66%)	Whorl on front of crown; strong brow fringe
cottoni (NE. Zaire, on R. bank of R. Zaire)	"	Short, reduced	Absent	Grizzled (50%)	—
cordieri (Zaire, E. of R. Lualaba)	Absent	"	"	Grizzled, with brownish tinge (90%)	—
prigoginei (from Verheyen, 1962: Zaire, Mt Kabobo, see below)	"	"	"	Yellowish white (90%)	Strong brow fringe; parting on front of crown; slight occipital crest
ruwenzorii (Mt Ruwenzori, Uganda)	Narrow	Long, profuse	Grizzled, broad (up to 170 by 100mm)	Grizzled (less than 5%)	Long thick coat; white cheek tufts long & bushy
adolfifriederici (E. Zaire, Rwanda, SW. Uganda)	"	Variable, less profuse than above	Grizzled, smaller than above (up to 100 by 80mm)	Grizzled (variable: 5-25%)	—
sharpei (Mts of SW. Tanganyika, N. Malawi)	Broad	Long, profuse	White, variable in size	Grizzled, grading into short white tip (50%); slightly bushy	Long thick coat
palliatus (Coast of S. Kenya, N. Tanganyika)	"	"	Narrow stripe (♂♂); (broader patch in ♀♀)	Grizzled, grading into variable white tip (25-75%)	Sparse white hairs on throat; whorl on nape; occipital crest

Colobus angolensis angolensis Sclater, 1860

Principal synonyms *Colobus abyssinicus sandbergi* Lönnberg, 1908:1; *Colobus (Colobus) palliatus weynsi* Matschie, 1913:207; *Colobus maniemae* Matschie, 1914:336; *Colobus benamakimae* Matschie, 1914:337.

Type localities Of *angolensis*, inland from Bembe, N. Angola; of *sandbergi*, near R. Lufizi (*c.* 13°S), R. Zambeze, Angola, corrected by Ansell (1959:170) to R. Lusiji, which rises *c.* 10°07′S 22°27′E and joins the upper R. Lulua at 10°08′S 22°54′E, southern Zaire; of *weynsi*, Lower Congo; of *maniemae*, between Kibombo and Pienimulamba (=Pene-Mapundu, 3°59′S 26°03′E) on middle Luala (= R. Lualaba), Zaire; of *benamakimae*, Bena Makima (4°58′S 21°06′E), Sankuru basin, Zaire.

Colobus angolensis adolfifriederici Matschie, 1914

Type locality Rugege Forest (*c.* 2°25′S 29°30′E), E. of L. Kivu, Rwanda.

Colobus angolensis cordieri Rahm, 1959

Type locality Captured alive at Kampunzu (= Kampungu), Pangi, Kivu Province, Zaire.

Colobus angolensis cottoni Lydekker, 1905

Principal synonyms *Colobus (Colobus) palliatus mawambicus* Matschie, 1913:205; *Colobus mawambicus nahani* Matschie, 1914:335.

Type localities Of *cottoni*, Qugo, corrected by Schwarz (1929:595) to Zokwa, between Mahagi and Irumu, Zaire. Powell-Cotton gives approximate co-ordinates 1°50′N 30°30′E; of *mawambicus*, North Pemba, between Irumu and Mawambi on upper R. Ituri, Zaire; of *nahani*, by Panga (1°52′N 26°23′E) on R. Aruwimi, Zaire.

Colobus angolensis palliatus Peters, 1868

Type locality East Africa opposite Zanzibar, restricted by Schwarz (1929:598) to R. Pangani. Moreau *et al.* (1946:406) note that this race occurs in evergreen forest down to the edge of the sea; Rodgers (1981) lists known records which extend to about 5000ft in Southern Highlands.

Colobus angolensis prigoginei Verheyen, 1959

Type locality Mt Kabobo (5°06′S 29°01′E), E. shore of L. Tanganyika, at 2400m, Zaire.

Colobus angolensis ruwenzorii Thomas, 1901

Type locality Bwamba Country, NW. flank of Mt Ruwenzori, Uganda, corrected by Moreau *et al.* (1946:407) to Ruwenzori, probably at high altitudes.

Colobus angolensis sharpei Thomas, 1902

Principal synonym *Colobus langheldi* Matschie, 1914:337.

Type localities Of *sharpei*, Fort Hill, Nyasa-Tanganyika plateau, corrected by Ansell (1959:170) to Nkuku Forest, Rungwe Mt (9°08′S 33°38′E), Tanganyika; of *langheldi*, purchased at Ujiji (E. shore of L. Tanganyika), from a caravan coming from Manyema country, Zaire.

Locality	Date of Collection	Sex	Age	Nature of Specimen	Accession no.	References, Remarks, etc.
Colobus (Colobus) angolensis angolensis Sclater 1860						
ANGOLA						
[About 300 miles from the coast], [N.Angola]	—	I	[Ad]	Skin	ZD.1860.7.23.1	Holotype *Colobus angolensis* Sclater 1860, Cited Sclater,P.L. 1860: 245, Cited Schwarz,E. 1929: 595
[ZAIRE]						
Bongandanga, Balolo, Congo, [1°30′N 21°03′E]	—	M	Ad	Skin, Skull	ZD.1895.4.1.2	
"Ibuka", Congo		M	Ad	Skin, Skull	ZD.1912.7.26.4	(*Colobus occidentalis*), Cited Schwarz,E. 1929: 595
" " "		M	Ad	Skin, Skull	ZD.1912.7.26.5	" " " "
Kiri, R.Lotoi, Lac Leopold II Dist., 1°40′S 19°E, 320m	Aug 1926	F	Ad	Skin, Skull	ZD.1927.3.1.3	
Kiri, R.Lotoi, Lac Leopold II Dist., Congo Belge, [1°40′S 19°E], 320m	29 Aug 1926	M	Ad	Skin, Skull	ZD.1927.3.1.1	
" " "	7 Sep 1926	M	Ad	Skin, Skull, Radiograph of penis	ZD.1927.3.1.2	Figured Colyer,J.F. 1936: 27-28, fig.31
Kongolo, Katanga, Belgian Congo, [5°20′S 27°E]	—	I	[Ad]	Skin	ZD.1913.6.9.1	(*Colobus sandbergi* Lönnberg)
St.Joseph de Luluabourg [= Mikalaie], Kasai, S.Congo, [6°01′S 22°19′E], 610m	8 Jun 1926	M	Ad	Skin, Skull, Radiograph of penis	ZD.1927.12.21.1	
St.Joseph de Luluabourg [= Mikalaie], Kasai, S.Congo, [6°01′S 22°19′E], 600[m]	8 Jan 1926	F	Ad	Skin, Skull	ZD.1926.11.1.16	
St.Joseph de Luluabourg [= Mikalaie], Kasai, S.Congo, [6°01′S 22°19′E]	—	[F]	Ad	Skin, Skull	ZD.1926.7.6.1	
St.Joseph de Luluabourg [= Mikalaie], Kasai, [6°01′S 22°19′E], 610m	14 Apr 1924	M	[Juv]	Skin	ZD.1926.7.6.2	(*[C.a.] sandbergi*)
St.Joseph de Luluabourg [= Mikalaie], [Kasai], [6°01′S 22°19′E], 610m	8 Jan 1926	M	[Inf]	Skin	ZD.1926.11.1.15	
Colobus (Colobus) angolensis adolfifriederici Matschie 1914						
[RWANDA]						
Rugege Forest, [E. of L.Kivu], [c.2°25′S 29°30′E]	—	M	Ad	Skin, Skull	ZD.1909.12.2.5	Cited Schwarz,E. 1929: 596
[TANGANYIKA]						
Bukoba, Tanganyika Territories, [1°19′S 31°49′E], 4000ft	—	[F]	[Ad]	Skin	ZD.1939.305	
Bukoba, [West Lake], Tanganyika Territories, [1°19′S 31°49′E], 4000ft	—	F	Inf	Skull	RCS(OM) A68.7	
" " "	—	M	Juv	Skull	RCS(OM) A68.71	
" " "	—	M	Ad	Skull	RCS(OM) G11.12	Figured Colyer,J.F. 1943: 114, fig.6
" " "	—	M	Ad	Skull	RCS(OM) G11.121	Figured Colyer,J.F. 1943: 112, fig.4
" " "	—	F	Ad	Skull	RCS(OM) G68.825	
" " "	—	F	Ad	Skull	RCS(OM) G68.8251	
UGANDA						
Kaiso Forest, Sango Bay, [0°58′S 31°32′E]	1949	M	[Ad]	Skin	ZD.1950.1459	
" " "	1949	M	[Ad]	Skin	ZD.1950.1460	
Katera, Minziro Forest, Masaka, [0°55′S 31°39′E], 3800ft	3 Nov 1928	M	Juv	Skin, Skull	ZD.1929.5.14.19	
" " "	3 Nov 1928	M	Juv	Skin, Skull	ZD.1929.5.14.20	
" " "	3 Nov 1928	F	Ad	Skin, Skull	ZD.1929.5.14.21	
" " "	3 Nov 1928	F	Ad	Skin, Skull	ZD.1929.5.14.22	
" " "	3 Nov 1928	F	Ad	Skin, Skull	ZD.1929.5.14.23	
" " "	3 Nov 1928	F	Ad	Skin, Skull	ZD.1929.5.14.24	
Malambigambo Forest, Sango Bay, S.Uganda [near Tanganyika border], Ankole Dist., 0°57′S 31°35′E, 4000ft	Jan 1949	M	Ad	Skin	ZD.1949.709	Associated with next specimen (1969.372)
Malambigambo Forest, Sango Bay, S.Uganda [near Tanganyika border], Ankole Dist., 0°57′S 31°35′E, [4000ft]	Jan 1949	M	Ad	Skull	ZD.1969.372	
Sango Bay Forest, Katera, Masaka, [c.0°54′S 31°43′E], 3700ft	12 Aug 1929	M	Ad	Skin, Skull, Skeleton	ZD.1930.8.1.13	
Sango Bay Forest, Masaka Dist., [c.0°54′S 31°43′E], 3500ft	[1945-47]	F	Ad	Skull	RCS(OM) A68.72	Cited Colyer,J.F. (unpubl.)a, Background Haddow,A.J. *et al.* 1951: 189-224, Correspondence Haddow,A.J. (unpubl.)
" " "	—	M	Ad	Skull	RCS(OM) G68.8252	
Tero Forest, S.E.Masaka, Victoria-Nyanza, [c.0°57′S 31°35′E]	27 Sep 1928	F	Ad	Skin, Skull	ZD.1929.5.14.18	
Colobus (Colobus) angolensis cordieri Rahm 1959						
[ZAIRE]						
Near village of Kibila between Rs.Tshoroba and Mandana, Congo, 2°59′S 26°15′E	10 Dec 1959	M	[Ad]	Skin	ZD.1969.404	Cited Rahm,U. and Christiaensen,A.R. 1960: 216

Locality	Date of Collection	Sex	Age	Nature of Specimen	Accession no.	References, Remarks, etc.
Colobus (Colobus) angolensis cottoni Lydekker 1905						
[ZAIRE]						
By Zokwa, between Mahagi and Irumu, Ituri Forest, c.1°50′N 30°30′E	15 Jun 1905	M	Ad	Skin, Skull	PCM Congo	
" " "	15 Jun 1905	M	Juv	Skin, Skull	PCM Congo	
Congo Forest, 3000ft	—	M	Juv	Skin, Skull	ZD.1914.6.2.4	
Ituri Forest, Congo, c.1°25′N 28°45′E	—	I	[Ad]	Skin	PCM Congo	(*Colobus palliatus mawambicus*)
Ituri Forest, W. of L.Edward	1 Jul 1931	I	[Juv]	Skin	ZD.1981.1	
" " "	1 Jul 1931	I	[Juv]	Skin	ZD.1981.2	
Jugu [=Djugu], [=by Zokwa, between Mahagi and Irumu], Ituri Forest, E.Congo Free State, [c.1°50′N 30°30′E]	[15 Jun 1905]	M	Ad	Skin, Skull	ZD.1906.6.2.1	Holotype *Colobus palliatus cottoni* Lydekker 1905, Cited Lydekker,R. 1905: 432 Cited Schwarz,E. 1929: 596, Figured Colyer,J.F. 1936: 210, fig.252
Loama, R.Uele [nr.Angu], [3°21′N 24°22′E]	—	F	Ad	Skin, Skeleton	ZD.1907.7.8.9	Cited Schwarz,E. 1929: 595
Nr. Pemba, Ituri Forest, 1°25′N 28°45′E, 2400ft	27 Jul 1905	M	Ad	Skin, Skull	PCM Congo	Paratype *Colobus* (*Colobus*) *palliatus mawambicus* Matschie 1913, Cited Matschie,P. 1913a: 205
R.Ituri (right bank), 50 miles S.W. of Irumu, Belgian Congo, [c.1°N 29°15′E], 3500ft	1 May 1930	M	Juv	Skin, Skull	ZD.1930.11.11.8	
" " "	1 May 1930	M	Ad	Skin, Skull	ZD.1930.11.11.9	
R.Ituri (right bank), 50 miles S.W. of Irumu, [c.1°N 29°15′E], 3500ft	29 Apr 1930	M	Juv	Skin, Skull	ZD.1930.11.11.10	
" " "	29 Apr 1930	F	Juv	Skin, Skull	ZD.1930.11.11.11	
NO DATA						
—	—	F	[Ad]	Skin	ZD.1923.12.7.3	(*Colobus polykomos adolfifriederici*)
Colobus (Colobus) angolensis cottoni? Lydekker 1908						
[ZAIRE]						
Congo Forest	[Feb 1914]	M	Juv	Skull	ZD.1914.6.2.4	Correspondence Blackburn-Maze,C.J. (unpubl.)
Colobus (Colobus) angolensis palliatus Peters 1868						
KENYA						
Ganda, S.E.Coast, [3°13′S 40°04′E]	—	M	Ad	Skull	RCS(OM) A68.8	
" " "	—	F	Ad	Skull	RCS(OM) A68.81	
6 miles from Kwale on Tanga Road, Coast Prov., 1000ft	18 Jan 1963	F	Juv	Skin, Skull, Skeleton	ZD.1972.154	
Mrima Hill, Kwale, Coast Prov., 4°29′S 39°16′E, 900ft	23 Sep 1959	M	Ad	Skin, Skull	ZD.1972.157	
Mrima Hill, Kwale, Coast Prov., 4°29′S 39°16′E, 200ft	—	F	Ad	Skull, Skeleton	ZD.1972.158	
Mrima Hill, Kwale, Coast Prov., [4°29′S 39°16′E], 200ft	18 Sep 1959	M	[Juv]	Skin	ZD.1972.155	
1 mile W. of Mrima, Kwale, Coast Prov., 4°29′S 39°16′E, 200ft	22 Sep 1959	F	Ad	Skin, Skull	ZD.1972.156	
[KENYA]						
Takaungu, B.E.A., [3°42′S 39°51′E], 70ft	—	F	Ad	Skin, Skull	ZD.1902.1.2.19	Cited Schwarz,E. 1929: 598
" " "	—	F	Ad	Skin, Skull	ZD.1902.1.2.20	Cited Schwarz,E. 1929: 598, Cited Colyer,J.F. 1936: 28
" " "	—	F	Ad	Skin, Skull	ZD.1902.1.2.21	Cited Schwarz,E. 1929: 598
" " "	—	I	[Inf]	Skin	ZD.1902.1.2.22	" " " "
" " "	—	I	[Inf]	Skin	ZD.1902.1.2.23	" " " "
Wasin, Coast, British East Africa, [4°40′S 39°22′E]	—			Skull, Skeleton	CMZ E7311A	(*Colobus (Guereza) angolensis*)
[TANGANYIKA]						
Amani, Tanganyika Territory, [5°09′S 38°36′E], 5000ft	1935	[M]	[Juv]	Skin	ZD.1935.10.18.1	
" " "	1935	I	[Juv]	Skin	ZD.1935.10.18.2	
E.Usambara Mts., (Near Amani), Tanganyika Territory, [c.5°09′S 38°36′E], 3000ft	14 Feb 1937	M	Ad	Skin, Skull, Skeleton	ZD.1937.8.18.1	
" " "	17 Feb 1937	F	Ad	Skin, Skull, Skeleton	ZD.1937.8.18.2	
E.Usambara Mts., [Tanga], Tanganyika Territory, [c.5°S 38°40′E], 3000ft	26 Oct 1937	F	[Ad]	Skin	ZD.1938.4.21.1	
" " "	27 Oct 1937	F	Ad	Skin, Skull, Skeleton	ZD.1938.4.21.2	
" " "	29 Oct 1937	M	Ad	Skin, Skull	ZD.1938.4.21.3	
E.Usambara, [Tanga], Tanganyika Territories, [c.5°S 38°40′E], 3000-4000ft	[1937]	M	Ad	Skull	RCS(OM) A68.82	Cited Colyer,J.F. (unpubl.): 20r
" " "	[1937]	F	Inf	Skull	RCS(OM) A68.83	" " " "
" " "	[1937]	F	Ad	Skull	RCS(OM) G54.4	Figured Colyer,J.F. 1943: 212, fig.33
" " "	[1937]	F	Ad	Skull	RCS(OM) G68.8261	Cited Colyer,J.F. (unpubl.): 20r
" " "	[1937]	F	Ad	Skull	RCS(OM) G68.8262	Cited Colyer,J.F. (unpubl.): 20r, Figured Colyer,J.F. 1943: 135, fig.11
" " "	[1937]	F	Ad	Skull	RCS(OM) G68.8263	Cited Colyer,J.F. (unpubl.): 19
" " "	[1937]	F	Ad	Skull	RCS(OM) G68.8264	
" " "	[1937]	M	Ad	Skull	RCS(OM) G68.8265	

Locality	Date of Collection	Sex	Age	Nature of Specimen	Accession no.	References, Remarks, etc.
Colobus (Colobus) angolensis palliatus *continued*						
Mafwemera [=Mafwemiro] Forest, Mpwapwa Dist., [6°54′S 36°34′E], 6200ft	24 Sep 1947	M	Ad	Skin, Skull	ZD.1964.2022	
" " "	24 Sep 1947	F	Ad	Skin, Skull	ZD.1964.2023	
Middle reaches of R.Umba, Lushoto Dist., Tanganyika Territories, [c.4°30′S 38°45′E]	11 Mar 1949	M	[Juv]	Skin, Skull	ZD.1951.362	
" " "	11 Mar 1949	F	Ad	Skin, Skull	ZD.1951.363	
R.Pangani Valley, [c.5°25′S]	1879	M	Ad	Skin, Skull	ZD.1879.11.12.1	Cited Gray,J.E. (unpubl.): 1722, Cited Schwarz,E. 1929: 598, Figured Colyer,J.F. 1936: 215, fig.257
Tanganyika Territory	—	I	[Juv]	Skin	ZD.1964.2024	
Colobus (Colobus) angolensis prigoginei Verheyen 1959						
NO SPECIMENS						
—	—	—	—	—	—	
Colobus (Colobus) angolensis ruwenzorii Thomas 1901						
UGANDA						
?Bwamba Forest, R.Semliki, W. of Ruwenzori	—	[M]	[Ad]	Skin	ZD.1927.7.3.5	
Kakuka, N. spur of Ruwenzori Mts., 0°35′N 29°59′E	Jan 1946	M	Ad	Skull	ZD.1969.374	Background Haddow,A.J. *et al.* 1951: 189-224
Kakuka, N. spur of Ruwenzori Mts., 0°35′N 30°04′E	Jan 1946	F	Ad	Skull	ZD.1969.373	" " " "
Kakuka, W. side of N. spur of Mt.Ruwenzori, Bwamba County, [c.0°35′N 30°E], 8500ft	—	M	Ad	Skin, Skull	ZD.1949.710	Background Haddow,A.J. *et al.* 1947: 677-700
" " "	—	M	Ad	Skin, Skull	ZD.1949.711	
N. spur of Mt.Ruwenzori, Toro Dist., [7000-8000ft]	[1951]	[M]	Inf	Skull	RCS(OM) A68.63	Cited Colyer,J.F. (unpubl.)a, Background Haddow,A.J. *et al.* 1951: 189-224, Correspondence Haddow,A.J. (unpubl.)
" " "	[1951]	M	Inf	Skull	RCS(OM) A68.631	" " " "
" " "	[1951]	M	Ad	Skull	RCS(OM) A68.632	" " " "
" " "	—	M	Ad	Skull	RCS(OM) G68.765	" " " "
[UGANDA/ZAIRE BORDER]						
Ruwenzori, ["probably at high altitudes"]	—	[M]	[Ad]	Skin	ZD.1901.8.9.13	Holotype *Colobus ruwenzorii* Thomas 1901, Cited Thomas,O. 1901a: 85, Cited Schwarz,E. 1929: 596, Cited Moreau,R.E. *et al.* 1946: 406-407
" " "	—	[M]	[Ad]	Skin	ZD.1901.8.9.14	Paratype *Colobus ruwenzorii* Thomas 1901, Cited Thomas,O. 1901a: 85, Cited Schwarz,E. 1929: 596, Cited Moreau,R.E. *et al.* 1946: 406-407
W.Ruwenzori	Feb 1930	[M]	[Ad]	Skin	ZD.1930.11.11.12	
Colobus (Colobus) angolensis sharpei Thomas 1902						
[MALAWI]						
N.Nyasaland	25 Oct 1897	M	[Ad]	Skin	ZD.1897.10.1.292	
Nyasaland	—	I	[Ad]	Skin	ZD.1892.2.6.1	
[MALAWI-TANGANYIKA BORDER]						
Nyasa-Tanganyika Plateau	—	I	[Ad]	Skin	ZD.1896.10.25.1	
[TANGANYIKA]						
Mt. ranges to E. of Fyfe station, [near Malawi/Tanganyika Border]	Jul 1896	I	[Inf]	Skin	ZD.1897.10.1.7	Cited Thomas,O. 1898: 927, Cited Schwarz,E. 1929: 597
" " "	Jul 1896	I	[Inf]	Skin	ZD.1897.10.1.8	" " " "
Mt.Rungwe, Tanganyika Territory, [9°08′S 33°38′E], c.7500ft	22 Nov 1949	F	Ad	Skin, Skull	ZD.1951.364	
Nkuka Forest, Rungwe Mt., [c.9°08′S 33°38′E]	—	F	Ad	Skin, Skull	ZD.1897.7.3.1	Holotype *Colobus sharpei* Thomas 1902, Cited Thomas,O. 1902: 118, Cited Schwarz,E. 1929: 597, Figured Colyer,J.F. 1936: 30-31, figs.35,37, Cited Ansell,W.F.H. 1959: 170
Colobus (Colobus) angolensis Sclater 1860						
TANGANYIKA						
Iringa?, Tanganyika Territory	—	[F]	[Ad]	Skin	ZD.1923.3.4.1	

COLOBUS (COLOBUS) GUEREZA Rüppell, 1835

Guereza, Abyssinian Colobus, Magistrate Colobus
(8 subspecies)

Taxonomic notes *Colobus guereza* Rüppell, 1835 replaces *Colobus abyssinicus* (Oken, 1816) as the valid name for the species since Oken's work was placed on the Official List of Rejected Works in Zoology (Opinion 417, ICZN, 1956). Both Schwarz (1929) and Rahm (1970) recognized 9 subspecies: *guereza, caudatus, dodingae, gallarum, kikuyuensis, matschiei, occidentalis, percivali* and *uellensis,* while Dandelot (1971) reduced the number to 6 by synonymising *percivali* with *guereza* and *uellensis* with *occidentalis* (and also, by inference, *dodingae*). The taxonomic arrangement of all three authors was based solely on external characters such as the length of the hair forming the white mantle and variations in the tail tuft. Following a craniometric study, Hull (1979) re-affirmed Rahm's arrangement, except that he synonymised *uellensis* with *occidentalis*. BM(NH) specimens confirm this view which has been adopted in the catalogue (Jenkins, personal communication, 1980).

Range From E. Nigeria through Cameroon, Congo (Brazzaville), N. Zaire and S. Sudan to Ethiopia; thence southwards to Kenya, Uganda and NW. Tanganyika (SE. to Mt Kilimanjaro).

Dimensions Adult specimens in the museum collection have the following field measurements (see p. 98 for comparative table):

h & b	20 ♂♂	615	(543-699)	:	22 ♀♀	576	(521-673)
tail	20 ♂♂	667	(521-826)	:	22 ♀♀	687	(528-797)
wt	8 ♂♂	9327	(7970-11 790)	:	12 ♀♀	7826	(5448-10 230)

Key to subspecies of *Colobus guereza*, adapted from Schwarz (1929), Rahm (1970) and Dandelot (1971)

Subspecies	White Mantle	Tail Base	White Tail Tuft
occidentalis (E. Nigeria, Cameroon and Zaire, N. of R. Zaire)	Short, just covering base of tail	Black	Sparse to full; 1/5 to 1/3 of tail length
dodingae (S. Sudan, N. Uganda)	Long, covering 1/4 to 1/3 of tail	Black	Full; 1/3 of tail length
guereza (W. Ethiopia)	Long, covering up to 1/3 of tail	Mixture of black and white hairs	Fairly full; 1/2 of tail length
gallarum (E. Ethiopia)	Long, covering base of tail; lateral fringe fairly short	Black	Fairly full; 1/4 to 1/2 tail length
percivali (Mt Warges, Kenya)	Long, covering 1/4 of tail	Black	Fairly full; 2/3 of tail length
matschiei (Kenya, W. of Rift Valley; Serengeti, NW. Tanganyika)	Long, covering 1/4 of tail	Black	Fairly full; 1/3 to 1/2 tail length
kikuyuensis (Kenya, E. of Rift Valley)	Very long, covering 1/3 to 1/2 of tail	Mixture of black and white hairs grading into black at base	Long and full; 2/3 of tail length
caudatus (Mt Kilimanjaro, Mt Meru and Kahe forest, N. Tanganyika)	Extremely long posteriorly, covering 1/2 of tail	Black or mixture of black and white hairs	Very long and full; 1/2 to almost entire length of tail

Colobus guereza guereza Rüppell, 1835

Principal synonyms *Guereza rueppellii* Gray, 1871:19; *Colobus abyssinicus poliurus* Thomas, 1901:800; *Colobus (Guereza) poliurus managaschae* Matschie, 1913:54.

Type localities Of *guereza,* S. and W. Abyssinia, restricted by Mertens (1925:37) to Damot (S. of L. Tsana within the arc of the R. Abbai in the province of Gojjam); of *rueppellii,* the same as for *guereza* (Mertens 1925:37); of *poliurus,* R. Omo, N. of L. Rudolf (= Turkana), Ethiopia; of *managaschae,* Managasha Forest, W. of Addis Ababa, in the region of the tributaries of the upper R. Awash, Ethiopia.

Colobus guereza caudatus Thomas, 1885

Principal synonym *Colobus albocaudatus* Lydekker, 1906:328, substitute name for *caudatus.*

Type locality Useri, NE. flank of Mt Kilimanjaro at 3000ft.

Colobus guereza dodingae Matschie, 1913

Type locality Dodinga (=Didinga) Mts (*c.* 4°10′N 33°42′E) at 5650ft, near tributaries of R. Kedef, Sudan.

Colobus guereza gallarum Neumann, 1902

Type locality Mountains in the headwaters of the Webbi Shebeli: the Gara Mulata near Harer (9°05′N 41°45′E) and the Djaffa (=Gasgar) Mts (7°40′N 40°10′E) in Arussi, Ethiopia.

Colobus guereza kikuyuensis Lönnberg, 1912

Principal synonyms *Colobus (Guereza) caudatus thikae* Matschie, 1913:56; *Colobus (Guereza) caudatus laticeps* Matschie, 1913:57.

Type localities Of *kikuyuensis,* Escarpment Station (1°01′S 36°37′E), Kenya; of *thikae,* W. flank of Mt Kenya; of *laticeps,* W. flank of Mt Kenya, at 7400ft.

Colobus guereza matschiei Neumann, 1899

Principal synonyms *Colobus abyssinicus roosevelti* Heller, 1913:5; *Colobus abyssinicus elgonis* Granvik, 1925:4, pl. 1, figs. 1-3.

Type localities Of *matschiei,* Kwa Kitoto (Kavirondo), Ugowe Bay, L. Victoria Nyanza, identified by Moreau *et al.* (1946:403) as being in Nyando Valley near Muhoroni (0°09′S 35°12′E), which is on the railway about 30 miles E. of Kisumu, Kenya; of *roosevelti,* Mau Forest, near Enjoro (= Njoro, 0°20′S 35°36′E), Kenya; of *elgonis,* eastern slopes of Mt Elgon, Kenya.

Colobus guereza occidentalis (Rochebrune, 1886-87)

Principal synonyms *Colobus (Guereza) matschiei uellensis* Matschie, 1913:47; *Colobus (Guereza) matschiei ituricus* Matschie, 1913:48; *Colobus (Guereza) matschiei dianae* Matschie, 1913:49; *Colobus (Guereza) matschiei brachychaites* Matschie, 1913:53; *Colobus abyssinicus terrestris* Heller, 1913:7; *Colobus (Guereza) escherichi* Matschie, 1914:342; *Colobus occidentalis rutschuricus* Lorenz, 1914:508.

Type localities Of *occidentalis,* Noki, Upper Congo. The suggestion by Schwarz (1929:590) that this may be Noqui on L. bank of lower Congo in Angola is considered improbable by Schouteden (1944:31) since Rochebrune stated explicitly 'Upper Congo' and in any case *Colobus* are not thought to occur in the area; of *uellensis,* Uele; of *ituricus,* Ituri; of *dianae,* Kissenge (=Kisenyi, near mouth of R. Niamweru, 0°19′S 29°51′E, on NE. shore of L. Edward, Uganda, according to Moreau *et al.* (1946:406); of *brachychaites,* by Modi (*c.* 4°N 31°45′E), between Kaya and Dufile (=Laropi), in the former Lado Enclave, Sudan; of *terrestris,* Rhino Camp (2°58′N 31°24′E), Lado Enclave, Uganda; of *escherichi,* Gombe (=?N'Gombe, 0°12′N 17°06′E), below Ikelemba, Sangha R., Congo (Brazzaville); of *rutschuricus,* R. Sassa, NE. edge of Rutshuru Plains, presumed by Moreau *et al.* (1946:404) to be R. Ishasha which here forms the boundary between Uganda and Zaire, SE. of L. Edward.

Colobus guereza percivali Heller 1913

Type locality Mt Varagess (=Warges), Matthews Range, Kenya.

Locality	Date of Collection	Sex	Age	Nature of Specimen	Accession no.	References, Remarks, etc.
Colobus (Colobus) guereza guereza Rüppell 1835						
ETHIOPIA						
Dello (Masslo), Dumal Valley, Bale Province, 6°40′N 39°45′E, 4900ft	—	—	—	Skin	ZD.1964.858	
" " "	—	—	—	Skin	ZD.1964.859	
" " "	—	—	—	Skin	ZD.1964.860	
N.W. of Lake Abaya, S.Ethiopia, [6°40′N 38°E]	18 Aug 1964	M	Ad	Skin, Skull	ZD.1964.2178	
Piccolo Abbai, N.E. of Dangila, N.Ethiopia, [11°02′N 37°08′E]	2 Sep 1964	F	Ad	Skin, Skull	ZD.1964.2179	
[ETHIOPIA]						
Abyssinia	—	M	Ad	Skin, Skull	ZD.1842.8.15.14	Paralectotype *Colobus guereza* Rüppell 1835, Cited Rüppell,W.P.E.S. 1835: 1, pl.1, Cited Mertens,R. 1925: 37
" " "	—	M	Ad	Skull	ZD.1848.1.27.47	Listed Gray,J.E. (unpubl.): 414
" " "	—	F	Ad	Skull	ZD.1919.6.1.1	Cited Colyer,J.F. 1936: 212
" " "	—	M	Ad	Skull	ZD.1922.3.4.1	
" " "	—	M	Ad	Skull	ZD.1922.3.4.2	
" " "	—	[F]	Ad	Skull	ZD.1922.3.4.3	
" " "	—	[F]	Ad	Skull	ZD.1922.3.4.4	
Dodgit, [W.Shoa], Abyssinia	26 Jun 1902	M	Ad	Skin, Skull	ZD.1902.9.9.1	Cited Thomas,O. 1903a: 308
" " "	26 Jun 1902	M	Ad	Skin, Skull	ZD.1902.9.9.2	Cited Thomas,O. 1903a: 308, Cited Colyer,J.F. 1936: 212
Fatam River, Big Abbai, 100 miles S. of Lake Tana, Abyssinia, [c.10°16′N 36°53′E], 6000ft	17 Mar 1927	M	Ad	Skin, Skull	ZD.1937.2.24.1	
Gumbela, Upper Sobat, 8°11′N 34°51′E	Jan 1900			Skin	ZD.1900.11.24.1	Cited Schwarz,E. 1929: 592
Halea Webbi, Bale, Abyssinia	20 Oct 1934	F	Juv	Skin, Skull	ZD.1936.5.20.1	
Managasha Forest, [Shoa], Abyssinia, c.9°N 38°25′E	6 Feb 1900	M	Ad	Skin, Skull	PCM Abys.I	Paratype *Colobus* (*Guereza*) *poliurus managaschae* Matschie 1913, Cited Matschie,P. 1913: 54
" " "	7 Feb 1900	F	Ad	Skin, Skull	PCM Abys.I	" " " "
" " "	10 Feb 1900	M	Ad	Skin, Skull	PCM Abys.I	Holotype *Colobus* (*Guereza*) *poliurus managaschae* Matschie 1913, Cited Matschie,P. 1913: 54, Cited Schwarz,E. 1929: 592
Metcha Forest, Shoa, 11,000ft	21 Nov 1906	M	Ad	Skin, Skull	ZD.1906.11.1.5	
R.Omo, [5°N 36°E]	—	M	[Ad]	Skin	ZD.1900.11.7.4	
R.Omo, [N. of] Lake Rudolf, [5°N 36°E]	21 Dec 1899	M	Ad	Skin, Skull	ZD.1900.11.7.2	Paratype *Colobus abyssinicus poliurus* Thomas 1901, Cited Thomas,O. 1901: 800
R.Omo, [N. of] L.Rudolf, [5°N 36°E]	21 Dec 1899	F	Ad	Skin, Skull	ZD.1900.11.7.3	Holotype *Colobus abyssinicus poliurus* Thomas 1901, Cited Thomas,O. 1901: 800
Welega Province, W.Abyssinia	—	M	Ad	Skull	RCS(OM) A68.44	
" " "	—	F	Ad	Skull	RCS(OM) A68.441	
" " "	—	M	Ad	Skull	RCS(OM) A68.442	
" " "	—	F	Ad	Skull	RCS(OM) G51.9	
Western Sidamo, Abyssinia, [c.7°40′N 39°45′E]	—	M	Ad	Skin, Skull	ZD.1909.6.1.2	Cited Schwarz,E. 1929: 592
SUDAN						
Towat Boma, S.E.Sudan, [c.6°12′N 34°25′E], 4000ft	8 Feb 1939	M	Ad	Skin, Skull	ZD.1940.11	
NO DATA						
—	18 Dec 1914	M	Ad	Skin, Skull	ZD.1939.1160	
Colobus (Colobus) guereza guereza? Rüppell 1835						
ETHIOPIA						
—	—		Juv	Skull	ZD.1976.12	
Colobus (Colobus) guereza caudatus Thomas 1885						
[KENYA/TANGANYIKA BORDER]						
Useri, N.E. flank of Mt.Kilimanjaro, 3000ft	28 Oct 1884	[F]	[Ad]	Skin	ZD.1885.1.17.3	Holotype *Colobus guereza caudatus* Thomas 1885, Figured Thomas,O. 1885: 219, pl.12, Cited Schwarz,E. 1929: 594
TANGANYIKA						
Kahe, [Kilimanjaro], [3°29′S 37°26′E]	—	M	Ad	Skull	DLPA 75.0.1	
" " "	—	F	Ad	Skull	DLPA 75.0.2	
[TANGANYIKA]						
E.Kilimanjaro, Tanganyika Territory, 6500ft	—	M	Ad	Skull	RCS(OM) G68.8237	Cited Colyer,J.F. (unpubl.): 20r
" " "	—	F	Ad	Skull	RCS(OM) G68.8238	" " " "
Handajega, Mbalangeti River Forest, Serengeti Closed Reserve, Tanganyika Territory, [2°17′S 34°04′E]	19 Aug 1932	I	[Ad]	Skin	ZD.1935.3.14.1	
Kilimanjaro, [2°45′-3°20′S 37°55′-38°40′E]	—	M	Ad	Skull	ZD.1892.10.18.5	
" " "	—	F	Ad	Skull	ZD.1892.10.18.6	
" " "	—	M	Ad	Skull	ZD.1892.10.28.6	
" " "	—	M	Ad	Skull	ZD.1892.10.28.7	

Colobus (Colobus) guereza caudatus *continued*

Locality	Date of Collection	Sex	Age	Nature of Specimen	Accession no.	References, Remarks, etc.
Kilimanjaro, Tanganyika Territory, [2°45′-3°20′S 37°55′-38°40′E]	—	I	[Ad]	Skin	ZD.1951.366	
" " "	—	[F]	[Ad]	Skin	ZD.1951.367	
" " "	—	I	[Ad]	Skin	ZD.1951.368	
Kilimanjaro?, Tanganyika Territory, [2°45′-3°20′S 37°55′-38°40′E]		I	[Ad]	Skin	ZD.1959.615	
" " "	—	I	[Ad]	Skin	ZD.1959.616	
" " "	—	I	[Juv]	Skin	ZD.1959.617	
" " "	—	[F]	[Ad]	Skin	ZD.1959.618	
" " "	—	I	[Ad]	Skin	ZD.1959.619	
" " "	—	[F]	[Ad]	Skin	ZD.1959.620	
" " "	—	I	[Ad]	Skin	ZD.1959.621	
" " "	—	I	[Juv]	Skin	ZD.1959.622	
" " "	—	[F]	[Ad]	Skin	ZD.1959.623	
" " "	—	[F]	[Ad]	Skin	ZD.1959.624	
" " "	—	I	[Inf]	Skin	ZD.1959.625	
" " "	—	I	[Juv]	Skin	ZD.1959.626	
" " "	—	I	[Ad]	Skin	ZD.1959.627	
" " "	—	I	[Ad]	Skin	ZD.1959.628	
" " "	—	I	[Juv]	Skin	ZD.1959.629	
" " "	—	I	[Ad]	Skin	ZD.1959.630	
" " "	—	I	[Juv]	Skin	ZD.1959.631	
" " "	—	I	[Juv]	Skin	ZD.1959.632	
" " "	—	I	[Ad]	Skin	ZD.1959.633	
Marangu, Kilimanjaro, [3°18′S 37°32′E], 7000ft	—	F	Ad	Skin, Skull, Skeleton	ZD.1897.3.14.1	
" " "	Sep	M	Ad	Skin, Skull	ZD.1897.3.14.2	Figured Colyer,J.F. 1936: 34-35, 593, figs.42,811, Cited Schwarz,E. 1929: 594
" " "	—	F	Ad	Skull	ZD.1897.3.14.3	
Mt.Kilimanjaro, [2°45′-3°20′S 37°55′-38°40′E]	—	F	Ad	Skull	ZD.1889.3.8.1	
" " "	—	F	Ad	Skull	ZD.1889.3.8.2	
Mt.Kilimanjaro, Tanganyika Territory, [3°04′S 37°22′E]	—	M	Ad	Skull	RCS(OM) G169.3	Cited Colyer,J.F. (unpubl.): 20r
[Mt.] Kilimanjaro	—	[M]	[Ad]	Skin	ZD.1939.3474	
Nr. Moshi, Kilimanjaro, [Kilimanjaro] Tanganyika Territory, [c.3°21′S 37°19′E]	—	M	Ad	Skull	RCS(OM) A68.62	Cited Colyer,J.F. (unpubl.): 20r
" " "	—	M	Ad	Skull	RCS(OM) A68.621	" " " "
" " "	—	M	Ad	Skull	RCS(OM) G68.824	" " " "
" " "	—	F	Ad	Skull	RCS(OM) G68.8241	" " " "
" " "	—	F	Ad	Skull	RCS(OM) G68.8242	" " " "
" " "	—	M	Ad	Skull	RCS(OM) G68.8243	" " " "
Old Mochi [=Old Moshi], Kilimanjaro, [3°18′S 37°25′E], 3000ft	—	I	[Ad]	Skin	ZD.1918.8.11.1	Cited Schwarz,E. 1929: 594
S.Kilimanjaro, Tanganyika Territory, 2500ft	—	M	Ad	Skull	RCS(OM) A68.6	Cited Colyer,J.F. (unpubl.): 20r
" " "	—	M	Ad	Skull	RCS(OM) A68.61	" " " "
" " "	—	F	Ad	Skull	RCS(OM) G68.683	" " " "
" " "	—	F	Ad	Skull	RCS(OM) G68.6831	" " " "
W.Kilimanjaro, Tanganyika Territory, 6000ft	—	F	Ad	Skull	RCS(OM) G68.8232	" " " "
" " "	—	M	Ad	Skull	RCS(OM) G68.8233	" " " "
" " "	—	M	Ad	Skull	RCS(OM) G68.8234	" " " "
" " "	—	M	Ad	Skull	RCS(OM) G68.8235	" " " "
W.Kilimanjaro, Tanganyika Territory, 7000ft	—	F	Ad	Skull	RCS(OM) G68.8236	" " " "
[W.Kilimanjaro], [5000ft]	—	F	Ad	Skull	RCS(OM) G68.672	Cited Colyer,J.F. (unpubl.): 20r, 307r
" " "	—	F	Ad	Skull	RCS(OM) G68.673	" " " "
" " "	—	M	Ad	Skull	RCS(OM) G68.674	" " " "

Colobus (Colobus) guereza dodingae Matschie 1913

SUDAN

Locality	Date of Collection	Sex	Age	Nature of Specimen	Accession no.	References, Remarks, etc.
Issore, Imatong Mountains, S.Sudan, [3°55′N 32°48′E], 4800ft	17 Feb 1939	M	Ad	Skin, Skull	ZD.1940.5	
Issore, Imatong Mountains, S.Sudan, [3°55′N 32°48′E]	—		Juv	Skin	ZD.1940.6	
Kipia, Imatong Mountains, S.Sudan, [3°57′N 32°50′E], 8800ft	10 Jan 1939	F	Ad	Skin, Skull	ZD.1940.1	
" " "	10 Jan 1939	M	Inf	Skin, Skull	ZD.1940.2	
Labone [=?Laboni], Imatong Mountains, S.Sudan, [?3°49′N 32°46′E], 3800ft	29 Jan 1939	M	Ad	Skin, Skull	ZD.1940.4	
Lomuleng, S.Sudan, 3°47′N 33°01′E, [7200ft]	16 Mar 1933	F	Ad	Skin, Skull	PCM Sudan I	(*Colobus polykomos dodingae*)
" " "	16 Mar 1933	F	Inf	Skin, Skull, Skeleton	PCM Sudan I	" " " "
" " "	16 Mar 1933	M	Ad	Skin, Skull, Skeleton	PCM Sudan I	" " " "
Lomuleng, Camp 5, [3°47′N 33°01′E], 7200ft	1 Feb 1933	F	Ad	Skin, Skull	ZD.1934.5.1.1	
Lotti Forest, Imatong Mountains, S.Sudan, 4°01′N 32°31′E, 2200ft	18 Jan 1938	M	Ad	Skin, Skull	ZD.1940.3	
Mt.Lotuke, Didinga Mountains, S.Sudan, [4°10′N 33°46′E], 6000-7000ft	4 Mar 1939	[F]	Ad	Skin, Skull	ZD.1940.8	
" " "	4 Mar 1939	F	Ad	Skin, Skull	ZD.1940.9	
" " "	4 Mar 1939	M	Juv	Skin	ZD.1940.10	

Locality	Date of Collection	Sex	Age	Nature of Specimen	Accession no.	References, Remarks, etc.
Colobus (Colobus) guereza dodingae *continued*						
Talanga Forest, Imatong Mountains, S.Sudan, 4°N 32°44′E	—	F	Ad	Skin, Skull	ZD.1940.7	
[SUDAN]						
Banks of R.Kinatey, N.Uganda, c.4°15′N 32°30′E	2 May 1903	M	Ad	Skin, Skull	PCM Uganda	Paratype *Colobus* (*Guereza*) *matschiei dodingae* Matschie 1913, Cited Matschie,P. 1913: 52, Cited Schwarz,E. 1929: 592
Nr. Kilio, N.Uganda, c.4°15′N 32°30′E, 2150ft	2 May 1903	M	Ad	Skin, Skull	PCM Uganda	,, ,, ,, ,,
,, ,, ,,	2 May 1903	F	Ad	Skin, Skull	PCM Uganda	,, ,, ,, ,,
[S.W.] Dodinga [=Didinga Mts.], N.Uganda, c.4°N 33°40′E	22 Mar 1903	F	Ad	Skin, Skull	PCM Uganda	,, ,, ,, ,,
[S.W.] Dodinga [=Didinga Mts.], N.Uganda, c.4°N 33°40′E, 5650ft	21 Mar 1903	M	Ad	Skin, Skull	PCM Uganda	Holotype *Colobus* (*Guereza*) *matschiei dodingae* Matschie 1913, Cited Matschie,P. 1913: 52, Cited Schwarz,E. 1929: 592
,, ,, ,,	21 Mar 1903	F	Ad	Skin, Skull	PCM Uganda	Paratype *Colobus* (*Guereza*) *matschiei dodingae* Matschie 1913, Cited Matschie,P. 1913: 52, Cited Schwarz,E. 1929: 592
[S.W.] Dodinga [=Didinga Mts.], N.Uganda, c.4°N 33°40′E, 5850ft	22 Mar 1903	F	Ad	Skin, Skull	PCM Uganda	,, ,, ,, ,,
,, ,, ,,	22 Mar 1903	F	Ad	Skin, Skull	PCM Uganda	,, ,, ,, ,,
,, ,, ,,	22 Mar 1903	F	Ad	Skin, Skull	PCM Uganda	,, ,, ,, ,,
,, ,, ,,	22 Mar 1903	M	Ad	Skin, Skull	PCM Uganda	,, ,, ,, ,,
Colobus (Colobus) guereza gallarum Neumann 1902						
[ETHIOPIA]						
Harer, Abyssinia	—	F	Ad	Skull	RCS(OM) G36.18	Associated with next specimen (1951.713)
,, ,, ,,	—	F	Ad	Skin	ZD.1951.713	
Koulobi [=K′ulubi], [Harer], Abyssinia, [9°28′N 41°40′E]	—	M	Ad	Skull	RCS(OM) G169.31	
Leucote, Sahatu Mts., Arussi, Abyssinia, c.8°N 39°E	11 Mar 1924	F	Ad	Skin, Skull	PCM Abys.II	
,, ,, ,,	11 Mar 1924	F	Juv	Skin, Skull	PCM Abys.II	Figured Colyer,J.F. 1936: 213, fig.255
Colobus (Colobus) guereza kikuyuensis Lönnberg 1912						
KENYA						
Aberdare Range, [c.0°10′-40′S 36°30′-50′E]	—	F	Juv	Skull	RCS(OM) A68.53	
,, ,, ,,	—	M	Juv	Skull	RCS(OM) A68.531	
,, ,, ,,	—	F	Ad	Skull	RCS(OM) A68.532	
,, ,, ,,	—	M	Ad	Skull	RCS(OM) G68.821	
,, ,, ,,	—	M	Ad	Skull	RCS(OM) G68.8211	
,, ,, ,,	—	F	Ad	Skull	RCS(OM) G169.32	
Kedong Escarpment, [c.1°10′S 36°30′E]		I	Inf	Skull	RCS(OM) A68.5	
,, ,, ,,	—	M	Inf	Skull	RCS(OM) A68.51	
,, ,, ,,	—	F	Juv	Skull	RCS(OM) A68.511	
,, ,, ,,	—	M	Ad	Skull	RCS(OM) A68.512	
,, ,, ,,	—	F	Ad	Skull	RCS(OM) A68.513	
,, ,, ,,	—	F	Ad	Skull	RCS(OM) G68.79	
,, ,, ,,	—	F	Juv	Skull	RCS(OM) G68.8215	
Kedong Escarpment, Central Province, [1°10′S 36°30′E], 7000-8000ft	9 Jan 1962	F	Juv	Skin, Skull, Skeleton	ZD.1972.136	
Kijabe, [0°56′S 36°34′E]	—	M	Ad	Skull	RCS(OM) A68.55	
,, ,, ,,	—	F	Ad	Skull	RCS(OM) A68.551	
,, ,, ,,	—	M	Ad	Skull	RCS(OM) G68.822	
,, ,, ,,	—	F	Ad	Skull	RCS(OM) G68.8221	
,, ,, ,,	—	M	Ad	Skull	RCS(OM) G68.8222	
,, ,, ,,	—	M	Ad	Skull	RCS(OM) G68.8223	
Lari	—	F	Ad	Skull	ZD.1900.1.3.11	
Mt.Kenya, [0°10′S 37°19′E]	—	[F]	[Ad]	Skin	ZD.1937.8.10.1	Cited Hull,D.B. 1978: 22
,, ,, ,,	—	M	Ad	Skull	ZD.1949.114	
,, ,, ,,	—	M	Ad	Skull	RCS(OM) A68.56	
,, ,, ,,	—	F	Ad	Skull	RCS(OM) G11.611	Cited Colyer,J.F. 1943: 138
,, ,, ,,	—	F	Ad	Skull	RCS(OM) G68.8216	Cited Colyer,J.F. 1943: 111, 137
,, ,, ,,	—	M	Ad	Skull	RCS(OM) G68.8217	,, ,, ,, ,,
Mt.Kenya, Nanyuki Dist., [0°10′S 37°19′E]	—	M	Ad	Skull	RCS(OM) A68.52	
Mt.Kenya, [0°10′S 37°19′E], 6700ft	14 Aug 1927	M	Juv	Skin, Skull	ZD.1934.10.28.1	
,, ,, ,,	14 Aug 1927	F	Juv	Skin, Skull	ZD.1934.10.28.2	
,, ,, ,,	1927	I	[Ad]	Skin	ZD.1934.10.28.3	
Mt.Kenya, [0°10′S 37°19′E], 7000ft	17 Sep 1899	M	Ad	Skin, Skull	ZD.1900.2.1.1	
Mt.Kenya, [0°10′S 37°19′E], 8000ft	17 Oct 1899	M	Ad	Skin, Skull	ZD.1900.2.1.2	Cited Thomas,O. 1900: 174
,, ,, ,,	14 Oct 1899	F	Ad	Skin, Skull	ZD.1900.2.1.3	Figured Colyer,J.F. 1936: 28, 213, fig.254, Cited Thomas,O. 1900: 174
N.Kinangop, Aberdare Mts., Central Province, [c.0°37′S 36°42′E], 9500ft	20 Nov 1962	M	Ad	Skin, Skull, Skeleton	ZD.1972.134	
,, ,, ,,	—	[M]	Ad	Skull, Skeleton	ZD.1972.135	
20 miles N. of Thompson's Falls [= Nyahururu Falls], [c.0°02′N 36°22′E]	—	I	[Ad]	Skin	ZD.1949.525	

Locality	Date of Collection	Sex	Age	Nature of Specimen	Accession no.	References, Remarks, etc.
Colobus (Colobus) guereza kikuyuensis *continued*						
Nanyuki, Mt.Kenya, [0°01′N 37°05′E], 8000-9000ft	—	[M]	[Ad]	Skin	ZD.1938.3.28.1	
,, ,, ,,	—	M	[Ad]	Skin	ZD.1938.3.28.2	
Nanyuki Dist.	—	M	Ad	Skull	RCS(OM) A68.522	
Nyeri, [0°26′S 36°56′E]	—	M	Ad	Skin, Skull	ZD.1904.11.5.1	Figured Colyer,J.F. 1936: 208, fig.249
Roromo [on Kikuyu scarp], [c.1°S 36°25′E]	Dec 1899	M	Ad	Skin, Skull	ZD.1900.6.21.52	
Solai, Mt.Kenya, [0°06′S 37°14′E], 8000ft	[8 Dec 1910]	M	Ad	Skin, Skull	ZD.1911.4.7.168	
,, ,, ,,	9 Dec 1910	M	Ad	Skin, Skull	ZD.1911.4.7.169	
,, ,, ,,	10 Dec 1910	M	Ad	Skin, Skull	ZD.1911.4.7.170	
,, ,, ,,	10 Dec 1910	M	Ad	Skin, Skull	ZD.1911.4.7.171	
,, ,, ,,	[6 Dec 1910]	F	Ad	Skin, Skull	ZD.1911.4.7.172	
,, ,, ,,	8 Dec 1910	F	Juv	Skin, Skull	ZD.1911.4.7.173	
,, ,, ,,	8 Dec 1910	F	Ad	Skin, Skull	ZD.1911.4.7.174	
,, ,, ,,	6 Dec 1910		Inf	Skin, Skull	ZD.1911.4.7.175	
Uplands, [1°03′S 36°39′E]	—	M	Ad	Skull	RCS(OM) A68.54	
,, ,, ,,	—	F	Ad	Skull	RCS(OM) G51.91	Cited Colyer,J.F. 1943: 235
,, ,, ,,	—	F	Ad	Skull	RCS(OM) G68.8212	
,, ,, ,,	—	F	Ad	Skull	RCS(OM) G68.8213	Cited Colyer,J.F. 1943: 159
,, ,, ,,	—	M	Ad	Skull	RCS(OM) G68.8214	
[KENYA]						
Fort Hall, B.E.A., [0°43′S 37°11′E]	—	M	Ad	Skull, Skeleton	ZD.1904.12.6.1	
,, ,, ,,	—	F	Juv	Skull, Skeleton	ZD.1904.12.6.2	
,, ,, ,,	—	[I]	Juv	Skull	ZD.1904.12.6.3	
Ithanga Hill, Kenya District, B.E.A., [c.1°S 37°25′E]	—	[F]	Ad	Skull	ZD.1917.8.21.1	
Mt.Kenya, 0°22′S 27°18′E, 6700ft	7 Aug 1927	M	Foetus	Wet,cadaver	ZD.1934.10.28.42	
Mutaragwa [=?Mutarakwa], Aberdares, B.E.A., [?0°47′S 35°14′E], 9000ft	—	M	Ad	Skull	ZD.1910.5.3.204	
Solai, Mt.Kenya, [0°06′S 37°14′E]	—	[F]	Ad	Skull	ZD.1911.12.8.1	
W. side of Mt.Kenya, Brit. East Africa, c.0°20′S 37°20′E, 7500ft	29 Mar 1902	F	Ad	Skin, Skull	PCM Uganda	Paratype *Colobus* (*Guereza*) *caudatus thikae* Matschie 1913, Cited Matschie,P. 1913: 56
,, ,, ,,	31 Mar 1902	F	Ad	Skin, Skull	PCM Uganda	,, ,, ,, ,,
W. side of Mt.Kenya, Brit. East Africa, c.0°20′S 37°20′E, 7400ft	1 Apr 1902	M	Ad	Skin, Skull	PCM Uganda	,, ,, ,, ,,
,, ,, ,,	1 Apr 1902	M	Ad	Skin, Skull	PCM Uganda	Holotype *Colobus* (*Guereza*) *caudatus laticeps* Matschie 1913, Cited Matschie,P. 1913: 57, Cited Schwarz,E. 1929: 594
NO DATA						
—	—	M	Juv	Skull, Skeleton	ZD.1972.160	
Colobus (Colobus) guereza kikuyuensis? Lönnberg 1912						
KENYA						
Central Province	—	[F]	Ad	Skull	ZD.1972.137	
Colobus (Colobus) guereza matschiei Neumann 1899						
KENYA						
Kaimosi, Kapsabet Road, Nandi, [0°08′N 34°47′E], 5000ft	28 Sep 1962	M	Ad	Skin, Skull, Skeleton, Radiograph of penis	ZD.1972.139	
,, ,, ,,	13 Oct 1962	F	Ad	Skin, Skull, Skeleton	ZD.1972.140	
Kakamega, Nyanza, [0°17′N 34°45′E], 5000ft	8 Sep 1962	F	Ad	Skin, Skull, Skeleton	ZD.1972.138	
Kaptagat, nr. Mt.Elgon, Uasin Gishu, 0°28′N 35°29′E, 8000ft	4 Nov 1959	M	Ad	Skin, Skull	ZD.1972.144	
,, ,, ,,	13 Nov 1959	M	Ad	Skin, Skull	ZD.1972.145	
,, ,, ,,	13 Nov 1959	M	Juv	Skin, Skull	ZD.1972.146	
,, ,, ,,	19 Nov 1959	F	Ad	Skin, Skull	ZD.1972.147	
,, ,, ,,	2 Jul 1960	F	Ad	Skin, Skull, Skeleton	ZD.1972.148	
,, ,, ,,	3 Jul 1960	M	Ad	Skin, Skull, Skeleton	ZD.1972.149	
,, ,, ,,	17 Jan 1961	F	Ad	Skin, Skull, Skeleton	ZD.1972.150	
Makutano, nr. Railway Bridge, 54 miles from Eldoret, [c.1°12′S 36°36′E], 8000ft	4 Feb 1960	M	Ad	Skin, Skull, Skeleton	ZD.1972.141	
Marindas Forest Reserve, Central Province, 9000ft	28 Jul 1960	M	Inf	Skin, Skull, Skeleton	ZD.1972.142	
,, ,, ,,	10 Aug 1960	M	Juv	Skin, Skull, Skeleton	ZD.1972.143	
Mt.Elgon, [c.1°07′N 34°35′E]		F	Ad	Skull	RCS(OM) A68.42	
,, ,, ,,		M	Ad	Skull	RCS(OM) A68.421	
W.Escarpment of Leikipia [=Laikipia], 0°25′N 36°08′E	Jun 1893	I	[Ad]	Skin	ZD.1893.12.2.1	Cited Schwarz,E. 1929: 592
,, ,, ,,	Jun 1893	I	[Ad]	Skin	ZD.1893.12.2.2	,, ,, ,, ,,
—	—		Ad	2 skulls	RCS(OM) S68.4(A)	
[KENYA]						
Kavirondo, N. of Nandi Road, British East Africa, c.0°10′N 35°E, 5250ft	5 Oct 1902	F	Ad	Skin, Skull	PCM Uganda	Cited Matschie,P. 1913: 51, Cited Schwarz,E. 1929: 592
,, ,, ,,	5 Oct 1902	F	Juv	Skin, Skull	PCM Uganda	,, ,, ,, ,,
,, ,, ,,	6 Oct 1902	F	Ad	Skin, Skull	PCM Uganda	,, ,, ,, ,,

Locality	Date of Collection	Sex	Age	Nature of Specimen	Accession no.	References, Remarks, etc.
Colobus (Colobus) guereza matschiei *continued*						
Kavirondo, N. of Nandi Road below Escarpment, B.E.A., c.0°10′N 35°E, 5250ft	5 Oct 1902	M	Juv	Skin, Skull	PCM Uganda	,, ,, ,, ,,
Mau Forest, [between Ravine Station and upper R.Nollosegelli], British East Africa, c.0°30′S 36°20′E, 7700ft	26 Jul 1902	M	Ad	Skin, Skull	PCM Uganda	,, ,, ,, ,,
,, ,, ,,	27 Jul 1902	F	Ad	Skin, Skull	PCM Uganda	,, ,, ,, ,,
,, ,, ,,	27 Jul 1902	M	Ad	Skin, Skull	PCM Uganda	,, ,, ,, ,,
,, ,, ,,	29 Jul 1902	F	Juv	Skin, Skull	PCM Uganda	,, ,, ,, ,,
,, ,, ,,	29 Jul 1902	M	Ad	Skin, Skull	PCM Uganda	,, ,, ,, ,,
,, ,, ,,	Jul 1902	I	[Ad]	3 skins	PCM Uganda	,, ,, ,, ,,
N.Nandi, below Escarpment, B.E.Africa, c.0°10′N 35°E, 5250ft	6 Oct 1902	M	Ad	Skin, Skull	PCM Uganda	,, ,, ,, ,,
,, ,, ,,	6 Oct 1902	M	Ad	Skin, Skull	PCM Uganda	,, ,, ,, ,,
UGANDA						
Mount Elgon, [1°09′N 34°29′E],9300ft	—	M	Ad	Skin, Skull	ZD.1931.11.1.1	
Mount Elgon, [1°09′N 34°29′E]	—			Skin	ZD.1934.6.2.1	
Namalu, at base of Kadan (Mt.Debasien), Karamoja District, 1°47′N 34°31′E	Nov 1951	M	Ad	Skull	ZD.1969.375	Cited Haddow,A.J. (unpubl.)a
,, ,, ,,	Nov 1951	[F]	Ad	Skull	ZD.1969.376	,, ,, ,, ,,
[Namalu], [at base of Kadan (Mt.Debasien)], Karamoja, [1°47′N 34°31′E]	[1951]		Ad	5 skulls	RCS(OM) S68.4(G)	,, ,, ,, ,,
Colobus (Colobus) guereza occidentalis (Rochebrune 1886-87)						
CAMEROON						
Assubam Village, Lomie Dist., 3°15′N 14°E	28 Mar 1937	M	Ad	Skin, Skull	PCM Merfield, RCS(OM) G68.811	Listed Merfield,F.G. (unpubl.)
,, ,, ,,	3 Feb 1937	M	Ad	Skin, Skull	PCM Merfield, RCS(OM) A68.33	,, ,, ,, ,,
Batouri Dist., 4°15′N 14°15′E	1932	I		3 skins	PCM Merfield	
Beri, [Batouri Dist.], [c.4°20′N 14°15′E]	Mar 1932	M		Skin	PCM CAM II	
Between Batouri and Lomie, 3°45′N 13°45′E	17 Apr 1935	F	Ad	Skin, Skull, Skeleton	PCM Merfield	Listed Merfield,F.G. (unpubl.)
,, ,, ,,	17 Apr 1935	F	Juv	Skin	PCM Merfield	,, ,, ,, ,,
Between Batouri and Lomie, Batouri Dist., 3°45′N 13°45′E	17 Apr 1935	F	Juv	Skull	RCS(OM) A68.32	,, ,, ,, ,, ,, ,, ,, ,,
Bitye, R.Dja, [3°01′N 12°22′E], 2000ft	—	M	Ad	Skin, Skull	ZD.1914.1.24.1	Figured Colyer,J.F. 1936: 214, fig.256
Doumé, Yoko-Tibati Road, 5°53′N 12°16′E	22 Jan 1932	I	[Ad]	Skin	ZD.1932.8.1.16	
,, ,, ,,	22 Jan 1932	I	[Ad]	Skin	ZD.1932.8.1.17	
Doumé, Yoko-Tibati Road, [5°53′N 12°16′E]	22 Jan 1932	I		8 skins	PCM CAM II	
Kanyol Village, (Mendjim Mey Tribe), Batouri Dist., 4°N 14°E	21 Jun 1932	F	Ad	Skull	RCS(OM) G68.8	Listed Merfield,F.G. (unpubl.)
Lelo Village, (Kaka Tribe), Batouri Dist., 4°15′N 14°15′E	11 Dec 1932	F	Juv	Skull	RCS(OM) A68.3	,, ,, ,, ,,
Mey Joss Village, (Mendjim Mey Tribe), Batouri Dist., 4°N 14°E	23 Jul 1932	M	Ad	Skull	RCS(OM) A68.35	,, ,, ,, ,,
[N′Dinga Village], Batouri Dist., 4°15′N 14°15′E	12 Mar 1933	F	Ad	Skin, Skull, Skeleton	PCM Merfield	,, ,, ,, ,,
Ndokofass, Yabassi Dist., 4°N 10°E	Jul 1939	M	Ad	Skin, Skull, Skeleton	PCM Merfield	,, ,, ,, ,,
Ndokyam Hills, Yabassi Dist., 4°15′N 14°15′E	Apr 1939	M	Ad	Skin, Skull	PCM Merfield, RCS(OM) G68.761	,, ,, ,, ,,
,, ,, ,,	12 May 1939	M	Ad	Skin, Skull	PCM Merfield	,, ,, ,, ,,
,, ,, ,,	12 May 1939	F	Ad	Skin, Skull	PCM Merfield, RCS(OM) A68.36	,, ,, ,, ,,
Near Beba, Lomie Dist., 3°15′N 13°30′E	14 Apr 1937	M	Ad	Skin, Skull	PCM Merfield, RCS(OM) A68.34	,, ,, ,, ,,
,, ,, ,,	14 Apr 1937	M	Ad	Skin, Skull, Skeleton	PCM Merfield	,, ,, ,, ,,
Nr. Koumarrie, N.Cameroons, c.8°N 13°E	26 Nov 1931	M	Ad	Skin, Skull	PCM CAM II	
Nr. Tibati, [6°25′N 12°33′E], 2900ft	—	M	Ad	Skin, Skull	ZD.1923.1.22.3	Cited Schwarz,E. 1929: 590
Obala Village, Batouri Dist., 4°15′N 14°15′E	17 Oct 1932	M	Ad	Skin, Skull, Skeleton	PCM Merfield	Listed Merfield,F.G. (unpubl.)
Obala Village, (Mendjim Mey Tribe), Batouri Dist., 4°15′N 14°15′E	22 Oct 1933	M	Juv	Skin, Skull	PCM Merfield, RCS(OM) A68.31	,, ,, ,, ,,
Obala Village, (Mendjim Mey Tribe), Batouri Dist., 3°45′N 14°15′E	Oct 1933	F	Ad	Skull	RCS(OM) G68.81	,, ,, ,, ,,
R.Nkan, Yabassi Dist., 4°N 10°E	24 Mar 1938	F	[Ad]	Skin	PCM Merfield	,, ,, ,, ,,
,, ,, ,,	1 Apr 1938	M	Ad	Skin, Skull, Skeleton	PCM Merfield	,, ,, ,, ,,
,, ,, ,,	1 Apr 1938	M	Juv	Skin, Skull	PCM Merfield	,, ,, ,, ,,
R.Nkan, Yabassi Dist., c.4°N 10°E	29 Dec 1938	M	Ad	Skin, Skull	PCM Merfield, RCS(OM) G11.6	Listed Merfield,F.G. (unpubl.), Cited Colyer,J.F. 1943: 113
Yabassi Dist., 4°N 10°E	9 Nov 1938	F	Ad	Skin, Skull	PCM Merfield	Listed Merfield,F.G. (unpubl.)
,, ,, ,,	—	M	[Ad]	Skin	PCM Merfield	,, ,, ,, ,,
,, ,, ,,	—	F	Ad	Skull	PCM Merfield	,, ,, ,, ,,
Yabassi Dist., c.4°15′N 10°15′E	—	F	Ad	Skin, Skull	PCM Merfield, RCS(OM) G68.76	Listed Merfield,F.G. (unpubl.),Cited Colyer,J.F. 1943: 113

Locality	Date of Collection	Sex	Age	Nature of Specimen	Accession no.	References, Remarks, etc.
Colobus (Colobus) guereza occidentalis *continued*						
Fotabong, Mamfe Division, [4000ft]	20 May 1933	I	[Ad]	Skin	ZD.1948.451	Listed Sanderson,I.T. (unpubl.), Cited Sanderson,I.T. 1940: 643
[CAMEROON]						
Ndop, Bamenda Highlands, British Cameroons, [6°N 10°25′E], 2000ft?	Apr 1942	I	[Ad]	Skin	ZD.1950.938	
N.foot of Mambila Plateau, British Cameroons, c.7°30′N 11°30′E, 2000ft	19 Mar 1948	M	[Ad]	Skin	ZD.1956.529	
,, ,, ,,	28 Mar 1948	F	[Ad]	Skin	ZD.1956.530	
[CENT. AFRICAN REP.]						
M'Pongu[?], [c.6°N 19°10′E]	11 [Oct] 1905	M	Inf	Skin, Skull	ZD.1907.7.8.13	Background Alexander,B. 1907: 215-217
R.Bamingui, [7°56′N 21°12′E to 8°31′N 19°02′E]	[1905]	M	Ad	Skin, Skull	ZD.1907.7.8.10	Cited Schwarz,E. 1929: 590
[CHAD/CENT. AFRICAN REP. BORDER]						
Banding, Dakoru, junction of Bahr Sara and Nana Barya, c.7°55′N 17°35′E	22 Jul 1925	I		Skin	PCM Chad	
Dakoru, junction of Bahr Sara and Nana Barya, c.7°55′N 17°35′E	24 Jul 1925	M	Ad	Skin, Skull	PCM Chad	
,, ,, ,,	24 Jul 1925	M	Inf	Skin, Skull	PCM Chad	
,, ,, ,,	24 Jul 1925	F	Ad	Skin, Skull	PCM Chad	
Nr. Daddig, Borro, [c.7°45′N 17°10′E]	22 Sep 1925	M	Ad	Skin, Skull	ZD.1928.11.11.1	
,, ,, ,,	22 Sep 1925	F	Ad	Skin, Skull	ZD.1928.11.11.2	
Nr. Daddig, by Borro, Ubangui-Chari, c.7°45′N 17°10′E	2 Sep 1925	M	Juv	Skin, Skull	PCM Chad	
,, ,, ,,	2 Sep 1925	F	Ad	Skin, Skull	PCM Chad	
,, ,, ,,	22 Sep 1925	F	Juv	Skin, Skull	PCM Chad	
,, ,, ,,	22 Sep 1925	F	Ad	Skin, Skull	PCM Chad	
Ubangui-Chari, c.7°45′N 17°10′E	Sep 1925	M		Skin	PCM Chad	
[CONGO (BRAZZAVILLE)]						
Dodina, French Congo, c.0°20′N 15°E	19 Jun 1927	I		Skin	PCM Fr.Congo	
M'Boko, French Congo, c.0°15′S 15°10′E	27 Mar 1927	M	[Ad]	Skin, Skull, Skeleton	PCM Fr.Congo	
,, ,, ,,	27 Mar 1927	F	[Ad]	Skin, Skull	PCM Fr.Congo	
Molémé, French Congo, [c.0°20′N 15°E]	8 Jun 1927	I		Skin	PCM Fr.Congo	
NIGERIA						
Gonda, Donga R., N.Nigeria, [7°15′N 10°33′E]	11 May 1923	[F]	Ad	Skin, Skull	ZD.1924.8.6.2	Cited Schwarz,E. 1929: 590
,, ,, ,,	11 May 1923	M	Ad	Skin, Skull	ZD.1924.8.6.3	,, ,, ,, ,,
N.Nigeria	5 Nov 1913	M	Inf	Skin, Skull, Casts of lower teeth	ZD.1913.12.10.1	
Samaki, Donga R., N.Nigeria	20 May 1923	F	Ad	Skin, Skull	ZD.1924.8.6.6	Cited Schwarz,E. 1929: 590
Turmi, Donga R., N.Nigeria, [c.7°-8°N 10°-11°E]	29 May 1923	[F]	Ad	Skin, Skull	ZD.1924.8.6.4	,, ,, ,, ,,
,, ,, ,,	29 May 1923	M	[Ad]	Skin	ZD.1924.8.6.5	,, ,, ,, ,,
[RWANDA]						
Kisenyi, Lake Kivu, [1°42′S 29°14′E]	—	I	[Juv]	Skin	ZD.1928.1.30.17	(*Colobus polykomos uellensis*)
[SUDAN]						
By Modi, between Kaya and Dufile [= Laropi], Lado Enclave, c.4°N 31°45′E	25 Mar 1905	M	Inf	Skin, Skull	PCM Congo	Syntype *Colobus* (*Guereza*) *matschiei brachychaites* Matschie 1913, Cited Matschie,P. 1913: 53
By Modi, between Kaya and Dufile [= Laropi], Lado Enclave, c.4°N 31°45′E, 2000ft	25 Mar 1905	F	Ad	Skin, Skull	PCM Congo	,, ,, ,, ,,
,, ,, ,,	25 Mar 1905	F	Ad	Skin, Skull	PCM Congo	,, ,, ,, ,,
,, ,, ,,	25 Mar 1905	F	Ad	Skin, Skull	PCM Congo	,, ,, ,, ,,
By Modi, between Kaya and Dufile [= Laropi], Lado Enclave, c.4°N 31°45′E, [2000ft]	25 Mar 1905	M	Ad	Skin, Skull	PCM Congo	Syntype *Colobus* (*Guereza*) *matschiei brachychaites* Matschie 1913, Cited Matschie,P. 1913: 53, Cited Schwarz,E. 1929: 591
Lalleika, nr. Lado	1882			Skin	HM	
UGANDA						
Allui, White Nile, Old Lado Enclave, 1600ft	7 Apr 1928	F	Ad	Skin, Skull	ZD.1928.9.8.1	(*Colobus polykomos uellensis*)
,, ,, ,,	7 Apr 1928	F	Ad	Skin, Skull	ZD.1928.9.8.2	,, ,, ,, ,,
,, ,, ,,	7 Apr 1928	F	Ad	Skin, Skull	ZD.1928.9.8.3	,, ,, ,, ,,
,, ,, ,,	7 Apr 1928	F	Ad	Skin, Skull	ZD.1928.9.8.4	,, ,, ,, ,,
Ankole District	1956	[M]	[Ad]	Skin	ZD.1966.3617	,, ,, ,, ,,
,, ,, ,,	1956	[F]	[Ad]	Skin	ZD.1966.3618	,, ,, ,, ,,
,, ,, ,,	1956	[F]	[Juv]	Skin	ZD.1966.3619	,, ,, ,, ,,
Bugoma Forest, Bunyoro Dist., [c.1°20′N 31°E], 4000ft	[1945-49]	M	Ad	Skull	RCS(OM) A68.389	(*Colobus guereza uellensis*), Cited Colyer,J.F. (unpubl.), Background Haddow,A.J. *et al* 1951: 189-224, Correspondence Haddow,A.J. (unpubl.), Cited Haddow,A.J. (unpubl.)a

Locality	Date of Collection	Sex	Age	Nature of Specimen	Accession no.	References, Remarks, etc.
Colobus (Colobus) guereza occidentalis *continued*						
Bugoma Forest, Bunyoro Dist., [c.1°20′N 31°E], 4000ft	[1942-49]	—	Ad	4 skulls	RCS(OM) S68.37(G)	(*Colobus guereza uellensis*), Cited Colyer,J.F. (unpubl.)a, Background Haddow,A.J. *et al* 1951: 189-224, Correspondence Haddow,A.J. (unpubl.)
Bunyoro, 3750ft	4 Jun 1930	M	Ad	Skin, Skull	ZD.1931.4.1.4	(*C.p.uellensis*)
Bwamba, Semliki Forest, Toro Dist., [c.0°48′N 30°07′E], 2500ft	1942-49			26 Skulls	RCS(OM) S68.37(A)	(*Colobus guereza uellensis*), Cited Colyer,J.F. (unpubl.)a, Background Haddow,A.J. *et al* 1947: 677-700, Background Haddow,A.J. *et al* 1951: 189-224, Background Haddow,A.J. 1952: 297-394, Correspondence Haddow,A.J. (unpubl.)
" " "	1942-49			25 skulls	RCS(OM) S68.37(G)	(*Colobus guereza uellensis*), Cited Colyer,J.F. (unpubl.)a, Background Haddow,A.J. *et al* 1947: 677-700, Background Haddow,A.J. *et al* 1951: 189-224, Correspondence Haddow,A.J. (unpubl.)
Bwamba Country, Toro, 0°50′N 30°03′E, 2500ft	30 Nov 1964	F	Ad	Skin, Skull	ZD.1972.153	(*Colobus guereza uellensis*)
Daro Forest [=Dura Forest], Toro, [0°15′N 30°20′E]	27 Oct 1911	F	Ad	Skin, Skull	ZD.1912.5.18.2	(*Colobus uellensis*)
Fort Portal, Mpanga Forest, N.Ruwenzori, [0°40′N 30°17′E], 5000ft	13 Sep 1906	F	Ad	Skin, Skull	ZD.1907.4.6.2	(*C.p.uellensis*)
" " "	14 Sep 1906	F	Ad	Skin, Skull	ZD.1907.4.6.3	" " " "
Kalinzu Forest, Ankole, [0°19′S 30°06′E], [4500ft]	1947	M	Ad	Skull	ZD.1954.762	(*Colobus abyssinicus ituricus*)
Kalinzu Forest, Ankole Dist., [c.0°19′S 30°06′E], 4500ft	[1942-49]	—	Ad	2 skulls	RCS(OM) S68.37(A)	(*Colobus guereza uellensis*), Cited Colyer,J.F. (unpubl.)a, Background Haddow,A.J. *et al* 1951: 189-224, Correspondence Haddow,A.J. (unpubl.), Cited Haddow,A.J. (unpubl.)a
Kasinga Channel, [0°36′S 29°44′E], 5000ft	—	F	Ad	Skull	ZD.1938.9.9.4	(*Colobus abyssinicus uellensis*)
[Kasunganyanja], Toro Dist., [0°25′N 30°12′E], 4000ft	[1951]	M	Inf	Skull	RCS(OM) A68.392	(*Colobus guereza uellensis*), Background Haddow,A.J. *et al* 1951: 189-224, Correspondence Haddow,A.J. (unpubl.), Cited Haddow,A.J. (unpubl.)a
" " "	[1951]	F	Juv	Skull	RCS(OM) A68.393	" " " "
" " "	[1951]	F	Ad	Skull	RCS(OM) A68.394	" " " "
" " "	[1951]	M	Ad	Skull	RCS(OM) A68.395	" " " "
" " "	[1951]	F	Ad	Skull	RCS(OM) G68.8178	" " " "
" " "	[1951]	F	Ad	Skull	RCS(OM) G68.8179	" " " "
" " "	[1951]	M	Ad	Skull	RCS(OM) G68.81791	" " " "
" " "	[1951]	M	Ad	Skull	RCS(OM) G68.81792	" " " "
Katunguru, Kazinga Channel, Ankole, [c.0°18′N 30°01′E]	19 Sep 1929	F	Ad	Skull	ZD.1930.8.1.12	(*Colobus polykomos uellensis*)
Kayonza Forest, Kigezi Dist., [c.1°S 29°35′E], [c.5000ft]	1945-47	M	Ad	Skull	RCS(OM) G169.312	(*Colobus guereza uellensis*), Cited Colyer,J.F. (unpubl.)a, Background Haddow,A.J. *et al* 1951: 189-224, Correspondence Haddow,A.J. (unpubl.), Cited Haddow,A.J. (unpubl.)a
Kibale Forest, Toro, 0°50′N 31°05′E, 5000ft	29 Sep 1964	M	Ad	Skin, Skull, Skeleton	ZD.1972.151	(*Colobus guereza uellensis*)
" " "	1 Oct 1964	M	Ad	Skin, Skull, Skeleton	ZD.1972.152	" " " "
Kibale Forest, Toro Dist., [0°30′N 30°24′E], 4000-5000ft	1942-47	F	Juv	Skull	RCS(OM) A68.3714	(*Colobus guereza uellensis*), Cited Colyer,J.F. (unpubl.)a, Background Haddow,A.J. *et al* 1951: 189-224, Correspondence Haddow,A.J. (unpubl.)
Kibale Forest, Toro Dist., [0°39′N 30°24′E], [5000ft]	[1945-49]	F	Ad	Skull	RCS(OM) G68.8169	(*Colobus guereza uellensis*), Cited Colyer,J.F. (unpubl.)a, Background Haddow,A.J. *et al* 1951: 189-224, Correspondence Haddow,A.J. (unpubl.),Cited Haddow,A.J. (unpubl.)a
Kibale Forest, Toro Dist., [0°39′N 30°24′E], 5000ft	[1945-49]		Ad	4 skulls	RCS(OM) S68.37(G)	" " " "
Kibale Forest, Toro Dist., [c.0°30′N 30°25′E], 5000ft	[1951]	F	Ad	Skull	RCS(OM) G169.313	(*Colobus guereza uellensis*), Cited Colyer,J.F. (unpubl.)a, Background Haddow,A.J. *et al* 1951: 189-224, Correspondence Haddow,A.J. (unpubl.)
Koboko, West Nile, [3°24′N 30°58′E], 4000ft	—	F	[Ad]	Skin	ZD.1951.533	(*Colobus abyssinicus uellensis*)
Masindi, Bunyoro, [1°41′N 31°42′E]	—	M	Ad	Skull	ZD.1930.12.1.5	(*Colobus uellensis*)
Masindi, Bunyoro, [1°41′N 31°42′E], 3750ft	3 Jun 1930	M	Ad	Skin, Skull	ZD.1931.4.1.5	(*C.p.uellensis*)
Masindi, Bunyoro, [1°41′N 31°42′E], 3700ft	1 Jun 1930	M	Ad	Skull	ZD.1931.4.1.6	" " " "

Locality	Date of Collection	Sex	Age	Nature of Specimen	Accession no.	References, Remarks, etc.
Colobus (Colobus) guereza occidentalis *continued*						
Mubende Area, Mubende Dist., [0°32′N 31°15′E], 3500ft	[1942-49]	M	Juv	Skull	RCS(OM) A68.388	(*Colobus guereza uellensis*), Cited Colyer,J.F. (unpubl.)a, Background Haddow,A.J. *et al* 1951: 189-224, Correspondence Haddow,A.J. (unpubl.), Cited Haddow,A.J. (unpubl.)a
" " "	[1942-49]	M	Ad	Skull	RCS(OM) A68.3881	" " " "
" " "	[1945-49]	F	Ad	Skull	RCS(OM) G68.815	" " " "
Muhangi Forest, Toro Dist., [c.0°52′N 30°48′E], 4500ft	[1942-49]	M	Juv	Skull	RCS(OM) A68.387	" " " "
8 miles upstream of Murchison Falls, Victoria Nile (left bank), [2°17′N 31°40′E]	19 Feb 1926	M	Ad	Skin, Skull	ZD.1926.11.18.2	(*Colobus matschiei uellensis*)
N. of Mbarara, 0°37′S 30°39′E, 5000ft	5 Feb 1938	M	Ad	Skin	ZD.1938.9.9.2	(*Colobus abyssinicus uellensis*)
R.Chambura, Kickwamba [Kichwamba], N.W.Ankole, [0°07′S 30°04′E], 3300ft	17 Sep 1929	F	Ad	Skin, Skull, Skeleton	ZD.1930.8.1.10	(*C.p.uellensis*)
" " "	17 Sep 1929	F	Juv	Skin, Skull, Skeleton	ZD.1930.8.1.11	" " " "
Rwansenge River, Semliki Forest, Bwamba County, c.0°49′N 30°04′E, 2400ft	Nov 1943	M	Ad	Skull	ZD.1969.377	(*Colobus abyssinicus ituricus*), Cited Haddow,A.J. (unpubl.)a
Semliki Forest, Bwamba County, 0°49′N 30°04′E, 2400ft	c.1946/47	F?	Ad	Skull	ZD.1969.378	(*Colobus abyssinicus ituricus*)
S. of Imatong Mts., nr. Sudan Border, North Acholi, [c.3°45′N 32°45′E]	[1949]	F	Ad	Skull	RCS(OM) A68.39	(*Colobus guereza uellensis*)
" " "	[1949]	F	Ad	Skull	RCS(OM) A68.391	" " " "
" " "	[1949]	F	Ad	Skull	RCS(OM) G96.617	" " " "
S. of Kasinga Channel [nr.] L.Edward, [0°36′S 29°44′E], 6000ft	4 Feb 1938	M	Inf	Skin, Skull	ZD.1938.9.9.3	(*Colobus abyssinicus uellensis*)
S. of Kasinga Channel, [0°36′S 29°44′E], 6000ft	4 Feb 1938	F	Ad	Skin, Skull	ZD.1938.9.9.5	" " " "
20 miles S. of Kasinga Channel, [c.0°30′S 30°E]	—	F	Ad	Skull	ZD.1938.9.9.6	" " " "
[UGANDA]						
Kibera's Village, Madi Country, B.E.A., 3°07′N 31°42′E, 2000ft	3 Nov 1901	M	Inf	Skin, Skull	ZD.1902.4.5.1	(*Colobus uellensis*), Cited Schwarz,E. 1929: 591
" " "	3 Nov 1901	F	Inf	Skin, Skull	ZD.1902.4.5.2	" " " "
Kissenye [=Kisenyi], [near mouth of R.Niamweru], N.E. shore of L.Edward, Congo, c.0°10′S 29°50′E	21 Sep 1906	F	Ad	Skin, Skull	PCM Congo	Paratype *Colobus* (*Guereza*) *matschiei dianae* Matschie 1913, Cited Matschie,P. 1913: 49
" " "	24 Sep 1906	M	Ad	Skin, Skull	PCM Congo	" " " "
" " "	24 Sep 1906	F	Ad	Skin, Skull	PCM Congo	" " " "
" " "	24 Sep 1906	F	Ad	Skin, Skull	PCM Congo	" " " "
" " "	24 Sep 1906	F	Juv	Skin, Skull	PCM Congo	" " " "
" " "	24 Sep 1906	F	Juv	Skin, Skull	PCM Congo	" " " "
" " "	24 Sep 1906	I	Ad	Skin, Skull	PCM Congo	" " " "
" " "	26 Nov 1906	F	Ad	Skull	PCM Congo	" " " "
Kissenye [=Kisenyi], [near mouth of R.Niamweru], N.E. of L.Edward, c.0°10′N 29°50′E	24 Sep 1906	M	Ad	Skin, Skull	PCM Congo	Holotype *Colobus* (*Guereza*) *matschiei dianae* Matschie 1913, Cited Matschie,P. 1913: 49, Cited Schwarz,E. 1929: 591, Cited Moreau,R.E. *et at.* 1946: 406
Kyehume [=?Kyetume], Chagwe [= Kiagwe]	—	[M]	Juv	Skin, Skull	ZD.1911.6.6.1	(*Colobus uellensis*)
Mpanga Forest [Kibale Forest], [0°33′N 30°24′E]	Sep 1900	[M]	Ad	Skull	ZD.1901.8.9.12	
N.E. of Lake George, Toro, [0° 30°12′E]	—	F	Juv	Skin	ZD.1911.6.6.2	(*Colobus uellensis*)
R.Wanka, Semliki Valley, [c.0°47′N 30°15′E], 3600ft	22 Mar 1930	M	Ad	Skin, Skull	ZD.1930.11.11.1	
Ruwenzori, B.E.A., [0°22′N 30°E]	—	M	Ad	Skin, Skull	ZD.1911.7.25.14	
" " "	—	M	Ad	Skin, Skull	ZD.1911.7.25.15	
S.E.Ruwenzori, [c.0°22′N 30°E], 3400ft	10 Jun 1906	M	Ad	Skin, Skull	ZD.1907.4.6.1	(*Colobus uellensis*)
S.W.Uganda	—	[M]	[Ad]	Skin	ZD.1934.8.10.7	(*Colobus polykomos uellensis*)
" " "	—	I	[Ad]	Skin	ZD.1934.8.10.8	" " " "
" " "	—	I	[Ad]	Skin	ZD.1934.8.10.9	" " " "
Toro, 4000ft	—	M	Ad	Skull	ZD.1901.8.9.15a	Cited Elliot,D.G. 1907: 195
—	[1947]	F	Ad	Skull	RCS(OM) G68.8112	
[UGANDA/ZAIRE BORDER]						
Sassa [=R.Ishasha], [S.E. of L.Edward], Congo, c.0°40′S 29°40′E	Nov 1906	M	Ad	Skin, Skull	PCM Congo	Paratype *Colobus* (*Guereza*) *matschiei dianae* Matschie 1913, Cited Matschie,P. 1913: 50, Cited Schwarz,E. 1929: 591
" " "	17 Nov 1906	M	Ad	Skin, Skull	PCM Congo	" " " "
" " "	17 Nov 1906	F	Ad	Skin, Skull	PCM Congo	" " " "
[ZAIRE]						
Aluta, W. of Beni, Belgian Congo, 3100ft	9 Jul 1930	M	Ad	Skin, Skull	ZD.1930.11.11.6	
(Beni), Ituri Forest, Belgian Congo, [0°29′N 29°29′E], 3100ft	31 Mar 1930	[F]	[Ad]	Skin	ZD.1930.11.11.7	

Locality	Date of Collection	Sex	Age	Nature of Specimen	Accession no.	References, Remarks, etc.
Colobus (Colobus) guereza occidentalis *continued*						
Between Irumu and Mawambe, Ituri Forest, Congo, c.1°25′N 28°45′E, 2500ft	26 Jul 1905	M	Ad	Skin, Skull	PCM Congo	Paratype *Colobus* (*Guereza*) *matschiei ituricus* Matschie 1913, Cited Matschie,P. 1913: 48, Cited Schwarz,E. 1929: 591
Between Irumu and Mawambe, Ituri Forest, Congo, c.1°25′N 28°45′E	25 Jul 1905	M	Ad	Skin, Skull	PCM Congo	,, ,, ,, ,,
By Maru Kifuku, Ituri Forest, Congo, c.1°25′N 29°50′E	25 Jul 1906	M	Ad	Skin, Skull	PCM Congo	,, ,, ,, ,,
By Mokoko, Ituri Forest, Congo, c.1°40′N 29°20′E, 2700ft	19 Jul 1905	M	Ad	Skin, Skull	PCM Congo	,, ,, ,, ,,
Ituri Forest, Belgian Congo, [c.1°N 29°30′E]	30 Mar 1930	F	Inf	Wet, cadaver	ZD.1930.11.11.5	
Ituri Forest, Belgian Congo	—	I	[Ad]	Skin	ZD.1939.3475	
Ituri Forest, S. of Irumu, Belgian Congo, 3400ft	30 Mar 1930	M	[Juv]	Skin	ZD.1930.11.11.2	
Ituri Forest, 40 miles S.W. of Irumu, Belgian Congo, 3600ft	30 Mar 1930	F	Ad	Skin, Skull	ZD.1930.11.11.4	
Opposite Mawambe, Ituri Forest, Congo, c.1°10′N 28°40′E	1 Sep 1905	F	Ad	Skin, Skull	PCM Congo	Paratype *Colobus* (*Guereza*) *matschiei ituricus* Matschie 1913, Cited Matschie,P. 1913: 48
,, ,, ,,	1 Sep 1905	F	Ad	Skin, Skull	PCM Congo	,, ,, ,, ,,
Penghe, 51 miles E. of Avakubi, Congo Forest, Congo, [1°18′N 28°35′E], 2000ft	8 Feb 1914	F	Ad	Skin, Skull	ZD.1914.6.2.3	
R.Ituri, 50 miles S.W. of Irumu, Belgian Congo, [1°N 29°15′E], 3500ft	4 Apr 1930	M	Ad	Skin, Skull	ZD.1930.11.11.3	
R.Meridi, Niam Niam, S. of Bahr el Ghazal, [4°40′N 29°33′E]		M	Ad	Skull	ZD.1905.7.25.1	
Sandeh [=Zande] Land, on the Nambiri, nr. Monbuttu, [c.3°40′N 28°32′E]	[20] Jul 1883	F	Juv	Skin, Skull	ZD.1887.12.1.4	(*Colobus polykomos uellensis*), Background Schweinfurth,G. *et al*, 1888: 449-452, Cited Thomas,O. 1888: 5, Cited Schwarz,E. 1929: 591
Upper Uele [R.], N.E.Congo, Congo, [c.3°37′N 28°33′E]	—	I	[Ad]	Skin	ZD.1916.6.29.1	(*Colobus uellensis*), Cited Schwarz,E. 1929: 591
,, ,, ,,	—	I	[Ad]	Skin	ZD.1916.6.29.2	,, ,, ,, ,,
NO DATA						
—	—	I	[Juv]	Skin	ZD.1971.489	
—	Died 29 Jan 1907	I	Juv	Skull	ZD.1981.264	
Colobus (Colobus) guereza occidentalis? (Rochebrune 1886-87)						
[UGANDA]						
White Nile District	16 Sep 1933	M	Ad	Skull	ZD.1934.6.3.3	(*Colobus uellensis?*)
Colobus (Colobus) guereza percivali Heller 1913						
KENYA						
Mount Urguess [=Warges Mts.], [0°56′N 37°24′E]	—	[M]	Juv	Skin, Skull	ZD.1912.7.1.1	Cited Schwarz,E. 1929: 593, Figured Colyer,J.F. 1936: 32, fig.39
Colobus (Colobus) guereza Rüppell 1835						
[ETHIOPIA]						
Abyssinia	—			Skin	CMZ E7314B	
,, ,, ,,	—			Skin	CMZ E7314C	
,, ,, ,,	—			Skin	SM A1930.9	
,, ,, ,,	—			Skin	GZD	
S.Abyssinia	Apr 1913			Skin	LM 36.13	
UGANDA						
Busesse, N.W.Ankole, 4000ft	Sep 1928	I		Skin	ZD.1929.5.14.25	
—	—			Skull	CMZ E7314A	
[ZAIRE]						
Nr. Beni, Semliki Forest, Congo, [c.0°29′N 29°29′E]	—	F	Juv	Skull	ZD.1920.10.17.5	
NO DATA						
—	—	M	[Ad]	Skin	ZD.1980.876	
—	—			2 skulls	EDA	(*Colobus guereza occidentalis*)
—	—	I	[Ad]	Skin	ZD.1981.181	
—	—	[M]	[Juv]	Skin	ZD.1981.182	
—	—	I	[Juv]	Skin	ZD.1981.183	
—	—	I	[Ad]	Skin	ZD.1981.184	
—	—	[M]	[Ad]	Skin	ZD.1981.185	
—	—	[F]	[Ad]	Skin	ZD.1981.186	
—	—	[F]	[Ad]	Skin	ZD.1981.187	
—	—	M	Ad	Skull	ZD.1981.260	
—	20 Dec 1914	M	Ad	Skull	ZD.1981.261	
—	—	I	[Ad]	Skin	ZD.1981.262	
Colobus (Colobus) guereza? Rüppell 1835						
UGANDA						
Mount Elgon, [1°09′N 34°29′E]	—			Skull	ZD.1931.11.1.2a	

COLOBUS (COLOBUS) SATANAS Waterhouse, 1838

Black Colobus

Principal synonyms *Semnopithecus anthracinus* Leconte, 1858:10; *Stachycolobus municus* Matschie, 1917:155; *Stachycolobus limbarenicus* Matschie, 1917:156; *Stachycolobus zenkeri* Matschie, 1917:157; *Colobus metternichi* Krumbiegel, 1942:311.

Type localities Of *satanas,* Fernando Póo: of *anthracinus,* Gabon; of *municus,* Wurminsog, S. of the mouth of Muni R. between 9°36′-9°48′E, Gabon; of *limbarenicus,* Limbareni (=Lambarené, 0°41′S 10°13′E), on Ogooué R., Gabon; of *zenkeri,* Bipindi (3°06′N 10°30′E), on R. Lokoundjé, S. Cameroon; of *metternichi,* Santa Isabel Mt, north slope, at 2000m, Fernando Póo.

Taxonomic notes Eisentraut (1973:54) has suggested that the mainland population with shorter glossier fur may be subspecifically distinct. The fur of the nominate race is longer and fluffier in texture, a distinction found in many primates from Fernando Póo. The earliest name for the mainland form is *Colobus satanas anthracinus. Colobus metternichi* Krumbiegel (1942:311), also from Fernando Póo, is considered by Eisentraut (1973:56) to be based on partial albinos.

Range Fernando Póo, S. Cameroon, Rio Muni and Gabon, reaching Congo (Brazzaville) at Odzala (0°45′N 14°45′E).

Dimensions An adult male and female in the museum collections have the following field measurements: h & b 670(♂), 635(♀); tail 800(♂), 800(♀).

Locality	Date of Collection	Sex	Age	Nature of Specimen	Accession no.	References, Remarks, etc.
Colobus (Colobus) satanas Waterhouse 1838						
CAMEROON						
Angon, Bipindi, [c.3°10′N 10°20′E]	1929	F	Ad	Skin, Skull	ZD.1930.12.15.2	
Ankong, c.2°50′N 10°30′E	1929	F	[Ad]	Skin, Skull	PCM Zenker I	
Bakoko, S.Cameroons, c.3°10′N 10°E	21 Feb 1931	I	[Ad]	Skin	PCM CAM II	
Batouri Dist., 4°15′N 14°15′E	11 Dec 1932	F	[Juv]	Skin, Skull	PCM Merfield	Listed Merfield,F.G. (unpubl.)
Between Ongue and N′yong [Rs.], S.Cameroons, c.3°30′N 10°E	May 1931	M	[Ad]	Skin, Skull	PCM CAM II	
„ „ „	May 1931	M	[Ad]	Skin, Skull	PCM CAM II	
„ „ „	May 1931	M	Ad	Skin, Skull	PCM CAM II	Figured Colyer,J.F. 1936: 27, fig.30
„ „ „	May 1931	M	[Ad]	Skin, Skull	PCM CAM II	
„ „ „	May 1931	M	[Ad]	Skin, Skull	PCM CAM II	
„ „ „	May 1931	F	[Ad]	Skin, Skull	PCM CAM II	
„ „ „	May 1931	F	[Ad]	Skin, Skull	PCM CAM II	
„ „ „	May 1931	F	[Ad]	Skin, Skull	PCM CAM II	
„ „ „	May 1931	F	[Ad]	Skin, Skull	PCM CAM II	
„ „ „	May 1931	F	[Ad]	Skin, Skull	PCM CAM II	
„ „ „	May 1931	F	[Ad]	Skin, Skull	PCM CAM II	
„ „ „	May 1931	F	[Ad]	Skin, Skull	PCM CAM II	
„ „ „	May 1931	F	[Ad]	Skin, Skull	PCM CAM II	
„ „ „	May 1931	F	[Ad]	Skin, Skull	PCM CAM II	
„ „ „	May 1931	M		17 skins and skulls	PCM CAM II	
„ „ „	May 1931	F		11 skins and skulls	PCM CAM II	
„ „ „	May 1931	?F		3 skins	PCM CAM II	
Between Ongue and N′yong [Rs.], S.Cameroon, c.3°30′N 10°E	May 1931	F	[Ad]	Skin, Skull	PCM CAM II	Cited Colyer,J.F. 1936: 28
„ „ „	May 1931	F	[Ad]	Skin, Skull	PCM CAM II	Figured Colyer,J.F. 1936: 29, fig.32
„ „ „	May 1931	F	[Ad]	Skin, Skull	PCM CAM II	Cited Colyer,J.F. 1936: 28
Bikiango, Bulu Bush, c.2°50′N 10°30′E	17 Feb 1930	M	[Juv]	Skin, Skull	PCM Zenker II	
Bipindi, 3°10′N 10°20′E	1929	M	Ad	Skin, Skull, Skeleton	ZD.1930.12.15.1	
„ „ „	7 Sep 1931	F	[Ad]	Skin, Skull	PCM Zenker V	
Bongan, S.Cameroons, c.3°10′N 10°E	21 Feb 1931	I	[Ad]	Skin	PCM CAM II	
Lelo Village, (Kaka Tribe), Batouri Dist., 4°15′N 14°15′E	11 Dec 1932	M	Juv	Skull	RCS(OM) A68.2	Listed Merfield,F.G. (unpubl.), Cited Colyer,J.F. 1943: 183-188
„ „ „	11 Dec 1932	F	Ad	Skull	RCS(OM) A68.22	Listed Merfield,F.G. (unpubl.)
„ „ „	11 Dec 1932	F	Ad	Skull	RCS(OM) G68.796	„ „ „ „
Lelo Village, Batouri Dist., 4°15′N 14°15′E	3 Feb 1933	F	[Ad]	Skin, Skull, Skeleton	PCM Merfield	„ „ „ „
Lomie Dist., 3°15′N 13°30′E	1937			2 skins	PCM Merfield	
Makav, near Bipindi, 3°10′N 10°20′E	16 Sep 1931	M	[Ad]	Skin, Skull	PCM Zenker V	
„ „ „	—	M	[Juv]	Skin	PCM Zenker VII	
Mouth of R.N′yong, S.Cameroons, c.3°20′N 9°50′E	19 Apr 1931	F	[Ad]	Skin, Skull	PCM CAM II	Cited Colyer,J.F. 1936: 28
Mouth of R.N′yong, S.Cameroons, [c.3°20′N 9°50′E]	18 Apr 1931	M	[Ad]	Skull, Skeleton	PCM CAM II	
Near Bipindi, S.Cameroons, c.3°10′N 10°20′E	20 Apr 1929			Skin	PCM CAM I	
Ongue Creek, N′yong R., S.Cameroons, c.3°30′N 10°E	Apr 1931	M	[Ad]	Skin, Skull	PCM CAM II	
„ „ „	Apr 1931	M	[Ad]	Skin, Skull	PCM CAM II	
Ongue Creek, S.Cameroon, 3°30′N 10°E	Mar 1932	M	Juv	Skin, Skull	ZD.1932.8.1.10	
„ „ „	Mar 1932	M	Ad	Skin, Skull	ZD.1932.8.1.11	Cited Colyer,J.F. 1936: 28
„ „ „	Mar 1932	M	Ad	Skin, Skull	ZD.1932.8.1.12	
„ „ „	Mar 1932	F	Juv	Skin, Skull	ZD.1932.8.1.13	
„ „ „	Mar 1932	F	Ad	Skin, Skull	ZD.1932.8.1.14	
„ „ „	Mar 1932	F	Ad	Skin, Skull	ZD.1932.8.1.15	
R.Bikiango, Bulu Bush, c.3°10′N 10°20′E	1930	M	Juv	Skin, Skull	ZD.1930.12.14.1	
Yabingon, S.Cameroons, c.3°10′N 10°E	21 Feb 1931	I	[Ad]	Skin	PCM CAM II	
—	—	[M]	[Ad]	Skin	ZD.1939.3476	
[CAMEROON]						
Bipindi Dist., S.Cameroons, c.3°N 10°30′E	c.1933	M	Juv	Skull	RCS(OM) A68.21	
Obala, Batouri Dist., French Cameroons, 3°45′N 14°15′E	28 Oct 1932	M	Ad	Skull	RCS(OM) G68.7961	Listed Merfield,F.G. (unpubl.)

Locality	Date of Collection	Sex	Age	Nature of Specimen	Accession no.	References, Remarks, etc.
Colobus (Colobus) satanas *continued*						
[CONGO (BRAZZAVILLE)]						
Odzala, French Congo, c.0°45′N 14°45′E	Apr 1927			Skin	PCM Fr. Congo	
FERNANDO PÓO						
Bubi Town, Bantabaré, [c.3°27′N 8°46′E], 500m	3 Feb 1904	I	[Ad]	Skin	ZD.1980.889	Background Thomas,O. 1905: 183-193, Cited Schwarz,E. 1929: 589
—	—	I	[Ad]	Skin	ZD.1841.1753	Syntype *Colobus satanas* Waterhouse 1838, Cited Waterhouse,G.R. 1838: 58. Listed Anon. (unpubl.)b: 81
—	—	I	[Ad]	Skin	ZD.1841.1754	,, ,, ,, ,,
—	Jun 1842	M	Juv	Skin	ZD.1847.1.18.74	
[FERNANDO PÓO]						
—	—	I	[Ad]	Skin	ZD.1855.12.24.404	Syntype *Colobus satanas* Waterhouse 1838, Cited Waterhouse,G.R. 1838: 58
[GABON]						
Gaboon	—	[F]	Ad	Skin, Skull	ZD.1871.5.27.7	
[RIO MUNI]						
Near Benito R., 25 miles from sea	11 Mar 1901	M	Ad	Skin, Skull	ZD.1901.11.21.1	Figured Elliot,D.G. 1913: Pl.17, Cited Schwarz,E. 1929: 590
,, ,, ,,	9 Mar 1901	F	Ad	Skin, Skull, Skeleton	ZD.1901.11.21.2	Cited Schwarz,E. 1929: 590
NO DATA						
—	—	M	Ad	Skin, Skull, Skeleton, Hyoid	ZD.1856.12.29.1	Listed Gray,J.E. (unpubl.): 1180
—	—	[F]	[Ad]	Skin	ZD.1921.1.3.1	

Subgenus: PILIOCOLOBUS Rochebrune, 1886-87

Red Colobus, Bay Colobus, Brown Colobus
(2 species: type species *Simia* (*Cercopithecus*) *badius* Kerr, 1792)

Principal synonym *Tropicolobus* Rochebrune, 1886-87 (type species: *Colobus rufomitratus* Peters, 1879).

Taxonomic note As Dandelot (1971) has stated, no definite answer can be given at present as to the number of species to be recognised. One of the characters that might be expected to influence speciation is the morphology of the external genitalia. Allen (1925), Pocock (1936), Hill (1958), Kuhn (1967, 1972) and Struhsaker (1975) note many differences in the female external genitalia and the size of the sexual swellings of *badius, waldroni, temminckii, preussi and tephrosceles.* Likewise the perineal organ of the male has been variously described as conspicuous in *C.b.badius*, the size of an apple in *C.b.brunneus* (=*oustaleti*), and as a small knob in *C.b.tephrosceles.* It is present in *C.b.kirkii* (Hill, 1958), but Struhsaker (1975) did not observe it during his field observations of *C.b.preussi, C.b.temminckii* and *C.b. rufomitratus.* Until detailed observations of these characters have been made and their significance fully elucidated, there will continue to be doubt about the number of species to be recognised. However, in all the most recent revisions, the same 14 populations are recognised, though arranged slightly differently, as follows:

Rahm, 1970	Dandelot, 1971	Groves & Dandelot (in prep.)
(1 species)	(8 species)	(4 species)
C.badius badius	*C.badius badius*	*C.badius badius*
C.badius temminckii	*C.badius temminckii*	*C.badius temminckii*
C.badius waldroni	*C.waldroni*	*C.badius waldroni*
C.badius preussi	*C.preussi*	*C.preussi*
C.badius pennantii	*C.pennantii pennantii*	*C.pennantii pennantii*
C.badius bouvieri	*C.pennantii bouvieri*	*C.pennantii bouvieri*
C.badius ellioti	*C.ellioti*	*C.pennantii ellioti*
C.badius tholloni	*C.tholloni*	*C.pennantii tholloni*
C.badius rufomitratus	*C.rufomitratus rufomitratus*	*C.pennantii oustaleti*
C.badius foai	*C.rufomitratus foai*	*C.pennantii foai*
C.badius tephrosceles	*C.rufomitratus tephrosceles*	*C.pennantii tephrosceles*
C.badius gordonorum	*C.rufomitratus gordonorum*	*C.pennantii gordonorum*
C.badius oustaleti	*C.rufomitratus oustaleti*	*C.pennantii kirkii*
C.badius kirkii	*C.kirkii*	*C.rufomitratus*

The catalogue follows the arrangement of Thorington & Groves (1970) which resembles that of Rahm (1970) except that *C. kirkii* is treated as a separate species.

Morphology Large slender red and brown (or red and black) monkeys with long untufted tails and paler underparts. Infants differ in colour from their parents (Struhsaker, 1975:13) but not so markedly as in *Colobus* (*Colobus*). Adult females have a large prominent clitoris and a sexual swelling at oestrus which is extremely large in *C.b.badius* and *C.b.preussi*. In *C.b.badius*, the female external genitalia is mimicked by the male perineal organ, even in such minute details as a 'pseudovaginal opening' and a 'pseudoclitoris' (Kuhn, 1972, pl. 1). The ischial callosities are well separated in both sexes. For skull characters distinguishing the subgenera, see *Colobus* (*Colobus*), morphology, p. 3. The intermembral index is considerably higher than in the black-and-white colobus (86 compared with 79), which is surprising, considering their rather similar ecology (see below). The stomach is divided into four compartments, as in *Procolobus*. Chromosome diploid no. 44 (*C. badius, C. kirkii*; Chiarelli, 1973).

Range Africa, south of the Sahara. There are three main centres of population: West Africa from Senegal to Ghana; E. Nigeria, Cameroon highlands and Fernando Póo; and the Zaire basin. Relict populations are also found in the Tana River flood plain, the montane forests of Tanganyika, and on Zanzibar Island.

Ecology Like *Colobus* (*Colobus*), *Colobus* (*Piliocolobus*) are found in rain forest, montane forest, gallery forest and forest-savanna mosaic. They range from sea level to over 5000ft (5350ft in the Ituri Forest, Zaire). The red colobus have a more varied diet than the black-and-white colobus and rarely eat mature leaves (Clutton-Brock, 1975). In spite of this, they can survive in seasonally dry areas such as the Mahali Mountains in Tanganyika (Kano, 1971), and in the Gambia (Struhsaker, 1975, pl. 31). In the latter instance, they frequently cross open ground between trees as do *C. guereza* in Sudan (see p. 3. *C. badius* frequently associate with other monkey species (*Cercocebus albigena, Cercopithecus ascanius, C. mitis*) whereas *C. guereza* tend to remain aloof (Struhsaker, 1975).

Social behaviour Group size is larger than in *Colobus* (*Colobus*), from 12 to 80 individuals. Smaller groups (up to 25 individuals) are found in *C.b.temminckii* in Senegal and *C.b.rufomitratus* in the Tana River. There is usually more than one fully adult male in each group, the adult sex ratio ranging from 1.5 to 3 females per male. A linear dominance hierarchy exists between the adult males of a group, the highest ranking male copulating most frequently. Home ranges are more than twice as big as those of *Colobus* (*Colobus*), averaging 35 hectares; and each home range almost completely overlaps those of adjacent groups. Core areas are not defended (Struhsaker & Oates, 1975); when one group supplants another, adult and subadult males engage in chasing, branch-shaking and leaping displays, accompanied by loud vocalisations.

Reproduction Little is know of the reproductive physiology of *Colobus* (*Piliocolobus*). The sexual swelling varies considerably in size, even between females of the same subspecies (Starin, personal communication, 1981).That of *C.b.waldroni* was described by Pocock (1936) as a trilobate cushion, subtriangular in shape, measuring 52mm from anus to the base of callosities and 59mm wide at anal level, diminishing to 38mm at mid-callosity level. Field notes state that the colour is white streaked with blue. In a female *C.b.preussi*, measured by Merfield (unpubl.), the swelling is recorded as 13½inches lengthwise, and 18½inches in circumference, which tends to support Struhsaker's statement (1975:58) that some *C.b.preussi* swellings appear to be equivalent to ¼ to ⅓ of the body volume of the female. *C.b.preussi* swellings are described as pink in colour. From a limited number of observations, Struhsaker (1975) concludes that the sexual swelling of the female is probably under hormonal control and associated with receptivity and copulation. The function of the male perineal organ is not clear. On the one hand, in *C.b.badius*, it mimicks the female sexual swelling, and Kuhn (1972) has suggested that it is probably of social significance, re-inforcing the appeasement gesture of presenting the hindquarters seen in young and halfgrown males. In *C.b.preussi*, on the other hand, where the female sexual swelling is largest, no male perineal organ has been observed (Struhsaker, 1975). Gestation is unknown; the interval between births is estimated at 20-24 months, as in *Presbytis entellus* (Struhsaker, 1975). He found little seasonality in births in Kibale Forest, Uganda (*C.b.tephrosceles*), but Merfield (unpubl.) notes that *C.b.preussi* births in Yabassi, Cameroon, were concentrated in May and June.

Species Two are recognised, following Thorington & Groves (1970). *C. kirkii* can be recognised by its slightly smaller size and characteristic long white brow fringes. The skulls can be separated in the following key (Verheyen, 1962):

Cranial capacity reaching at most 62cc .... *C. kirkii*
Cranial capacity equal to or greater than 62cc .... *C. badius*

C.b.rufomitratus (not studied by Verheyen) is however as small as *C. kirkii*.

COLOBUS (PILIOCOLOBUS) BADIUS (Kerr, 1792)

Red Colobus, Bay Colobus, Brown Colobus
(13 subspecies, see subgenus *Piliocolobus*, taxonomic notes, p. 22).

Range W. Africa from Senegal, the Gambia, Sierra Leone, Ivory Coast, Liberia and Ghana; Yabassi district, Cameroon highlands; Fernando Póo; Congo (Brazzaville), Zaire, Rwanda, Burundi, Uganda and Tanganyika; isolated populations in the R. Tana flood plain, Kenya (*rufomitratus*), and the Uzungwa Mountains, Tanganyika (*gordonorum*).

Dimensions Adult specimens in the museum collections have the following field measurements (see p. 98 for comparative table):

h & b	30 ♂♂	569	(470-690)	:	20 ♀♀	528	(470-584)
tail	30 ♂♂	665	(540-800)	:	20 ♀♀	665	(515-750)
wt	7 ♂♂	9348	(6300-11 340)	:	7 ♀♀	6770	(5500-9072)

Subspecies The difficulty of providing a key to the subspecies of *C. badius* has been noted by Dandelot (1971:33). They are most readily understood when divided geographically, first into western and eastern groups, and second into regions within those groups. The dividing line between east and west is not completely clear. *C.b.bouvieri* and *C.b.tholloni* of Dandelot's western group are found west of the Zaire/Ubangi and south of the R. Zaire. *C.b.oustaleti, ellioti* and *foai* of the eastern group are found north and east of the Zaire/Lualaba and south of the Ubangi. *C.b.oustaleti* is also said to be found on the right bank of the Zaire in *C.b.bouvieri* country. The following key and descriptive notes, based on pelage colour, may help in identification:

1 Crown and back the same colour; little individual variation; (no whorl behind ears) (Western subspecies) 2
1a Crown and back different colours; much individual variation (Eastern subspecies) 3
2 Cheeks the same colour as limbs .... Western subspecies group A
badius
temminckii
waldroni
preussi
2a Cheeks contrasting with limbs .... Western subspecies group B
pennantii
bouvieri
tholloni
3 Predominantly black with red crown (sometimes red back as well) and white underparts; no whorl behind ears
Eastern subspecies group C
gordonorum
3a Predominantly brown or red brown; whorl behind ears 4
4 Predominantly red brown with some brown or fawn tones, particularly on legs and underparts .... Eastern subspecies group D
oustaleti
ellioti
foai
4a Predominantly brown or greyish brown .... Eastern subspecies group E
tephrosceles
rufomitratus

Note: most *C. badius* subspecies have a prominent black brow band which extends to the ears or spreads on to the cheeks. The whorls behind the ears of the eastern subspecies deflect the hair forwards, forming a crest at the sides of the crown.

Group A from W. and West Central Africa.
C.b.badius (Sierra Leone, Liberia, W. Ivory Coast): head and back black, extending to outer arms and thighs; tail mainly black; all other parts of body deep maroon red.
C.b.waldroni (E. Ivory Coast, Ghana): more erythristic, with forehead and thighs also deep maroon.
C.b.temminckii (Senegal, Gambia and Guinea): a washed-out version of *C.b.badius* with black and maroon replaced by blue-grey and orange.
C.b.preussi (between Rs Cross and Sanaga): head and back black ticked with orange; cheeks, limbs and tail bright red.

Group B shows increasing erythrism from Fernando Póo to West and South Congo basin.
C.b.pennantii (Fernando Póo): head, back and tail dark red brown; cheeks white; limbs red brown.
C.b.bouvieri (Congo (Brazzaville)): head, dorsal stripe and tail red brown; cheeks white; flanks and limbs bright red.
C.b.tholloni (S. of R. Zaire): head, upper back and tail deep dark red; cheeks reddish, speckled with black; lower back and limbs bright red.

Group C (Mts of Central Tanganyika):
C.b.gordonorum: black above, white below: lower back often suffused with dark red; lower leg greyish white; crown deep chestnut red. Colour pattern approaching *C. kirkii* (p. 32).

Group D (between Rs Ubangi and Zaire/Lualaba, bounded to the east by the western Rift Valley): specimens have whorls behind their ears and show much individual variation in colour.
C.b.oustaleti (between Rs Ubangi & Zaire): crown dark red brown contrasting with fawn speckled cheeks; back varies from dark mahogany to red brown, olive brown or light brown; limbs generally lighter and browner than back; tail light brown to black.
C.b.ellioti (Ituri Forest): crown and cheeks deep red; back dark mahogany; body and limbs suffused with red anteriorly, brown posteriorly; tail black.
C.b.foai (between R. Lualaba and L. Tanganyika): crown and cheeks deep red (long hairs at front of crown); upper back dark mahogany; lower back, limbs and tail red brown.

Group E E. of western Rift Valley: specimens have whorls behind their ears and show much less individual variation in colour.
C.b.tephrosceles (Uganda, Rwanda, Burundi, W. Tanganyika): red crown, fawn speckled cheeks; sharp contrast between dark brown back, upper arms, hands, feet and tail and light fawn rump, forearms and legs.
C.b.rufomitratus (Tana R., Kenya): Light red crown with black-tipped crests which almost meet on vertex; fawn speckled cheeks; rest of body more uniform than *tephrosceles*, light greybrown, slightly darker on back and tail.

Colobus (Piliocolobus) badius badius (Kerr, 1792)

Principal synonyms *Simia ferruginea* Shaw, 1800:59: *Colobus ferruginosus* E. Geoffroy, 1812:92; *Colobus rufoniger* Ogilby, 1838:273.

Type localities Of *badius*, Sierra Leone; of *ferruginea*, Sierra Leone; of *ferruginosus*, La Guinée; of *rufoniger*, Sierra Leone.

Colobus (Piliocolobus) badius bouvieri (Rochebrune, 1886-87)

Principal synonym *Colobus (Piliocolobus) likualae* Matschie, 1914: 346.

Type localities Of *bouvieri*, doubtful. Bouvier's specimen was sent to Rochebrune from Gambia, hence Allen's (1939:170) type locality selection. However the type specimen, collected by De Brazza and figured by Rochebrune, is labelled: 'Mongo, right bank of Congo, 2 days journey by boat above the mouth of the R. Alima' (Schwarz, 1928:97: Moreau *et al*., 1946:407); of *likualae*, on R. Sangha opposite the mouth of R. Likouala aux Herbes, Congo (Brazzaville).

Colobus (Piliocolobus) badius ellioti Dollman, 1909

Principal synonyms *Piliocolobus anzeliusi* Matschie, 1914:339; *Piliocolobus ellioti melanochir* Matschie, 1914:339; *Colobus variabilis* Lorenz, 1914:383; *Colobus multicolor* Lorenz, 1914:385; *Colobus badius langi* J A Allen, 1925:443.

Type localities Of *ellioti*, 90km W. of S. end of L. Albert Edward (=L. Edward), Zaire; of *anzeliusi*, upper Ituri Province, Zaire; of *melanochir*, between Beni and Irumu, upper Ituri region, Zaire; of *variabilis*, Ituri forest, Zaire; of *multicolor*, by Mawambi (1°04′N 28°34′E), on R. Ituri, Zaire; of *langi*, Risimu, between Stanleyville and Bafwaboli, Zaire.

Colobus (Piliocolobus) badius foai Pousargues, 1899

Principal synonyms *Colobus graueri* Dollman, 1909:474; *Piliocolobus kabambarei* Matschie, 1914:338; *Piliocolobus lulindicus* Matschie, 1914:338.

Type localities Of *foai*, Ouroua country between the SW. shore of L. Tanganyika and the upper Congo (=Lualaba), Zaire; of *graueri*, Wabembeland, 80km W. of N. end of L. Tanganyika, Zaire; of *kabambarei*, Kabambare, between Baraka on NW. shore of L. Tanganyika and Kasonga on R. Lualaba, Zaire; of *lulindicus*, on R. Lulindi in the neighbourhood of its confluence with R. Lualaba (4°39′S 26°48′E), near Kasongo, Zaire.

Colobus (Piliocolobus) badius gordonorum (Matschie, 1900)

Type locality Uzungwa Mts (8°20′S 38°45′E), Uhehe, Tanganyika.

Colobus (Piliocolobus) badius oustaleti Trouessart, 1906

Principal synonyms *Colobus badius nigrimanus* Trouessart, 1906:444; *Colobus (Piliocolobus) powelli* Matschie, 1913:61; *Colobus (Tropicolobus) umbrinus* Matschie, 1914:343; *Colobus (Tropicolobus) schubotzi* Matschie, 1914:345; *Colobus (Piliocolobus) brunneus* Lönnberg, 1919:112.

Type localities Of *oustaleti*, 'Youmba country, Oubangui basin'. Moreau *et al*. (1946:407) note that Youmba is on the R. bank of the lower Ubangi at about 0°30′N 17°50′E in Congo (Brazzaville); of *nigrimanus*, Liranga, on R. bank of R. Zaire in Congo (Brazzaville) as noted by Moreau *et al*. (1946:407); of *powelli*, by Zokwa, between Mahagi and Irumu, NE. Zaire, at 5350 ft; of *umbrinus*, Bungi, on R. Sangha, between Ouesso and Ikelemba. This locality is in Congo (Brazzaville) rather than Zaire, as originally cited (Moreau *et al*., 1946:407); of *schubotzi*, Koloka, between Rs Likati and Bima, tributaries of the Itimbiri and Uele respectively, Zaire; of *brunneus*, Sili, upper Uele District, Zaire.

Colobus (Piliocolobus) badius pennantii Waterhouse, 1838

Type locality Fernando Póo.

Colobus (Piliocolobus) badius preussi (Matschie, 1900)

Type locality Barombi, Elephant Lake, N. Cameroon.

Colobus (Piliocolobus) badius rufomitratus Peters, 1879

Type locality Muniuni, Tana River, Kenya.

Colobus (Piliocolobus) badius temminckii Kuhl, 1820

Principal synonyms *Colobus fuliginosus* Ogilby, 1835:97; *Colobus rufofuliginus* Ogilby, 1838:270, substitute for *fuliginosus*.

Type localities Of *temminckii*, none cited; of *fuliginosus* and *rufofuliginus*, the Gambia.

Colobus (Piliocolobus) badius tephrosceles Elliot, 1907

Principal synonym *Tropicolobus gudoviusi* Matschie, 1914:340.

Type localities Of *tephrosceles*, Ruahara River, Toro, E. side of Mt Ruwenzori, Uganda at 4000ft; of *gudoviusi*, between Ussuwi and Ihangiro, SSE. of L. Burigi, W. of L. Victoria, NW. Tanganyika.

Colobus (Piliocolobus) badius tholloni Milne Edwards, 1886

Principal synonym *Colobus (Piliocolobus) lovizettii* Matschie, 1913:207.

Type localities Of *tholloni*, none cited (collected by De Brazza); of *lovizettii*, Kutu, L. Leopold II, Zaire.

Colobus (Piliocolobus) badius waldroni Hayman, 1936

Type locality Goaso, Ashanti, Ghana.

Locality	Date of Collection	Sex	Age	Nature of Specimen	Accession no.	References, Remarks, etc.
Colobus (Piliocolobus) badius badius (Kerr 1792)						
IVORY COAST						
Bakanda, W. bank of Bandama R., S. of Tiassalé	7 Jul 1953	F	Inf	Skin, Skull	ZD.1956.367	(*Procolobus badius badius*), Listed Grubb,P. and Booth,A.H. (unpubl.): 3
Guiglo, [6°40′N 7°28′W]	18 Jul 1953	F	Ad	Skin, Skull	ZD.1956.365	" " " "
" " "	Aug 1953	M	[Inf]	Skin	ZD.1956.366	(*Procolobus badius badius*)
Pont de Nzi, between Bandama and Nzi rivers, just N. of confluence, [c.5°58′N 4°52′W]	28 Mar 1954	F	Ad	Skin	ZD.1956.364	" " " "
Pont de Nzi, just N. of Bandama-Nzi confluence, E. of Bandama and Nzi rivers, [c.5°58′N 4°51′W]	27 Mar 1954	F	Inf	Skin, Skull	ZD.1956.362	(*Procolobus badius badius*), Listed Grubb,P. and Booth,A.H. (unpubl.): 3
" " "	27 Mar 1954	F	Ad	Skin, Skull	ZD.1956.363	" " " "
Tiassalé, West of Bandama R., [5°53′N 4°57′W]	5 Jul 1953	F	Ad	Skin, Skull	ZD.1956.368	" " " "

Locality	Date of Collection	Sex	Age	Nature of Specimen	Accession no.	References, Remarks, etc.
Colobus (Piliocolobus) badius badius *continued*						
W. edge of Bouaflé Forest Reserve, near Daloa	15 Jul 1953	M	Ad	Skin, Skull	ZD.1956.369	" " " "
—	—	F	Ad	Skull	ZD.1971.2375	(*Procolobus badius badius*), Listed Grubb,P. and Booth,A.H. (unpubl.): 3, Listed Brandon-Jones,D. (unpubl.)
—	—	F	Ad	Skull	ZD.1971.2376	" " " "
—	—	F	Juv	Skull	ZD.1971.2377	" " " "
—	—	F	Ad	Skull	ZD.1971.2378	" " " "
—	—	F	Ad	Skull	ZD.1971.2379	" " " "
—	—	F	Ad	Skull	ZD.1971.2380	" " " "
LIBERIA						
50 miles inland from Monrovia	—	F	Ad	Skin, Skull	ZD.1904.6.2.2	(*Colobus ferrugineus*)
Nanna Kru, [4°50′N 8°44′W]	10 Jan 1911	M	Ad	Skin, Skull	ZD.1911.11.3.1	
St.Paul's River	—	I	[Ad]	Skin	ZD.1919.7.11.1	(*Colobus ferrugineus*)
—	—	I	[Juv?]	Skin	ZD.1910.4.18.1	(*Colobus ferrugineus*)
SIERRA LEONE						
Kenema, [c.7°51′N 11°11′W], 495ft	13 Apr 1936	F	Ad	Skin, Skull	ZD.1936.8.14.1	
Makari	21 Feb 1946	M	Ad	Skin, Skull	ZD.1950.1867	
Near Potoru, Barri Chiefdom, [c.7°31′N 11°28′W]	30 Aug 1950	F	Inf	Skin, Skull	ZD.1952.23	
Near Rokupr, [c.8°40′N 12°23′W]	1945	I	[Ad]	Skin	ZD.1950.1868	
Near Zimmi, Makpele Chiefdom, [c.7°19′N 11°18′W]	1 Aug 1948	M	Ad	Skin, Skull	ZD.1950.1865	
Potoru, Barri Chiefdom, [7°31′N 11°29′W]	30 Aug 1950	F	Juv	Skin, Skull	ZD.1952.22	
Sangbaia, Sangbaia Chiefdom, [8°58′N 11°40′W]	15 Jun 1946	M	Juv	Skin, Skull	ZD.1946.849	
Zimmi, Makpele Chiefdom, [Pujehu], [c.7°19′N 11°18′W]	1 Aug 1948	M	Ad	Skin, Skull	ZD.1950.1866	
—	Mar 1922	[F]	[Ad]	Skin	ZD.1923.9.3.1	(*Colobus rufoniger*)
—	—	I	[Ad]	Skin	ZD.1923.9.15.1	" " " "
—	—	[I]	[Ad]	Skin	ZD.1923.9.15.2	" " " "
?SIERRA LEONE						
—	—	I	[Ad]	Skin	ZD.1842.11.4.11	
NO DATA						
—	—	[F]	[Ad]	Skin	ZD.1855.12.24.407	Syntype *Colobus rufoniger* Ogilby 1838, Cited Ogilby,W. 1838a: 273
—	—	[F]	Juv	Skull	RCS(OM) A67.1	
Colobus (Piliocolobus) badius bouvieri (Rochebrune 1886-87)						
[CONGO (BRAZZAVILLE)]						
Butando, French Congo, [c.1°S 16°30′E]	29 Dec 1926	F	Ad	Skin, Skull	ZD.1928.11.11.3	
Butando, French Congo, c.1°S 16°30′E	30 Dec 1926	F	Ad	Skin, Skull	PCM Fr.Congo	
Nola, French Congo, 3°30′N 16°E	Sep 1927	M	[Ad]	Skull	PCM Fr.Congo	
Colobus (Piliocolobus) badius ellioti Dollman 1909						
[RWANDA]						
Kisenyi, Lake Kivu, [1°42′S 29°15′E]	—	I	[Ad]	Skin	ZD.1928.1.30.15	
" " "	—	I	[Ad]	Skin	ZD.1928.1.30.16	
" " "	—	I	Ad	Skin	ZD.1929.4.27.4	
[ZAIRE]						
Aluta, W. of Beni, 3100ft	9 Apr 1930	F	Ad	Skin, Skull	ZD.1930.11.11.27	Cited Colyer,J.F. 1936: 28
" " "	12 Apr 1930	M	[Ad]	Skin	ZD.1930.11.11.28	
" " "	9 Apr 1930	F	[Ad]	Skin	ZD.1930.11.11.29	
" " "	9 Apr 1930	F	[Ad]	Skin	ZD.1930.11.11.30	
" " "	12 Apr 1930	F	[Ad]	Skin	ZD.1930.11.11.31	
Ambalango, Mawambe-Beni Road, Ituri, Congo, c.0°40′N 29°10′E	31 Jul 1906	M	Ad	Skin, Skull	PCM Congo	Cited Matschie,P. 1913: 59
Beni, [0°29′N 29°29′E]		M	Ad	Skull	ZD.1981.98	
" " "		M	Ad	Skull	ZD.1981.99	
" " "		M	Ad	Skull	ZD.1981.100	
Between Kasanga and Kasubi, Ituri Forest, Congo, c.1°10′N 28°30′E	19 Jul 1906	M	[Ad]	Skin	PCM Congo	Cited Matschie,P. 1913: 59
" " "	19 Jul 1906	F	Ad	Skin, Skull	PCM Congo	" " " "
" " "	19 Jul 1906	M	Ad	Skin, Skull	PCM Congo	" " " "
" " "	19 Jul 1906	M	Ad	Skin, Skull	PCM Congo	" " " "
" " "	19 Jul 1906	F	Ad	Skin, Skull	PCM Congo	" " " "
" " "	19 Jul 1906	M	Ad	Skin, Skull	PCM Congo	" " " "
" " "	19 Jul 1906	M	Ad	Skin, Skull	PCM Congo	" " " "
" " "	19 Jul 1906	M	Ad	Skin, Skull	PCM Congo	" " " "
" " "	19 Jul 1906	F	Ad	Skin, Skull	PCM Congo	" " " "
" " "	19 Jul 1906	M	Ad	Skin, Skull	PCM Congo	" " " "
" " "	19 Jul 1906	F	Ad	Skin, Skull	PCM Congo	" " " "
Between Kasanga and Kasubi, Mawambe-Avakubi Road, Congo, [c.1°10′N 28°30′E]	19 Jul 1906	M	Ad	Skin, Skull, Radiograph of penis	ZD.1930.11.7.2	

Locality	Date of Collection	Sex	Age	Nature of Specimen	Accession no.	References, Remarks, etc.
Colobus (Piliocolobus) badius ellioti *continued*						
Between Mawambe and Lenda, Ituri Forest, Congo, c.1°10′N 28°30′E	14 Sep 1905	F	Ad	Skin, Skull	PCM Congo	Cited Matschie,P. 1913: 59
„ „ „	14 Sep 1905	F	Juv	Skin, Skull	PCM Congo	„ „ „ „
Bwanasura, E.Ituri Forest, Irumu-Beni Rd., [1°05′N 29°37′E], 3700ft	27 Mar 1930	M	Ad	Skin, Skull	ZD.1930.11.11.17	
„ „ „	27 Mar 1930	F	Ad	Skin, Skull	ZD.1930.11.11.18	
„ „ „	27 Mar 1930	F	Ad	Skin, Skull	ZD.1930.11.11.19	
„ „ „	27 Mar 1930	F	Ad	Skin, Skull	ZD.1930.11.11.20	
„ „ „	27 Mar 1930	M	Inf	Skin, Skull	ZD.1930.11.11.21	
„ „ „	27 Mar 1930	M	Inf	Skin, Skull	ZD.1930.11.11.22	
By Makala, Ituri Forest, Congo, c.0°30′N 28°E	31 Jul 1906	I	[Ad]	Skin	PCM Congo	Cited Matschie,P. 1913: 59
„ „ „	31 Jul 1906	I	[Ad]	Skin	PCM Congo	„ „ „ „
By Moro Kifuku, Ituri Forest, Congo, c.1°20′N 29°50′E	25 Jul 1906	M	Ad	Skin, Skull	PCM Congo	„ „ „ „
Fundi Katama, 30 miles E. of Avakubi, Congo Forest, 2500ft	4 Feb 1914	M	Juv	Skin, Skull	ZD.1914.6.2.2	(*Colobus nigrimanus*)
Ituri, 50 miles S.W. of Irumu, [c.1°N 29°15′E], 3500ft	1 Apr 1930	M	Ad	Skin, Skull	ZD.1930.11.11.24	
„ „ „	1 Apr 1930	M	Ad	Skin, Skull	ZD.1930.11.11.25	
„ „ „	4 Apr 1930	M	Ad	Skin, Skull	ZD.1930.11.11.26	
Ituri Forest near Bwanasura, [c.1°05′N 29°37′E], 3500ft	29 Mar 1930	M	Ad	Skin, Skull	ZD.1930.11.11.23	
Kulu Kulu, Ituri Forest, Congo, c.1°N 28°E	4 Oct 1905	M	Ad	Skin, Skull	PCM Congo	Cited Matschie,P. 1913: 59
„ „ „	4 Oct 1905	F	Ad	Skin, Skull	PCM Congo	„ „ „ „
90km W. of S. end of Lake Edward	—	M	[Ad]	Skin	ZD.1909.12.2.3	Holotype *Colobus ellioti* Dollman 1909, Cited Dollman,G. 1909: 475
90km W. of S. end of Lake Edward, 1600m	Feb 1908	F	Ad	Skin, Skull	ZD.1909.12.2.4	Paratype *Colobus ellioti* Dollman 1909, Cited Dollman,G. 1909: 475
Lindi River, S. Ituri Forest, [c.0°30′S 28°30′E]	—	[F]	[Ad]	Skin	ZD.1923.8.1.2	
4 hours W. of Mawambe, Ituri Forest, Congo, c.1°15′N 28°35′E, 2499ft	30 Aug 1905	M	Ad	Skin, Skull	PCM Congo	Cited Matschie,P. 1913: 58
6 hours W. of Mawambe, Ituri Forest, Congo, c.1°15′N 28°35′E, 2454ft	15 Aug 1905	M	Ad	Skin, Skull	PCM Congo	„ „ „ „
„ „ „	15 Aug 1905	M	Ad	Skin, Skull	PCM Congo	„ „ „ „
„ „ „	15 Aug 1905	F	Ad	Skin, Skull	PCM Congo	„ „ „ „
„ „ „	15 Aug 1905	F	Ad	Skin, Skull	PCM Congo	„ „ „ „
„ „ „	15 Aug 1905	F	Ad	Skin, Skull	PCM Congo	„ „ „ „
„ „ „	15 Aug 1905	F	Inf	Skin, Skull	PCM Congo	„ „ „ „
Mgalese, 30 miles E. of Stanleyville, Congo, 1600ft	3 Mar 1914	F	Ad	Skin, Skull	ZD.1914.6.2.1	
Near Beni, Semliki Valley, [Kivu], Belgian Congo, [c.0°29′N 29°29′E]	—	M	Ad	Skin, Skull	ZD.1920.10.17.1	
„ „ „	—	M	Ad	Skin, Skull	ZD.1920.10.17.2	
Nr. Fundi, Mawambe-Beni Road, Ituri, Congo, c.1°N 29°E	28 Jul 1906	F	Ad	Skin, Skull	PCM Congo	Cited Matschie,P. 1913: 59
„ „ „	29 Jul 1906	M	Ad	Skin, Skull	PCM Congo	„ „ „ „
„ „ „	29 Jul 1906	F	Ad	Skin, Skull	PCM Congo	„ „ „ „
„ „ „	29 Jul 1906	I	[Inf]	Skin	PCM Congo	„ „ „ „
Nr. Kokomegle, Mawambe- Beni Road, Congo, c.0°40′N 29°10′E	3 Aug 1906	M	Ad	Skin, Skull	PCM Congo	„ „ „ „
„ „ „	3 Aug 1906	F	Ad	Skin, Skull	PCM Congo	„ „ „ „
R.Luna, E.Ituri Forest, [1°17′N 29°41′E], 3400ft	21 Mar 1930	M	Ad	Skin, Skull	ZD.1930.11.11.13	
„ „ „	21 Mar 1930	M	Ad	Skin, Skull	ZD.1930.11.11.14	
„ „ „	21 Mar 1930	F	Ad	Skin, Skull	ZD.1930.11.11.15	
„ „ „	21 Mar 1930	M	Ad	Skin, Skull	ZD.1930.11.11.16	
Colobus (Piliocolobus) badius foai Pousargues 1899						
[ZAIRE]						
Kabimba, L.Tanganyika, Congo Belge, [5°33′S 29°21′E]	1946-47	[F]	[Ad]	Skin	ZD.1953.13	(*Colobus badius graueri*)
Vicinity of Albertville [=Kalémié], Belgian Congo, [c.5°57′S 29°10′E]	—	I	[Juv]	Skin	ZD.1936.2.28.3	
Wabembeland, 80km W. of N. end of L.Tanganyika	—	M	Ad	Skin, Skull	ZD.1909.12.2.1	Holotype *Colobus graueri* Dollman 1909, Cited Dollman,G. 1909: 474
„ „ „	—	M	Ad	Skin, Skull	ZD.1909.12.2.2	Paratype *Colobus graueri* Dollman 1909, Cited Dollman,G. 1909: 474
Colobus (Piliocolobus) badius gordonorum (Matschie 1900)						
[TANGANYIKA]						
Iringa, Tanganyika Territory, [7°49′S 35°39′E]	—	I	[Ad]	Skin	ZD.1923.3.4.2	
Colobus (Piliocolobus) badius gordonorum? (Matschie 1900)						
TANGANYIKA						
—	—	I	[Ad]	Skin	ZD.1925.12.12.1	(*C.badius tephrosceles?*)

Locality	Date of Collection	Sex	Age	Nature of Specimen	Accession no.	References, Remarks, etc.
Colobus (Piliocolobus) badius oustaleti Trouessart 1906						
SUDAN						
S.W.Sudan	—	I	[Ad]	Skin	ZD.1934.9.6.1	
[ZAIRE]						
Loama, R.Uele, [3°21′N 24°22′E]	11 Feb 1906	F	Juv	Skin, Skull	ZD.1907.7.8.8	(*Colobus badius powelli*)
R.Ituri (W. bank), 50 miles S.W.of Irumu, 3500ft	30 Apr 1930	M	Ad	Skin, Skull	ZD.1930.11.11.32	(*Colobus powelli*)
,, ,, ,,	30 Apr 1930	M	Ad	Skin, Skull	ZD.1930.11.11.33	,, ,, ,, ,,
,, ,, ,,	30 Apr 1930	F	Ad	Skin, Skull	ZD.1930.11.11.34	,, ,, ,, ,,
,, ,, ,,	30 Apr 1930	F	Juv	Skin, Skull	ZD.1930.11.11.35	,, ,, ,, ,,
R.Ituri (W.bank), 50 miles S.W. of Irumu, 3500ft	1 May 1930	F	Ad	Skin, Skull	ZD.1930.11.11.36	,, ,, ,, ,,
,, ,, ,,	1 May 1930	F	Ad	Skin, Skull	ZD.1930.11.11.37	,, ,, ,, ,,
,, ,, ,,	1 May 1930	F	Juv	Skin, Skull	ZD.1930.11.11.38	,, ,, ,, ,,
,, ,, ,,	1 May 1930	F	Ad	Skin, Skull	ZD.1930.11.11.39	,, ,, ,, ,,
,, ,, ,,	30 Apr 1930	F	Inf	Skin, Skull	ZD.1930.11.11.40	,, ,, ,, ,,
Upper Uele or Ituri [Rivers], [N.E.Congo]	—	F	[Ad]	Skin	ZD.1916.6.29.3	,, ,, ,, ,,
Zakwa, Ituri Forest, Congo, c.1°30′N 30°30′E, 5350ft	13 Jun 1905	M	Ad	Skin, Skull	PCM Congo	Holotype *Colobus* (*Piliocolobus*) *powelli* Matschie 1913, Cited Matschie,P. 1913: 61-64
,, ,, ,,	13 Jun 1905	M	Ad	Skin, Skull	PCM Congo	Paratype *Colobus* (*Piliocolobus*) *powelli* Matschie 1913, Cited Matschie,P. 1913: 61-64
,, ,, ,,	13 Jun 1905	F	Ad	Skin, Skull	PCM Congo	,, ,, ,, ,,
,, ,, ,,	13 Jun 1905	F	Ad	Skin, Skull	PCM Congo	,, ,, ,, ,,
,, ,, ,,	13 Jun 1905	F	Ad	Skin, Skull	PCM Congo	,, ,, ,, ,,
,, ,, ,,	13 Jun 1905	F	Ad	Skin, Skull	PCM Congo	,, ,, ,, ,,
Zakwa, Ituri Forest, [c.1°30′N 30°30′E]	—	F	Ad	Skin, Skull	ZD.1930.11.7.1	,, ,, ,, ,,
NO DATA						
—	—	I	[Juv]	Skin	ZD.1935.8.26.1	
—	—	I	[Ad]	Skin	ZD.1935.8.26.3	(*Colobus badius powelli*)
—	—	I	[Ad]	Skin	ZD.1935.8.26.4	,, ,, ,, ,,
—	—	I	[Ad]	Skin	ZD.1935.8.26.5	,, ,, ,, ,,
—	—	[F]	[Ad]	Skin	ZD.1935.8.26.6	,, ,, ,, ,,
—	—	I	[Juv]	Skin	ZD.1935.8.26.7	,, ,, ,, ,,
Colobus (Piliocolobus) badius pennantii Waterhouse 1838						
FERNANDO PÓO						
—	—	[F]	[Ad]	Skin	ZD.1841.1751	Lectotype *Colobus badius pennantii* Waterhouse 1838, Cited Waterhouse,G.R. 1838: 57, Listed Anon. (unpubl.)b: 80, Cited Schwarz,E. 1928: 97
—	—	[F]	[Ad]	Skin	ZD.1841.1752	Paralectotype *Colobus badius pennantii* Waterhouse 1838, Cited Waterhouse,G.R. 1838: 57, Listed Anon. (unpubl.)b: 80, Cited Schwarz,E. 1928: 97
—	—	I	[Ad]	Skin	ZD.1842.11.4.10	
Colobus (Piliocolobus) badius preussi (Matschie 1900)						
CAMEROON						
[Ndokyam Hills], Yabassi Dist., 4°15′N 10°15′E	Apr 1939	M	[Ad]	Skin	ZD.1947.1553	Listed Merfield,F.G. (unpubl.)
[Ndokyam Hills], Yabassi Dist., [4°15′N 10°15′E]	18 Apr 1939	M	Ad	Skin, Skull, Skeleton	RSM 1948/89/2	,, ,, ,, ,,
,, ,, ,,	18 Apr 1939	M	Ad	Skin, Skull, Skeleton	RSM 1948/89/3	,, ,, ,, ,,
N.E. of Yabassi, Ndokofass Dist., 4°30′N 10°30′E	May 1939	M	Ad	Skin	ZD.1947.1552	,, ,, ,, ,,
,, ,, ,,	[Jun] 1939	F	Ad	Skin, Skeleton	ZD.1947.1554	,, ,, ,, ,,
,, ,, ,,	[Jun] 1939	F	[Ad]	Skin	ZD.1947.1555	,, ,, ,, ,,
,, ,, ,,	Jun 1939	M	Ad	Skin, Skull, Skeleton	ZD.1949.90	,, ,, ,, ,,
,, ,, ,,	May 1939	F	Ad	Skin, Skull, Skeleton	ZD.1949.91	,, ,, ,, ,,
,, ,, ,,	Jun 1939	M	Ad	Skin, Skull, Skeleton	PCM Merfield	,, ,, ,, ,,
,, ,, ,,	Jun 1939	M	Ad	Skin, Skull, Skeleton	PCM Merfield	,, ,, ,, ,,
,, ,, ,,	Jun 1939	F	Ad	Skin, Skull, Skeleton	PCM Merfield	,, ,, ,, ,,
,, ,, ,,	Jun 1939	F	Juv	Skin, Skull, Skeleton	PCM Merfield	,, ,, ,, ,,
,, ,, ,,	Jun 1939	F	Ad	Skin, Skull, Skeleton	PCM Merfield	,, ,, ,, ,,
,, ,, ,,	Jun 1939	F	Ad	Skin, Skull, Skeleton	PCM Merfield	,, ,, ,, ,,
,, ,, ,,	Jun 1939	M	Ad	Skin, Skull, Skeleton	PCM Merfield	,, ,, ,, ,,
,, ,, ,,	Jun 1939	M	[Ad]	Skin	PCM Merfield	,, ,, ,, ,,
,, ,, ,,	Jun 1939	F	Ad	Skin, Skull	PCM Merfield	,, ,, ,, ,,
,, ,, ,,	Jun 1939	F	[Ad]	Skin	PCM Merfield	,, ,, ,, ,,
,, ,, ,,	Jun 1939	F	Ad	Skin, Skull	PCM Merfield	,, ,, ,, ,,
,, ,, ,,	Jun 1939	F	Ad	Skin, Skull, Skeleton	PCM Merfield	,, ,, ,, ,,
,, ,, ,,	Jun 1939	F	Ad	Skin, Skull, Skeleton	PCM Merfield	,, ,, ,, ,,
,, ,, ,,	Jun 1939	M	Ad	Skin, Skull, Skeleton	PCM Merfield	,, ,, ,, ,,
,, ,, ,,	Jun 1939	F	Ad	Skin, Skull, Skeleton	PCM Merfield	,, ,, ,, ,,
,, ,, ,,	Jun 1939	F	Ad	Skin, Skull, Skeleton	PCM Merfield	,, ,, ,, ,,
,, ,, ,,	Jun 1939	F	Ad	Skin, Skull, Skeleton	PCM Merfield	,, ,, ,, ,,
,, ,, ,,	Jun 1939	F	Ad	Skin, Skull, Skeleton	PCM Merfield	,, ,, ,, ,,

Locality	Date of Collection	Sex	Age	Nature of Specimen	Accession no.	References, Remarks, etc.
Colobus (Piliocolobus) badius preussi *continued*						
,, ,, ,,	Jun 1939	F	[Ad]	Skin	PCM Merfield	,, ,, ,, ,,
,, ,, ,,	Jun 1939	F	Ad	Skin, Skull	PCM Merfield	,, ,, ,, ,,
,, ,, ,,	Jun 1939	F	[Ad]	Skin	PCM Merfield	,, ,, ,, ,,
,, ,, ,,	Jun 1939	F	Ad	Skin, Skull, Skeleton	PCM Merfield	,, ,, ,, ,,
,, ,, ,,	Jun 1939	F	[Ad]	Skin	PCM Merfield	,, ,, ,, ,,
,, ,, ,,	Jun 1939	M	[Ad]	Skin	PCM Merfield	,, ,, ,, ,,
,, ,, ,,	Jun 1939	F	[Ad]	Skin, Skull, Skeleton	PCM Merfield	,, ,, ,, ,,
,, ,, ,,	Jun 1939	F	[Ad]	Skin, Skull, Skeleton	PCM Merfield	,, ,, ,, ,,
,, ,, ,,	Jun 1939	F	[Ad]	Skin, Skull, Skeleton	PCM Merfield	,, ,, ,, ,,
,, ,, ,,	Jun 1939	F	[Ad]	Skin, Skull	PCM Merfield	,, ,, ,, ,,
,, ,, ,,	Jun 1939	M	Ad	Skin, Skull, Skeleton	PCM Merfield	,, ,, ,, ,,
,, ,, ,,	May 1939	M	[Ad]	Skin, Skull, Skeleton	RSM 1948/89/1	,, ,, ,, ,,
,, ,, ,,	[Jun 1939]	F	Ad	Skin, Skull, Skeleton	RSM 1948/89/4	,, ,, ,, ,,
,, ,, ,,	[Jun 1939]	F	[Ad]	Skin, Skull, Skeleton	RSM 1948/89/5	,, ,, ,, ,,
,, ,, ,,	[Jun 1939]	F	Ad	Skin, Skull, Skeleton	RSM 1948/89/6	,, ,, ,, ,,
Ndokyam Hills, Yabassi Dist., 4°15′N 10°15′E	1 Apr 1939	M	Ad	Skin, Skull, Skeleton	PCM Merfield	,, ,, ,, ,,
,, ,, ,,	Apr 1939	F	Ad	Skin, Skull, Skeleton	PCM Merfield	,, ,, ,, ,,
,, ,, ,,	Apr 1939	F	Ad	Skin, Skull, Skeleton	PCM Merfield	,, ,, ,, ,,
,, ,, ,,	Apr 1939	F	Ad	Skin, Skull, Skeleton	PCM Merfield	,, ,, ,, ,,
,, ,, ,,	Apr 1939	F	Ad	Skin, Skull, Skeleton	PCM Merfield	,, ,, ,, ,,
,, ,, ,,	Apr 1939	F	Ad	Skin, Skull, Skeleton	PCM Merfield	,, ,, ,, ,,
,, ,, ,,	Apr 1939	F	Ad	Skin, Skull, Skeleton	PCM Merfield	,, ,, ,, ,,
,, ,, ,,	Apr 1939	F	[Ad]	Skin	PCM Merfield	,, ,, ,, ,,
,, ,, ,,	Apr 1939	F	Ad	Skin, Skull, Skeleton	PCM Merfield	,, ,, ,, ,,
,, ,, ,,	Apr 1939	F	Foetus	Wet, cadaver	PCM Merfield	,, ,, ,, ,,
,, ,, ,,	Apr 1939	F	Ad	Skin, Skull, Skeleton	PCM Merfield	,, ,, ,, ,,
Yabassi Dist., 4°15′N 10°15′E	Apr 1939	M	Ad	Skin, Skull, Skeleton	ZD.1940.108	,, ,, ,, ,,
,, ,, ,,	Apr 1939	F	Ad	Skin, Skull, Skeleton	ZD.1940.109	,, ,, ,, ,,
,, ,, ,,	1938	M	[Ad]	Skin	ZD.1968.525	
,, ,, ,,	Mar 1939	F?	[Ad]	Skin	PCM Merfield	
[CAMEROON]						
Manjeme, British Cameroons, c.5°20′N 9°40′E, 1500ft	12 Dec 1948	M	Juv	Skin	ZD.1956.531	
Ndokyam Hills, Yabassi Dist., 4°15′N 10°15′E	Apr-May 1939	F	Ad	Skeleton	ZD.1977.217	Listed Merfield,F.G. (unpubl.)
,, ,, ,,	1939	F	Ad	Skin, Skull	PCM Merfield, RCS(OM) G11.7	Listed Merfield,F.G. (unpubl.), Cited Colyer,J.F. 1940: 757-768, Cited Colyer,J.F. 1943: 112, 114
,, ,, ,,	1939	F	Ad	Skin, Skull	PCM Merfield, RCS(OM) G68.771	Listed Merfield,F.G. (unpubl.), Cited Colyer,J.F. 1940: 757-768
,, ,, ,,	1939	F	Ad	Skin, Skull	PCM Merfield, RCS(OM) G68.775	,, ,, ,, ,,
,, ,, ,,	1939	F	Ad	Skin, Skull	PCM Merfield, RCS(OM) G68.779	,, ,, ,, ,,
,, ,, ,,	Apr-May 1939	F	Ad	Skull	RCS(OM) G68.792	Listed Merfield,F.G. (unpubl.), Figured Colyer,J.F. 1940: Figs.7a-b, Figured Colyer,J.F. 1943: 207, fig.28
,, ,, ,,	1939	F	Ad	Skin, Skull	PCM Merfield, RCS(OM) G68.793	Listed Merfield,F.G. (unpubl.), Figured Colyer,J.F. 1940: Fig.6, Figured Colyer,J.F. 1943: 185, Fig.26
N.E. of Yabassi, Ndokofass Dist., 4°30′N 10°30′E	Jun 1939	F	Ad	Skeleton	ZD.1977.218	Listed Merfield,F.G. (unpubl.)
,, ,, ,,	Jun 1939	F	Ad	Skeleton	ZD.1977.219	,, ,, ,, ,,
,, ,, ,,	1939	F	Ad	Skin, Skull	PCM Merfield, RCS(OM) G68.77	Listed Merfield,F.G. (unpubl.), Cited Colyer,J.F. 1940: 757-768
,, ,, ,,	1939	F	Ad	Skin, Skull	PCM Merfield, RCS(OM) G68.772	Listed Merfield,F.G. (unpubl.), Figured Colyer,J.F. 1940: Fig.2
,, ,, ,,	1939	F	Ad	Skin, Skull	PCM Merfield, RCS(OM) G68.773	Listed Merfield,F.G. (unpubl.), Cited Colyer,J.F. 1940: 757-768
,, ,, ,,	1939	F	Ad	Skin, Skull	PCM Merfield, RCS(OM) G68.774	,, ,, ,, ,,
,, ,, ,,	1939	F	Ad	Skin, Skull	PCM Merfield, RCS(OM) G68.776	,, ,, ,, ,,
,, ,, ,,	1939	F	Ad	Skin, Skull	PCM Merfield, RCS(OM) G68.777	,, ,, ,, ,,
,, ,, ,,	1939	F	Ad	Skin, Skull	PCM Merfield, RCS(OM) G68.778	Listed Merfield,F.G. (unpubl.) Figured Colyer,J.F. 1940: Fig.3, Figured Colyer,J.F. 1943: 184, fig.23
N.E. of Yabassi, Ndokofass Dist., French Cameroons, 4°30′N 10°30′E	1939	F	Ad	Skull	RCS(OM) G68.78	Listed Merfield,F.G. (unpubl.), Cited Colyer,J.F. 1940: 757-768
N.E. of Yabassi, Ndokofass Dist., 4°30′N 10°30′E	1939	F	Ad	Skull	RCS(OM) G68.781	Listed Merfield,F.G. (unpubl.), Figured Colyer,J.F. 1940: Fig.5, Figured Colyer,J.F. 1943: 185, fig.25
,, ,, ,,	1939	F	Ad	Skull	RCS(OM) G68.782	Listed Merfield,F.G. (unpubl.) Figured Colyer,J.F. 1943: 185, fig.24 Figured Colyer,J.F. 1940: Figs.4,9,
,, ,, ,,	1939	F	Juv	Skull	RCS(OM) G68.79	Listed Merfield,F.G. (unpubl.), Cited Colyer,J.F. 1940: 757-768, Figured Colyer,J.F. 1943: 213, fig.40
,, ,, ,,	1939	M	Ad	Skull	RCS(OM) G68.791	Listed Merfield,F.G. (unpubl.), Cited Colyer,J.F. 1940: 757-768
,, ,, ,,	1939	F	Ad	Skin, Skull	PCM Merfield, RCS(OM) G68.794	,, ,, ,, ,,

Locality	Date of Collection	Sex	Age	Nature of Specimen	Accession no.	References, Remarks, etc.
Colobus (Piliocolobus) badius preussi *continued*						
„ „ „	1939	F	Ad	Skull	RCS(OM) G169.2	Listed Merfield,F.G. (unpubl.), Cited Colyer,J.F. 1940: 757-768 Figured Colyer,J.F. 1943: 233, fig.42
„ „ „	1939	F	Ad	Skin, Skull	PCM Merfield, RCS(OM) G169.21	Listed Merfield,F.G. (unpubl.), Cited Colyer,J.F. 1940: 757-768, Cited Colyer, J.F. 1943: 112
Colobus (Piliocolobus) badius rufomitratus Peters 1879						
KENYA						
Tana R.	28 Mar 1923	M	Ad	Skin, Skull, Skeleton	ZD.1923.10.15.1	Cited Schwarz, E. 1928: 97, Cited Colyer, J.F. 1936: 28
„ „ „	28 Mar 1923	F	Juv	Skin, Skull, Skeleton	ZD.1923.10.15.2	„ „ „ „
„ „ „		F	Ad	Skull	ZD.1923.10.15.5	Cited Colyer, J.F. 1936: 28
[KENYA]						
[Tana R.]		F	Ad	Skull	ZD.1936.12.30.1	
„ „ „		F	Ad	Skull	ZD.1936.12.30.2	
„ „ „		F	Ad	Skull	ZD.1936.12.30.3	
„ „ „		F	Ad	Skull	ZD.1936.12.30.4	
„ „ „		F	Juv	Skull	ZD.1936.12.30.5	
Colobus (Piliocolobus) badius temminckii Kuhl 1820						
GAMBIA						
Bakalari, [13°25′N 16°26′W]	24 Dec 1928	F	Ad	Skin, Skull	ZD.1931.12.1.6	
Bakkendik, [13°27′N 16°27′W]	19 Dec 1928	F	Ad	Skin, Skull	ZD.1931.12.1.5	
„ „ „	17 Dec 1928	F	Ad	Skin, Skull	ZD.1931.12.1.4	
„ „ „	17 Dec 1928	M	Inf	Skin, Skull	ZD.1931.12.1.3	
„ „ „	—	M	Ad	Skin, Skull, Radiograph of penis	ZD.1931.12.1.2	
„ „ „	17 Dec 1928	M	Ad	Skin, Skull	ZD.1931.12.1.1	
10 miles S. of Cape St. Mary, [c.13°30′N 16°40′W]	5 Feb 1946	[F]	Ad	Skin, Skull	ZD.1946.133	
Seringmass	30 Dec 1928	F	Ad	Skin, Skull	ZD.1931.12.1.7	
—	—	F	Ad	Skin, Skull	ZD.1922.12.11.3	(*Colobus ferrugineus*)
—	—	I	Inf	Skin, Skull	ZD.1924.12.14.1	„ „ „ „
—	23 Apr 1945	F	[Inf]	Skin, Skull in skin	ZD.1945.149	
[GAMBIA]						
—	—	I	[Inf]	Skin	ZD.1922.12.11.4	(*Colobus ferrugineus*)
—	—	[F]	[Ad]	Skin	ZD.1837.7.15.355	Holotype *Colobus fuliginosus* Ogilby 1835, Cited Ogilby,W. 1835: 97, Figured Ogilby,W. 1838a: 270, Listed Anon. (unpubl.)b: 67
[GUINEA-BISSAU]						
Crocoli Camp, Portuguese Guinea, c.12°10′N 13°55′W	27 Jun 1911	M	[Ad]	Skin, Skull	PCM Guinea	(*Colobus fuliginosus*)
„ „ „	27 Jun 1911	F	[Ad]	Skin, Skull	PCM Guinea	„ „ „ „
Kiu Camp, Portuguese Guinea, c.12°10′N 14°50′W	14 Apr 1911	F	[Juv]	Skin, Skull	PCM Guinea	„ „ „ „
„ „ „	19 Apr 1911	F	[Ad]	Skin, Skull	PCM Guinea	„ „ „ „
Kopulan Camp, Portuguese Guinea, c.12°14′N 13°58′W	24 Apr 1911	F	[Inf]	Skin	PCM Guinea	„ „ „ „
„ „ „	24 Apr 1911	M	[Ad]	Skin, Skull	PCM Guinea	„ „ „ „
NO DATA						
—	—	[F]	Juv	Skin, Skull	ZD.1839.12.26.13	Listed Anon. (unpubl.)b: 67
—	—	M	Ad	Skull	ZD.1840.1.18.12	„ „ „ „
—	—	[F]	Ad	Skull, Skeleton	ZD.1850.11.13.1	Listed Gray,J.E. (unpubl.): 778
Colobus (Piliocolobus) badius temminckii? Kuhl 1820						
NO DATA						
—	—	[F]	[Ad]	Skin	ZD.1981.111	
—	—	[F]	[Ad]	Skin, Skull in skin	ZD.1981.113	
Colobus (Piliocolobus) badius tephrosceles Elliot 1907						
TANGANYIKA						
Biharamulo, S.W. of L.Victoria, Tanganyika Territory	—	M	Ad	Skin	ZD.1939.303	
„ „ „	—	I	Juv	Skin	ZD.1939.304	
Biharamulo, S.W. of L.Victoria, Tanganyika Territory, 4500ft	[c.1939]	F	Juv	Skull	RCS(OM) A67.3	
„ „ „	[c.1939]	M	Ad	Skull	RCS(OM) A67.31	
Biharamulo, [4500ft]	—	M	Ad	Skeleton	ZE.1968.7.25.1	
„ „ „	—	F	Juv	Skeleton	ZE.1968.7.25.2	
Mbizi Forest, Ufipa Plateau E. of Sumbawanga	9 Jul 1957	M	Ad	Skin, Skull	ZD.1958.352	
Near Ruiga Bay, W. coast of L.Victoria, 3750ft	21 Apr 1949	F	Ad	Skin, Skull	ZD.1951.365	

Locality	Date of Collection	Sex	Age	Nature of Specimen	Accession no.	References, Remarks, etc.
Colobus (Piliocolobus) badius tephrosceles *continued*						
[TANGANYIKA]						
E. side of Mahari [=Mahali] Mts., Kigoma Dist., Tanganyika Territory, 5000ft	—	I	Juv	Skull	RCS(OM) A67.32	
,, ,, ,,	—	F	Ad	Skull	RCS(OM) A67.33	
Lupa Forest, 10 miles S. of Busenga Village, near upper Niamanzi R., Tanganyika Territory, 5°50′S 30°55′E	Oct 1943	[M]	Ad	Skin, Skull	ZD.1944.41	Correspondence Moreau,R.E. (unpubl.)
Mahare [=Mahali] Mts., S.Kigoma Dist., Tanganyika Territory	—	I	[Ad]	Skin	ZD.1936.2.28.4	(*Colobus badius gudoviusi?*)
Mahari [=Mahali] Mts., Kigoma Dist., Tanganyika Territory, 5000ft	Dec 1945	F	[Ad]	Skin	ZD.1964.2025	
Near Biharamulo, Lake Province, Tanganyika Territory	[1938]	I	[Ad]	Skin	ZD.1947.1550	
,, ,, ,,	[1938]	I	[Ad]	Skin	ZD.1947.1551	
UGANDA						
Busesse, N.W.Ankole	Sep 1928	M	Ad	Skin, Skull	ZD.1929.5.14.26	
,, ,, ,,	Sep 1928	F	Ad	Skin, Skull	ZD.1929.5.14.27	
,, ,, ,,	Sep 1928	F	Inf	Skin, Skull	ZD.1929.5.14.28	
,, ,, ,,	—	M	Ad	Skull	ZD.1929.5.14.34	
Katebe, N.W.Ankole, [0°31′S 30°37′E], 4000ft	28 Aug 1929	M	Ad	Skin, Skull, Skeleton	ZD.1930.8.1.1	
,, ,, ,,	28 Aug 1929	M	Ad	Skin, Skull, Skeleton	ZD.1930.8.1.2	
,, ,, ,,	28 Aug 1929	M	Ad	Skin, Skull	ZD.1930.8.1.3	
,, ,, ,,	28 Aug 1929	M	Ad	Skin, Skull	ZD.1930.8.1.4	Figured James,W.Warwick 1960: 240-241, figs.61a-c
,, ,, ,,	28 Aug 1929	M	Ad	Skin, Skull	ZD.1930.8.1.5	
,, ,, ,,	28 Aug 1929	F	Ad	Skin, Skull, Skeleton	ZD.1930.8.1.6	
,, ,, ,,	28 Aug 1929	F	Juv	Skin, Skull, Skeleton	ZD.1930.8.1.7	
,, ,, ,,	28 Aug 1929	F	Juv	Skin, Skull, Skeleton	ZD.1930.8.1.8	
,, ,, ,,	28 Aug 1929	I	Inf	Skin, Skull, Skeleton	ZD.1930.8.1.9	
,, ,, ,,	28 Mar 1929	F	Foetus	Wet cadaver	ZD.1930.12.1.4	
Kibale Forest, Toro Dist., 0°30′N 30°25′E	Mar 1946	M	Ad	Skull	ZD.1969.379	Cited Haddow,A.J. (unpubl.)a
,, ,, ,,	Mar 1946	F	Ad	Skull	ZD.1969.380	,, ,, ,, ,,
Kibale Forest, Toro Dist., [0°30′N 30°25′E]	1946	F	Inf	Skull	RCS(OM) A67.4	Cited Colyer,J.F. (unpubl.), Background Haddow,A.J. *et al* 1951: 189-224, Correspondence Haddow,A.J. (unpubl.), Cited Haddow,A.J. (unpubl.)a
,, ,, ,,	1942-47	F	Juv	Skull	RCS(OM) A67.41	,, ,, ,, ,,
,, ,, ,,	1942-47	M	Juv	Skull	RCS(OM) A67.42	,, ,, ,, ,,
Kibale Forest, Toro Dist., [0°30′N 30°25′E], 4000ft	1942-49	M	Ad	Skull	RCS(OM) A67.43	,, ,, ,, ,,
Kibale Forest, Toro Dist., [0°30′N 30°25′E], 5000ft	1942-49	F	Ad	Skull	RCS(OM) A67.44	,, ,, ,, ,,
Kibale Forest, Toro Dist., [0°30′N 30°25′E]	1942-47	F	Ad	Skull	RCS(OM) G11.1211	,, ,, ,, ,,
,, ,, ,,	1942-47	[F]	Ad	Skull	RCS(OM) G68.7951	,, ,, ,, ,,
,, ,, ,,	1942-47	F	Juv	Skull	RCS(OM) G68.7952	,, ,, ,, ,,
,, ,, ,,	1942-47	F	Ad	Skull	RCS(OM) G80.14	,, ,, ,, ,,
,, ,, ,,	1942-49	F	Ad	Skull	RCS(OM) G167.52	,, ,, ,, ,,
Near Kanyawara, Kibale Forest, W.Uganda, 0°35′N 30°20′E, 4800ft	1 Apr 1971	M	Ad	Skin, Skull	ZD.1971.2064	Correspondence Struhsaker,T.T. (unpubl.)
,, ,, ,,	1 Apr 1971	M	Ad	Skin, Skull	ZD.1971.2065	,, ,, ,, ,,
R.Mpanga, 5 miles N. of Rusangwe, Toro Dist., 4000ft	Feb 1930	M	Juv	Skin, Skull	ZD.1930.12.1.3	
Toro Dist., 0°50′N 31°05′E, 5000ft	1 Oct 1964	M	Ad	Skin, Skull, Skeleton, Radiograph of penis	ZD.1972.133	
[UGANDA]						
Ruahara R., Ruwenzori, [Toro]	—	F	Ad	Skin, Skull	ZD.1901.8.9.130	
,, ,, ,,	—	F	Inf	Skin, Skull	ZD.1901.8.9.131	
Ruahara R., Toro Dist., 4000ft	Aug 1900	M	[Ad]	Skin	ZD.1901.8.9.129	Holotype *Colobus tephrosceles* Elliot 1907, Cited Elliot,D.G. 1907: 195, Cited Thomas,O. 1901a: 86
Ruahara R., [Toro], 4000ft	Aug 1900	M	Ad	Skin, Skull	ZD.1901.8.9.15	
,, ,, ,,	Aug 1900	M	Ad	Skull, Skeleton	ZD.1901.8.9.46	Paratype *Colobus tephrosceles* Elliot 1907, Cited Elliot,D.G. 1907: 195, Figured Elliot,D.G. 1913: Pl.15, Cited Thomas,O. 1901a: 86
Colobus (Piliocolobus) badius tholloni Milne Edwards 1886						
[ZAIRE]						
Bopania, Lac Leopold II, Congo Belge, [1°12′S 18°59′E], c.320m	16 Sep 1926	M	Ad	Skin, Skull	ZD.1927.3.1.4	Figured Colyer,J.F. 1936: 208, fig.251
Lukallila, Congo, 1°30′S	Sep 1894	I	[Ad]	Skin	ZD.1898.1.5.1	
Nkolongo, Congo	—	M	Ad	Skin, Skull	ZD.1912.7.26.6	
,, ,, ,,	—	F	Ad	Skin, Skull	ZD.1912.7.26.7	

Locality	Date of Collection	Sex	Age	Nature of Specimen	Accession no.	References, Remarks, etc.
Colobus (Piliocolobus) badius waldroni Hayman 1936						
GHANA						
Kwama Bi Krom, W.Ghana, [6°38′N 3°07′W]	May 1969	F	[Ad]	Skin	ZD.1970.2624	
Tano Lodge, Sefwi Wiawso, [c.6°22′N 2°29′W]	—	M	Ad	Skull	ZD.1971.2381	(*Procolobus badius waldroni*)
,, ,, ,,	—	F	Juv	Skull	ZD.1971.2382	,, ,, ,, ,,
,, ,, ,,	—	F	Ad	Skull	ZD.1971.2383	,, ,, ,, ,,
,, ,, ,,	—	F	Ad	Skull	ZD.1971.2384	,, ,, ,, ,,
,, ,, ,,	—	F	Ad	Skull	ZD.1971.2385	,, ,, ,, ,,
,, ,, ,,	—	F	Ad	Skull	ZD.1971.2386	,, ,, ,, ,,
,, ,, ,,	—	F	Ad	Skull	ZD.1971.2387	,, ,, ,, ,,
,, ,, ,,	—	F?	Ad	Skull	ZD.1971.2388	(*Procolobus badius?*)
,, ,, ,,	—	F	Ad	Skull	ZD.1971.2389	,, ,, ,, ,,
,, ,, ,,	—	F	Ad	Skull	ZD.1971.2390	,, ,, ,, ,,
,, ,, ,,	—	F	Ad	Skull, Dental casts	ZD.1971.2391	,, ,, ,, ,,
,, ,, ,,	—	F	Ad	Skull	ZD.1971.2392	,, ,, ,, ,,
[GHANA]						
Ahafo-Mim, 20 miles N. of Goaso, Gold Coast	6 Jul 1952	F	Juv	Skin, Skull	ZD.1956.375	Listed Grubb,P. and Booth,A.H. (unpubl.)
,, ,, ,,	25 Dec 1952	F	Juv	Skin, Skull	ZD.1956.376	,, ,, ,, ,,
Bibianaha, Gold Coast, [6°30′N 2°08′W], 720[ft]	16 Apr 1912	F	Inf	Skin, Skull	ZD.1912.6.20.1	(*Colobus ferrugineus*)
Bibianaha, Gold Coast, [6°30′N 2°08′W]	—	[M]	Inf	Skin, Skull	ZD.1912.4.10.2	,, ,, ,, ,,
Bibiani, Sefwi, Gold Coast, [6°28′N 2°20′W]	Died 29 Jul 1952	F	Inf	Skin, Skull	ZD.1956.372	Listed Grubb,P. and Booth,A.H. (unpubl.)
Gambia [=Kenyasi Mohu], 30 miles N. of Goaso, Gold Coast, [7°04′N 2°39′W]	27 Dec 1952	F	Juv	Skin, Skull	ZD.1956.370	,, ,, ,, ,,
Gambia [=Kenyasi Mohu], 30 miles N. of type locality [Goaso], Gold Coast, [7°04′N 2°39′W]	6 [Sep] 1953	F	Ad	Skin, Skull	ZD.1956.371	,, ,, ,, ,,
Goaso, Ashanti, Gold Coast, [6°49′N 2°27′W]	23 Dec 1933	M	Ad	Skin, Skull	ZD.1935.1.30.3	Paratype *Colobus badius waldroni* Hayman 1936, Cited Hayman,R.W. 1936: 915
,, ,, ,,	23 Dec 1933	M	Ad	Skin, Skull	ZD.1935.1.30.4	,, ,, ,, ,,
,, ,, ,,	29 Dec 1933	M	Juv	Skin, Skull	ZD.1935.1.30.5	,, ,, ,, ,,
,, ,, ,,	29 Dec 1933	M	Ad	Skin, Skull	ZD.1935.1.30.6	,, ,, ,, ,,
Goaso, Ashanti, [6°49′N 2°27′W]	31 Dec 1933	M	Ad	Skin, Skull	ZD.1935.1.30.7	Holotype *Colobus badius waldroni* Hayman 1936, Figured Hayman,R.W. 1936: 915, pl.1
,, ,, ,,	26 Dec 1933	F	Ad	Skin, Skull	ZD.1935.1.30.8	Paratype *Colobus badius waldroni* Hayman 1936, Cited Hayman,R.W. 1936: 915, Figured James,W.Warwick 1960: 242-243, figs.62a-c
,, ,, ,,	27 Dec 1933	F	Ad	Skin, Skull	ZD.1935.1.30.9	Paratype *Colobus badius waldroni* Hayman 1936, Cited Hayman,R.W. 1936: 915, Listed Lowe,W.P. (unpubl.)b
,, ,, ,,	30 Dec 1933	F	Ad	Skin, Skull	ZD.1935.1.30.10	Paratype *Colobus badius waldroni* Hayman 1936, (*Colobus badius ferrugineus*), Cited Hayman,R.W. 1936: 915, Listed Lowe,W.P. (unpubl.)b
,, ,, ,,	16 Dec 1934	M	Ad	Skin, Skull	ZD.1935.10.22.1	Paratype *Colobus badius waldroni* Hayman 1936, Cited Hayman,R.W. 1936: 915
,, ,, ,,	26 Dec 1934	M	Ad	Skin, Skull	ZD.1935.10.22.2	
,, ,, ,,	—	F	[Ad]	Wet,cadaver	ZD.1935.10.22.3	Paratype *Colobus badius waldroni* Hayman 1936, Cited Hayman,R.W. 1936: 915, Listed Lowe,W.P. (unpubl.)b
Mim, 20 miles N. of type locality, Goaso, Gold Coast, [6°54′N 2°34′W]	25 Dec 1952	M	Ad	Skin, Skull	ZD.1956.379	Listed Grubb,P. and Booth,A.H. (unpubl.)
IVORY COAST						
Pont de Nzi, just N. of Bandama-Nzi confluence	27 Mar 1954	F	[Ad]	Skin	ZD.1956.373	,, ,, ,, ,,
NO DATA						
—	—	I	[Juv]	Skin	ZD.1935.8.26.8	
Colobus (Piliocolobus) badius waldroni? Hayman 1936						
GHANA						
E.Region	—	F	Ad	Skull	ZD.1981.59	
Sefwi, Wiawso, W.Ghana, [6°12′N 2°19′W]	1972	I	[Ad]	Skin	ZD.1976.1792	
Colobus (Piliocolobus) badius (Kerr 1792)						
NO DATA						
—	—	F	Ad	Skull	ZD.1981.89	

COLOBUS (PILIOCOLOBUS) KIRKII Gray, 1868

Kirk's Colobus

Type locality Zanzibar

Taxonomic notes As mentioned on p. 22, *C. kirkii* is retained as a separate species, following Thorington & Groves (1970) and Dandelot (1971). Lectotype, designated by Elliot (1913:128), a young adult female cadaver in 1% phenoxytol, BM(NH) 1868.3.16.1.

Range Zanzibar Island and, according to BM(NH) specimens, the East African coast opposite Zanzibar.

Dimensions Adult specimens in the museum collections have the following field measurements (see p. 98 for comparative table:)

h & b	1♂ 450	: 3♀♀ 490 (465-506)
tail	1♂ 595	. 3♀♀ 671 (644-715)
wt	1♂ 5800	: 3♀♀ 5463 (5100-6100)

Locality	Date of Collection	Sex	Age	Nature of Specimen	Accession no.	References, Remarks, etc.
Colobus (Piliocolobus) kirkii Gray 1868						
[TANGANYIKA]						
Dar-es-Salaam	—	M	[Inf]	Wet, cadaver	ZD.1874.4.20.5	
East African Coast opposite Zanzibar near Pangani	—	F	Ad	Skin, Skull, Skeleton	ZD.1874.4.20.1	Listed Gray,J.E. (unpubl.): 1648
ZANZIBAR						
Between Jembiani and Mkunduchi, East Coast	5 Aug 1918	F	[Ad]	Skin, Skull	ZD.1918.11.1.1	
Jozani Forest, [6°15′S 39°24′E], Sea level	7 Dec 1959	M	Inf	Skin, Skull	ZD.1972.131	
Kitogani, Muongoni, [6°17′S 39°26′E]	29 Jul 1954	M	Ad	Skin, Skull	ZD.1964.1006	
" " "	29 Jul 1954	F	Ad	Skin, Skull	ZD.1964.1007	
" " "	29 Jul 1954	F	Inf	Skin, Skull	ZD.1964.1008	
Mkunduchi, [6°25′S 39°33′E]	—	F	Ad	Skin, Skull	ZD.1924.12.15.2	Figured Colyer,J.F. 1936: 31, fig.38
Mkunduchi, [6°25′S 39°33′E], 100ft	11 Apr 1919	M	Juv	Skin, Skull	ZD.1919.6.9.1	
[Pangani], [Village on Zanzibar]	—	F	[Juv]	Wet,cadaver	ZD.1868.3.16.1	Lectotype *Colobus kirkii* Gray 1868, Figured Gray,J.E. 1868: 180, pl.15
Pete, 17 miles S.E. of Zanzibar City, Pete, [6°17′S 39°25′E]	15 Jul 1954	F	Juv	Skin, Skull	ZD.1964.1009	
" " "	17 Jul 1954	F	Ad	Skin, Skull	ZD.1964.1010	
—	—	M	[Ad]	Skin, Skull, Skeleton	ZD.1906.6.5.1	
—	—	[F]	[Ad]	Skin, Skull	ZD.1906.6.5.2	
—	—	[F]	[Ad]	Skin, Skull	ZD.1906.6.5.3	Figured Elliot,D.G. 1913: Pl.16
—	—	I	Juv	Skin, Skull	ZD.1906.6.5.4	
—	—	I	[Inf]	Skin	ZD.1906.6.5.5	
—	—	F	Ad	Skin, Skull	ZD.1937.5.10.1	
Sea level	5 Apr 1968	F	Ad	Skin, Skull, Skeleton	ZD.1972.132	
[ZANZIBAR]						
—	—	[F]	Juv	Skull, Skeleton	ZD.1869.1.26.1	Listed Gray,J.E. (unpubl.): 1648

Locality	Date of Collection	Sex	Age	Nature of Specimen	Accession no.	References, Remarks, etc.
Colobus sp. indet.						
[KENYA]						
Kanam East, [S. of Kavirondo Gulf], [0°21′S 34°30′E]	1937	I	Juv	Mandibular fragment, P3-4 erupting, Cast of same	BM(NH)A M15911	(Lower Pleistocene), Figured Szalay,F.S. and Delson,E. 1979: 386, fig.193, H-J

PROCOLOBUS Rochebrune, 1886-87

Olive Colobus
(1 species: type species *Colobus verus* Van Beneden, 1838)

Principal synonym *Lophocolobus* Pousargues, 1895 (type species: *Colobus verus* Van Beneden, 1838).

Taxonomic note Although resembling *Colobus (Piliocolobus)* in a number of characters, the Olive Colobus has been given generic status, following Thorington & Groves (1970).

Morphology Smallest colobine genus with the characteristic features of a complex stomach which resembles that of the subgenus *Piliocolobus* rather than that of *Colobus* (Kuhn, 1964), and the absence of cheek pouches (Hill, 1952). The thumb is reduced to a tubercle. With its olive-brown back, greyish limbs and greyish white underparts, *Procolobus* is as subdued in appearance as *Colobus* is flamboyant. Its only adornment is the insignificant longitudinal compressed crest on the crown. Its position in relation to *Colobus* is comparable to that of the talapoin *(Miopithecus)* to the larger more spectacular forest guenons. A sexual swelling appears in females before puberty and is more or less permanent once developed (Hill, 1952). There is also a perineal swelling in newborn and juvenile males, but not in adults (Kuhn, 1967). *Procolobus* differs from *Colobus* in many details of its visceral anatomy and male genitalia (e.g. relatively large testes) (Hill, 1952). Unlike *Colobus*, sexual dimorphism in body size is minimal. Characters separating the skull from that of the genus *Colobus* are cited from Verheyen (1962) on p. 3.

Range From Sierra Leone and SE. Guinea to Benin (Booth, 1958). Recently an isolated population has been located east of the R. Niger near Ayangba (7°30′N 7°18′E) (Menzies, 1970).

Ecology Apart from Booth's observations (1957), little is known of the life of *Procolobus*. It has never survived in captivity for more than a few days. In natural conditions, its olive-brown cryptic coloration and retiring disposition make it difficult to observe. *Procolobus* often associates with other monkey species *(Cercopithecus mona, C. campbelli* and *C. petaurista)*, and makes good use of their warning cries to conceal itself from danger. It lives at low levels of the forest where it feeds at a height of not more than 20 ft from the ground, only climbing higher to sleep or to escape observation. It has a preference for riverbanks and swamps; its native name in Togo signifies 'water monkey'. Unlike *Colobus*, it has never been observed to come to the ground. Diet, based on the examination of the contents of 33 stomachs, consists almost exclusively of leaves; neither insects nor fruits were found (Booth, 1957).

Social behaviour *Procolobus* lives in groups of 6-20 individuals. More than one adult male may be present.

Reproduction Little is known of reproduction. The sexual swelling of the female appears to be reduced in pregnant and lactating females, and is probably under oestrogenic control. The colour of the newborn scarcely differs from that of the adult. The olive colobus is unique among the Anthropoidea in the method by which the infant is transported. For several weeks after birth, the infant is carried in the mother's mouth. Later it clings by its own efforts to its mother's abdomen.

PROCOLOBUS VERUS Van Beneden, 1838

Olive Colobus

Principal synonyms *Semnopithecus (Colobus) olivaceus* Wagner, 1840:309, a substitute for *Colobus verus; Colobus cristatus* Gray, 1866:77. *(Colobus ?? chrysurus* Gray, 1866: 77, a furriers's skin received from W. Africa with the holotype of *Colobus cristatus*, has been identified as a howler monkey).

Type localities Of *verus* and *olivaceus*, 'Afrique'; of *cristatus*, West Africa.

Range As for the genus.

Dimensions Adult specimens in the museum collections have the following field measurements (see p. 98 for comparative table):

h & b	6 ♂♂	480 (455-510)	:	14 ♀♀	466	(380-546)
tail	6 ♂♂	559 (500-610)	:	14 ♀♀	573	(430-710)
wt	1 ♂	4000	:	2 ♀♀	3500	(2900, 4100)

Locality	Date of Collection	Sex	Age	Nature of Specimen	Accession no.	References, Remarks, etc.
Procolobus verus (Van Beneden 1838)						
GHANA						
Sukusuku, W.Ghana, [6°34′N 3°14′W]	9 Jul 1969	[M]	[Ad]	Skin	ZD.1970.2626	
„ „ „	Nov 1970	I	[Ad]	Skin	ZD.1976.1794	
Tano Lodge, Sefwi Wiawso, [c.6°22′N 2°29′W]	—	M	Ad	Skull	ZD.1971.2396	Listed Grubb,P. and Booth,A.H. (unpubl.)
„ „ „	—	F	Juv	Skull	ZD.1971.2397	„ „ „ „
„ „ „	—	F.	Ad	Skull	ZD.1971.2398	„ „ „ „
„ „ „	—	F	Ad	Skull	ZD.1971.2399	„ „ „ „
„ „ „	—	M	Ad	Skull	ZD.1971.2400	(*Procolobus ?verus*), Listed Grubb,P. and Booth,A.H. (unpubl.)
„ „ „	—	F	Ad	Skull	ZD.1971.2401	„ „ „ „
Trans-Bia, W.Ghana	Oct 1970	F	Ad	Skull	ZD.1976.1795	
[GHANA]						
Akroso, [5°46′N 0°46′W]	9 Nov 1945	F	Ad	Skin, Skull	ZD.1946.365	
?Apapaso, Afram plains	—	I	[Ad]	Skin	ZD.1937.8.16.1	
Atimpoku Island, Near Kpong, Volta R., [6°14′N 0°06′E]	30 Mar 1955	F	[Ad]	Skin	ZD.1956.341	
Bekwai, Ashanti, Gold Coast, 600[ft]	9 Feb 1947	F	Inf	Skin, Skull	ZD.1950.1500	
Goaso, Ashanti, [6°49′N 2°27′W]	23 Dec 1933	F	Juv	Skin, Skull	ZD.1935.1.30.1	
„ „ „	18 Dec 1934	F	Ad	Skin, Skull	ZD.1935.10.22.4	Cited Lowe,W.P. (unpubl.)b
„ „ „	30 Dec 1933	F	Ad	Skin, Skull	ZD.1935.1.30.2	
Gold Coast	1927	F	[Ad]	Skin	ZD.1927.6.22.2	(*Colobus cristatus*)
Mankrong, Afram River, Gold Coast, [6°42′N 0°15′W]	8 Apr 1952	M	[Juv]	Skin, Radiograph of penis	ZD.1956.339	
„ „ „	3 Feb 1953	F	Juv	Skin, Skull	ZD.1956.340	Listed Grubb,P. and Booth,A.H. (unpubl.)
IVORY COAST						
Daloa-Man Road, near Daloa, [c.6°56′N 6°28′W]	15 Jul 1953	F	Juv	Skin, Skull	ZD.1956.337	
25km W. of Daloa, [c.6°48′N 6°40′W]	30 Mar 1954	F	[Juv]	Skin	ZD.1956.338	
„ „ „	30 Mar 1954	M	Ad	Skull	ZD.1956.338A	Listed Grubb,P. and Booth,A.H. (unpubl.)

Locality	Date of Collection	Sex	Age	Nature of Specimen	Accession no.	References, Remarks, etc.
Procolobus verus *continued*						
Guiglo, [6°40′N 7°28′W]	9 Apr 1954	M	Inf	Skin, Skull	ZD.1956.332	
,, ,, ,,	4 Apr 1954	M	Ad	Skin, Skull, Radiograph of penis	ZD.1956.333	Listed Grubb,P. and Booth,A.H. (unpubl.)
,, ,, ,,	20 Jul 1953	F	Ad	Skin, Skull	ZD.1956.334	,, ,, ,, ,,
,, ,, ,,	9 Apr 1954	F	[Ad]	Skin	ZD.1956.335	
Tiassalé, [5°53′N 4°57′W]	5 Jul 1953	F	Ad	Skin, Skull	ZD.1956.336	Listed Grubb,P. and Booth,A.H. (unpubl.)
[Tiassalé], [5°53′N 4°57′W]	[5 Jul 1953]	M	Juv	Skull	ZD.1971.2393	,, ,, ,, ,,
NIGERIA						
Ayangba, 46 miles N.E. of Idah [E. of R.Niger], 7°32′N 7°15′E	17 Feb 1967	F	Ad	Skin, Skull	ZD.1967.1085	Correspondence Menzies,J.I. (unpubl.), Cited Menzies,J.I. 1970: 83
SIERRA LEONE						
Baoma, Koya Chiefdom	15 Nov 1948	F	[Juv]	Skin	ZD.1952.21	
Baoma, Koya Chiefdom, S.W.Sierra Leone	15 Oct 1948	M	[Ad]	Skin	ZD.1950.1873	
,, ,, ,,	15 Oct 1948	F	[Ad]	Skin	ZD.1950.1874	
Futa, Peje Chiefdom, [7°34′N 11°34′W]	30 Aug 1950	M	Ad	Skin, Skull	ZD.1962.450	
Futa, Peje Chiefdom, S.W.Sierra Leone, [7°34′N 11°34′W]	10 Aug 1949	M	Ad	Skin, Skull	ZD.1950.1869	
,, ,, ,,	11 Aug 1949	F	Ad	Skin, Skull	ZD.1950.1870	
,, ,, ,,	11 Aug 1949	[F]	[Ad]	Skin, Skull	ZD.1952.14	
,, ,, ,,	30 Aug 1950	F	Ad	Skin, Skull	ZD.1952.15	
,, ,, ,,	30 Aug 1950	F	Ad	Skin, Skull	ZD.1952.16	
,, ,, ,,	30 Aug 1950	F	Ad	Skin, Skull	ZD.1952.17	
,, ,, ,,	30 Aug 1950	F	Juv	Skin, Skull	ZD.1952.18	
,, ,, ,,	30 Aug 1950	F	Ad	Skin, Skull	ZD.1952.19	
Gegbwema, Tunkia Chiefdom, [7°35′N 11°10′W]	15 Dec 1948	F	Ad	Skin, Skull	ZD.1950.1872	
Near Potoru, [c.7°31′N 11°28′W]	14 Aug 1952	F	Juv	Skin, Skull	ZD.1953.127	
Near Potoru, Barri Chiefdom, [c.7°31′N 11°28′W]	30 Aug 1950	F	[Ad]	Skin	ZD.1962.451	
Near Potoru, Pujehun District, [c.7°31′N 11°28′W]	—	M	Inf	Skull	ZD.1962.452	
,, ,, ,,	—	I	Juv	Skull	ZD.1962.453	
,, ,, ,,	—	I	Inf	Skull	ZD.1962.454	
,, ,, ,,	—	I	Juv	Skull	ZD.1962.455	
,, ,, ,,	—	I	Juv	Skull	ZD.1962.456	
,, ,, ,,	—	I	Juv	Skull	ZD.1962.457	
,, ,, ,,	—	I	Juv	Skull	ZD.1962.458	
Near Zimmi, Makpele Chiefdom, S.W.Sierra Leone, [c.7°19′N 11°18′W]	1 Aug 1948	F	Ad	Skin, Skull	ZD.1950.1871	
Poturu, Barri Chiefdom, [7°31′N 11°28′W]	30 Aug 1950	M	Ad	Skin, Skull	ZD.1952.11	
Potoru, Barri Chiefdom, S.W.Sierra Leone, [7°31′N 11°28′W], 200ft	31 Aug 1950	M	Ad	Skin, Skull	ZD.1950.2079	
,, ,, ,,	31 Aug 1950	M	Ad	Skin, Skull	ZD.1950.2080	
,, ,, ,,	31 Aug 1950	F	Ad	Skin, Skull, Dental casts	ZD.1950.2081	
,, ,, ,,	31 Aug 1950	F	Juv	Skin, Skull	ZD.1950.2082	
,, ,, ,,	31 Aug 1950	F	Ad	Skin, Skull	ZD.1950.2083	
,, ,, ,,	31 Aug 1950	F	Ad	Skin, Skull	ZD.1950.2084	
Potoru, Barri Chiefdom, S.W.Sierra Leone, [7°31′N 11°28′W]	30 Aug 1950	F	Juv	Skin, Skull	ZD.1952.12	
Talia, Yawme Chiefdom, Bonthe District	24 Sep 1950	F	Ad	Skin, Skull	ZD.1952.20	
[SIERRA LEONE]						
Near Potoru, Barri Chiefdom, [c.7°31′N 11′28′W]	30 Aug 1950	F	Juv	Skin, Skull	ZD.1952.13	
NO DATA						
—	—	M	[Ad]	Skin, Skull	ZD.1861.6.1.2	Holotype *Colobus cristatus* Gray 1866, Cited Gray,J.E. 1866a: 77
—	—	[M]	[Ad]	Skin	ZD.1935.8.26.2	
—	—	M	Ad	Skull	ZD.1971.2395	Listed Grubb, P. and Booth, A.H. (unpubl.)

LIBYPITHECUS Stromer, 1913 (extinct)

(1 species: type species *Libypithecus markgrafi* Stromer, 1913)

Morphology About the size of *Colobus badius*, with typical colobine dentition, *Libypithecus* is known from a single but very complete male cranium. Damage during life has lead to asymmetry of the anterior teeth, the left incisors having been displaced mesially. Significant features are marked sagittal and nuchal crests, narrow inter-orbital region and long nasal bones. The original orientation of the maxilla, which makes contact with the cranial bones at only two points, is questionable, and a pictorial restoration is shown in Szalay & Delson (1979, fig. 174) to whom reference should be made for further details. They consider *Libypithecus* closer to present-day *Colobus* than to the south and east African *Paracolobus-Cercopithecoides* assemblage.

Range Known only from the Wadi Natrun, Northern Egypt. *Age* Late Miocene (late Turolian, estimated at 6 MYA by Cooke & Maglio, 1972).

Ecology The shores of Natrun Lake were probably wooded at this time, but surrounded by open country which was inhabited by a species of macaque, as well as ostrich, mastodon and *Hipparion* (Simons & Delson, 1978:109).

LIBYPITHECUS MARKGRAFI Stromer, 1913 (extinct)

Type locality Gar Maluk, Wadi Natrun, northern Egypt. *Age* Late Miocene (*c*. 6 MYA).

Referred material The type skull lacks its mandible but with it, at the same site, was found a single worn isolated lower first molar which is probably referable here (Simons & Delson, 1978:108). Known only from the type site.

Dimensions Estimated to be of medium size, similar to *Colobus badius* (Simons & Delson, 1978).

Locality	Date of Collection	Sex	Age	Nature of Specimen	Accession no.	References, Remarks, etc.
Libypithecus markgrafi Stromer 1913 (extinct)						
EGYPT						
Wadi el Natrùn [Gar Maluk], [c.30°20′N 30°15′E]	—	[M]	Ad	Cast of skull	BM(NH)A M14776	Holotype *Libypithecus markgrafi* Stromer 1913, Figured Stromer, E. 1913: 351, pls.7-8, Figured Simons,E.L. 1972: 200, fig.83, Cited Stromer,E. 1920: 349, Cited Hill,W.C.Osman 1970: 635
[EGYPT]						
[Wadi el Natrùn, Gar Maluk], [c.30°20′N 30°15′E]	—	M	Ad	Cast of upper teeth	BM(NH)A M32192	Holotype *Libypithecus markgrafi* Stromer 1913, Figured Stromer,E. 1913: 351, pls.7-8, Listed Delson,E. (unpubl.): 12
,, ,, ,,	—	I		Cast of upper P3-M1	BM(NH)A M32193	(*Libypithecus*), Figured Stromer,E. 1913: 349, fig.4, Listed Delson,E. (unpubl.): 12

CERCOPITHECOIDES Mollett, 1947 (extinct)

(1 species: type species *Cercopithecoides williamsi* Mollett, 1947)

Taxonomic notes A second species from the Koobi Fora formation, East Turkana, has been described by Leakey & Leakey (1973) and will be named by M G Leakey (in press). Other finds from Olduvai and Makapansgat may also be referred here (Szalay & Delson, 1979:392).

Morphology Large-sized colobines with strong sexual dimorphism in cranial shape; male skulls are long and narrow while females have a short facial region and globular braincase. There is a pronounced depression (ophryonic groove) behind the brow-ridge, but sagittal and nuchal crests are absent. The teeth are typically colobine and interorbital region is broad. *Cercopithecoides* appears to be most closely allied to *Paracolobus* but differs in its smaller size, very short muzzle with straight profile, and the shape of the mandible, the corpus of which is of constant depth, with the ascending ramus sloping slightly backwards at an angle, as in many cercopithecines. No post-cranial bones have as yet been assigned to *Cercopithecoides*.

Range One of the most widely ranging African fossil monkeys, *Cercopithecoides* has been found from sites in South and East Africa. *Age* Late Pliocene to middle Pleistocene.

Ecology Associated with both large and small baboons and hominids in the South African breccias, and thus thought to have lived in an open country habitat with varying amounts of tree and bush cover (see Vrba, 1975).

CERCOPITHECOIDES WILLIAMSI Mollett, 1947 (extinct)

Principal synonyms *Parapapio coronatus* Broom and Robinson, 1950:242; *Cercopithecoides molletti* Freedman, 1957:232.

Type localities Of *williamsi*, Makapan Valley; of *coronatus*, Kromdraai australopithecine site, Transvaal, South Africa; of *molletti*, Swartkrans, Transvaal, South Africa.

Taxonomic notes In 1950, Broom and Robinson proposed the name *Parapapio coronatus* for a calvarium and two mandibular fragments found in a surface pocket in the Kromdraai breccia from which specimens of *Paranthropus robustus* had been recovered. Its affinities were considered by Freedman (1957:241) to be uncertain, but later (1972:6) he and Brain re-assigned it to *Cercopithecoides williamsi*. In 1957, Freedman proposed the name *C. molletti* for some larger *Cercopithecoides* material recovered from Swartkrans, but in 1961 relegated it to the synonymy of *C. williamsi* on the grounds that further material had diminished, and in many cases bridged, the size gap between the two. This view was later confirmed by Maier (1971:74)

Referred material South African cave deposits including Makapansgat, Sterkfontein, Swartkrans, Bolt's Farm, Cooper's, Kromdraai 'ape-man' site, Taung and Graveyard (Freedman, 1970). *Age* Late Miocene to middle Pleistocene.

Dimensions Large-sized colobines; the cranium of *C. williamsi* is larger than that of any living species of *Colobus* (Leakey & Leakey, 1973:129).

Locality	Date of Collection	Sex	Age	Nature of Specimen	Accession no.	References, Remarks, etc.
Cercopithecoides williamsi Mollett 1947 (extinct)						
S.AFRICA						
Makapansgat, Transvaal, [c.24°08'S 29°12'E]	—	[M]	Ad	Cast of cranium	BM(NH)A M32261	Figured Freedman,L. 1961: 29, figs.10-13
" " "	—	[M]	Ad	Cast of mandible	BM(NH)A M32262	Figured Freedman,L. 1957: 230, 232, pl.54,fig.117, pl.55,fig.119
Makapansgat, [Transvaal], [c.24°08'S 29°12'E]	—		Ad	Cast of upper L P4-M3	BM(NH)A M32263	
Makapansgat, Transvaal, [c.24°08'S 29°12'E]	—	M	Ad	Cast of skull in 3 parts	BM(NH)A M32264	Holotype *Cercopithecoides williamsi* Mollett 1947, Figured Mollett,O.D.v.d.S. 1947: 298, figs.5-7, Figured Freedman,L. 1957: 226, pls.52-53, figs.112,114
[S.AFRICA]						
Bolt's Farm, [Krugersdorp], [Transvaal], [26°02'S 27°52'E]	—	[F]	Ad	Cast of cranium	BM(NH)A M32259	
" " "	—	[F]	Ad	Cast of mandible	BM(NH)A M32260	
[Kromdraai "ape-man" site], [2km E. of Sterkfontein], [Transvaal], [25°58'S 27°47'E]	—	[F]	[Ad]	Cast of lower L P3-M2	BM(NH)A M32198	(*"Parapapio" coronatus*), Figured Broom,R. and Robinson,J.T. 1950: 244, fig.5, Figured Freedman,L. 1957: 238-240, pl.58, figs.128-129, Cited Freedman,L. and Brain,C.K. 1972: 6
" " "	—	I	Ad	Cast of lower M2-M3	BM(NH)A M32199	" " " "
Makapansgat, Transvaal, [c.24°08'S 29°12'E]	—	M	Ad	Cast of cranium	BM(NH)A M16837	[Holotype] [*Cercopithecoides williamsi* Mollett 1947], Figured Mollett,O.D.v.d.S. 1947: 298, figs.5-7
[Makapansgat], [Transvaal], [c.24°08'S 29°12'E]	—	[M]	[Ad]	Cast of R C	BM(NH)A M32258	Cited Freedman,L. 1957: 233
[Swartkrans], [6 miles N.W. of Krugersdorp], [Transvaal], [26°01'S 27°42'E]	—	[M]	[Ad]	Cast of L maxilla with C-M3	BM(NH)A M32294	Holotype *Cercopithecoides molletti* Freedman 1957, Figured Freedman,L. 1957: 232, 235-237, pl.55, fig.121, pl.56, fig.122

PARACOLOBUS R E F Leakey, 1969 (extinct)

(1 species: type species *Paracolobus chemeroni* R E F Leakey, 1969)

Taxonomic note Eck (1977) described a mandibular fragment of a large colobine from the Shungura Formation (Member G), R. Omo, Ethiopia, as Colobinae gen. et. sp. indet. (C). Recently however Szalay & Delson (1979:396) have assigned it to the genus *Paracolobus* as cf. *Paracolobus* sp. A number of other colobine fragments from Hadar, Omo and E. Turkana may well be referable to *Paracolobus*, but at present only one species can be recognised with any certainty.

Morphology Based on an almost complete individual from the Chemeron Beds, W. of L. Baringo, Kenya, *Paracolobus* is a large colobine with a short broad face, broad inter-orbital region and typical colobine dentition. It differs from *Cercopithecoides* in its larger size, marked post-orbital constriction, broad rounded muzzle, and in the shape of the mandible, the corpus of which is deeper at M3 level than anteriorly. The mandibular ramus is at right angles to the corpus. The post-cranial skeleton is very complete but has not yet been described. The limb bones, particularly the femur, are robust and the forelimb is almost as long as the hind limb, suggesting a terrestrial way of life, though the foot, with its long phalanges, seems better adapted for arboreal life. The hand is unfortunately missing. Szalay & Delson (1979) consider that *Paracolobus* may somewhat resemble a larger version of *Nasalis*, a colobine which has the limb proportions of a terrestrial monkey, yet spends most of its time in trees.
The even larger mandible described and figured by Eck (1977), if correctly referred here, would indicate that *Paracolobus* exhibits gigantism similar to that of many East African Pliocene and Pleistocene mammals (Simons & Delson, 1978).

Range Chemeron Beds, W. of L. Baringo, Kenya. *Age* Pliocene. Material from the Shungura Formation, Omo River, and other East African sites may also be referred here.

PARACOLOBUS CHEMERONI R E F Leakey, 1969 (extinct)

Type locality Basal deposits of the Chemeron Beds, W. of L. Baringo, Kenya, Site JM90. *Age* Originally stated to be lower Pleistocene, but Simons & Delson (1978:109) note that the skeleton is about 4 million years old.

Dimensions Similar in proportions to *Nasalis* but slightly larger.

Locality	Date of Collection	Sex	Age	Nature of Specimen	Accession no.	References, Remarks, etc.
Paracolobus chemeroni Leakey,R. 1969 (extinct)						
KENYA						
[S.W. of] L.Baringo, Chemeron beds, Site JM90, [c.0°30'N 36°E]	—	M	Ad	Cast of skull	BM(NH)A M32297	Holotype *Paracolobus chemeroni* Leakey,R. 1969, Figured Leakey,R.E.F. 1969: 54, pl.1

Locality	Date of Collection	Sex	Age	Nature of Specimen	Accession no.	References, Remarks, etc.
Cf. Paracolobus sp. (extinct)						
ETHIOPIA						
[R.] Omo, Shungura Formation, Member G5, [c.5°N 36°E]	—	I	Ad	Cast of L mandibular fragment with M2-3	BM(NH)A M35366	Cited Eck,G. 1976: 336, Figured Eck,G. 1977: 58, pl.1, Figured Szalay,F.S. and Delson,E. 1979: 396

Locality	Date of Collection	Sex	Age	Nature of Specimen	Accession no.	References, Remarks, etc.
Colobinae gen. et sp. indet.						
ETHIOPIA						
[R.] Omo, Shungura Formation, Member G, [c.5°N 36°E]	—	I	Ad	Cast of mandibular corpora with Cs-M3s	BM(NH)A M35369	Cited Eck,G. 1976: 334, Cited Szalay,F.S. and Delson,E. 1979: 396
KENYA						
Gamble's Cave II, Elmenteita, Gt. Rift Valley, [c.0°30′S 36°05′E]	—	I	Ad	3 lower M3s	BM(NH)A M20626	(Upper Pleistocene *(Gamblian)*)
Rawi, [S.E. of] Kavirondo Gulf, L.Victoria, [c.0°30′S 34°45′E]	—		[Ad]	Fragment of R mandible with M3	BM(NH)A M18945	(Middle Pleistocene), Listed Delson,E. (unpubl.): 11, Listed Delson,E. (unpubl.)c: 5
[TANGANYIKA]						
Garusi, Vogel River, [Northern], [3°13′S 35°10′E]	Feb 1939	F	Ad	Cast of upper R C-P3	BM(NH)A M32215	(*Papio (Simopithecus) serengetensis*), Background Dietrich,W.O. 1942:43-133, Listed Delson,E. (unpubl.): 11, Correspondence Delson,E. (unpubl.)b
" " "	Feb 1939			Cast of upper L P3-M1	BM(NH)A M32224	(*Papio (Simopithecus) serengetensis*), Background Dietrich,W.O. 1942:43-133, Listed Delson,E. (unpubl.): 10, Correspondence Delson,E. (unpubl.)b

Asian Colobinae

PRESBYTIS Eschscholtz, 1821

Langurs, Leaf Monkeys
(4 species groups: 17 species: type species *Presbytis mitrata* Eschscholtz, 1821 = *Simia melalophos* Raffles, 1821, see taxonomic notes).

Principal synonyms *Semnopithecus* Desmarest, 1822 (type species: *Simia entellus* Dufresne, 1797); *Trachypithecus* Reichenbach, 1862 (type species: *Semnopithecus pyrrhus* Horsfield, 1832); *Kasi* Reichenbach, 1862 (type species: *Cercopithecus johnii* Fischer, 1829); *Entellus* Gray, 1871 (type species: *Simia entellus* Dufresne, 1797, by tautonomy); *Corypithecus* Trouessart, 1879 (type species: *Semnopithecus frontatus* Müller, 1838); *Lophopithecus* Trouessart, 1879 (type species: *Semnopithecus rubicundus* Müller, 1838); *Presbypithecus* Trouessart, 1879 (substitute for *Presbytis* Reichenbach, 1862, type species: *Cercopithecus kephalopterus* Zimmermann, 1780).

Taxonomic notes *Type species: Presbytis mitrata* Eschscholtz, 1821, the type of *Presbytis* by monotypy, is probably a synomym of *Presbytis melalophos* rather than the Javan Leaf Monkey, previously known as *P. aygula* (see under *P. comata*, taxonomic notes). Brandon-Jones (personal communication, 1979) noted its resemblance to *Presbytis melalophos fuscomurina* Elliot, 1906, though there are one or two striking differences (e.g. facial skin almost black in *mitrata*, pale in *fuscomurina*; hands and feet reddish brown in *mitrata*, whitish in *fuscomurina*). The particular character described by Eschscholtz that makes *P. mitrata* positively identifiable as *P. melalophos* is the crown pattern: a black band running from brow to ear turns at right angles to cross the top of the head in front of the occipital crest. Together with the dark eyebrows, this black band forms a 'D'-shape which surrounds a semicircle of yellowish fur on the front of the crown. This character is seen very well in Müller & Schlegel's illustration of *P. melalophos* (1843, pl. 18bis, fig. 2). It is not seen in the Javan Leaf Monkey in which the entire crown is completely black. Eschscholtz states that the type specimen was bought from the inhabitants of Sumatra as the ship on which he was sailing passed through the Straits of Sunda. Thus on the basis of locality and general coloration it is possible that *fuscomurina* is a synonym of *mitrata*. However, lacking an up-to-date revision of the Sumatran subspecies of *P. melalophos*, *P. melalophos mitrata* is kept separate.

If this identification is accepted, *Simia melalophos* becomes the type species of *Presbytis* rather than *Simia aygula* Linnaeus, 1758. As the latter name is no longer available for the Javan Leaf Monkey (see *Presbytis comata*, taxonomic notes), the species group previously known as the *P. aygula* group is henceforth referred to as the *P. melalophos* group.

Species groups Pocock (1928) divided the genus into three groups, based on the colour of the newborn: the *P. entellus* group (including all Indian and Sri Lankan Leaf Monkeys) with black (*sic*) young, the *P. cristata* group with orange young, and the *P. aygula* group (now the *P. melalophos* group) with white young with the dark 'cruciger' pattern (see p. 39). Later, in 1934, Hill pointed out that the newborn of the purple-faced leaf monkeys of Sri Lanka were not black but grey with white cheeks. He therefore separated them from the *P. entellus* group, uniting them with the southwest Indian species, *P. johnii*. He gave the groups generic status: *Presbytis* for the *P. melalophos* group; *Kasi* for the *P. vetulus* [=*senex*] group (see *P. vetulus*, taxonomic notes, p. 72); *Semnopithecus* for the *P. entellus* group; and *Trachypithecus* for the *P. cristata* group. By 1935 Pocock had independently reached the same conclusion and continued this usage in 1939 in the first volume of his *Fauna of British India*. However Chasen (1940) and Miller (1942) continued to treat *Presbytis* as a single genus, as did Simpson (1945) and Ellerman & Morrison-Scott (1951). More recently Hooijer (1962) has reverted to generic status for the two leaf monkey groups in the Sunda Islands, while Medway (1965) uses subgenera to express their inter-relationships. Here however the species groups of Ellerman & Morrison-Scott (1951) are retained, following Thorington & Groves (1970). The following names are available for the groups if generic status is desired:

Presbytis Eschscholtz, 1921. Type species by monotypy *Presbytis mitrata* Eschscholtz, 1821 = *Simia melalophos* Raffles, 1821.
Trachypithecus Reichenbach, 1862. Type species designated by Pocock (1935) *Semnopithecus pyrrhus* Horsfield, 1832 = *Simia cristata* Raffles, 1821.
Kasi Reichenbach, 1862. Type species by monotypy *Semnopithecus cucullatus* I. Geoffroy, 1830 = *Simia johnii* Fischer, 1829.
Semnopithecus Desmarest, 1822. Type species designated by Lereboullet (1838) *Simia entellus* Dufresne, 1797.

Morphology Large but slender monkeys with short muzzles, long limbs and long tails of uniform thickness. The hairs of the crown often form a crest or cap. Hands and feet are long and slender with short big toes and extremely short thumbs (Jouffroy & Lessertisseur, 1977). Newborn infants are strikingly different in colour from adults (see under species, p. 39). Ischial callosities are variable in size and shape, contiguous in males but separate in females (Hill, 1936:121, pl. 2). Cheek pouches are absent and salivary glands enlarged. The large stomach is tripartite with marked dilatations, an adaptation to the ruminant-like digestion of a bulk diet of mature leaves (Bauchop & Martucci, 1968; Bauchop, 1978). The skull is orthognathous, particularly in the *P. melalophos* group, with broad interorbital region. The dentition is typically colobine, with high pointed molar cusps and sharply defined shearing crests (Kay & Hylander, 1978). The skulls of the *P. melalophos* and *cristata* groups can be distinguished as follows:

P. melalophos group	*P. cristata group*
Brow ridges weak, upper edge of orbit straight, horizontal	Brow ridges stronger, swept back and arched over orbits
Nasal profile convex	Nasal profile concave
Face short (prognathism index 100-115 (30♂♂); 103-111 (23♀♀)*	Face longer (prognathism index 115-125 (20♂♂); 111-122 (28♀♀)*
Deep consistent underbite†	Edge-to-edge bite†
Homomorphic incisors+	Heteromorphic incisors§
Lower M3 bears 4 cusps+	Lower M3 bears 5 cusps+

* Prognathism index $= \frac{\text{prosthion — basion}}{\text{nasion — basion}} \times 100$ (Washburn, 1944)

† Zingeser, 1970
+Weitzel (1978)
§ Swindler & Sirianni (1969)

The skulls of the *P. vetulus* group are difficult to distinguish from these of the *P. cristata* group, but *P. entellus* group skulls can be recognised by their heavy horizontal brow ridges with marked depression posteriorly (ophryonic groove). In the latter group, Colyer (1936:218) noted underbite in only 11.8% of specimens (n = 289).

Limb proportions vary, as mentioned by Washburn (1944), the most notable difference being the brachial indices of the *P. cristata* and *P. melalophos* groups. The former has the radius slightly shorter than the humerus (brachial index 94, the same as in *Colobus* (*Piliocolobus*)), whereas the radius in the *P. melalophos* group is considerably longer than the humerus (brachial index 109, see Appendix 3, p. 100). On the other hand the intermembral index is lower in the *P. melalophos* group: 76 compared with 82 in the *P. cristata* group. These and other anatomical differences are reflected in the locomotor behaviour of the two groups in Malaya, *P. melalophos* leaping more than the mainly quadrupedal *P. obscura*. Bimanual progression by armswinging was only seen in *P. melalophos* (Fleagle, 1977, 1977*a* & 1978).

Chromosome diploid no. 44 (*P. entellus, P. vetulus, P. obscura*; Chiarelli, 1973).

Range From Kashmir, Pakistan, Nepal, S. Tibet, Bhutan and Burma, throughout the Indian subcontinent and Sri Lanka; Indo-China with limited incursions into China in Yunnan and Kwangsi; the Malay Peninsula, Borneo, Sumatra, Java, Bali, Lombok, the Mentawai Islands and many other offshore islands.

Ecology The ecology of *Presbytis* is probably even more varied than that of *Colobus*. *P. entellus* lives at high altitudes in the Himalayas (8-12,000 ft in Kashmir, BMNH 1910. 1.18.65), *P. cristata* in tidal mangrove swamps in Malaya (Furuya, 1963), and *P. francoisi* in the rocky limestone hills of N. Vietnam (Lowe, cited in Osgood, 1932:204). *P. entellus* infiltrates towns and villages in India where it is considered sacred and given some degree of protection in spite of the fact that it raids crops (Jay, 1965). Because of its ability to survive on a diet consisting solely of mature leaves, *P. entellus* can live in many dry zones of India (e.g. Mount Abu in Rajasthan) where even the ubiquitous Rhesus macaque is absent (Hrdy, 1977). *Presbytis* eat mainly leaves but also fruits and seeds. Almost no animal protein is eaten, except larvae adventitiously attached to foliage. Some earth is occasionally ingested, presumably for its mineral content (Hladik, 1977). When representatives of two species groups are sympatric, some differences in their niches are observable. In Sri Lanka *P. entellus thersites* eats considerably more fruit than *P. vetulus philbricki* (Hladik & Hladik, 1972); it is much more terrestrial in habit (Ripley, 1977; Hladik, 1978), and its home ranges are larger. Similarly in Malaya, *P. melalophos* eats more fruit than does *P. obscura,* and while neither species is terrestrial, they utilise the forest in different ways. *P. melalophos* is more mobile, feeding at all levels of the forest from ground level to the emergent layer while *P. cristata* feeds mainly in the middle and upper canopy (Curtin, 1977). Their home ranges are about the same size (*P. melalophos* 16 hectares: *P. cristata* 17 hectares) (Curtin & Chivers, 1978).

Social behaviour For a bibliography of field studies of *Presbytis,* see Baldwin *et al.* (1975). The extreme variability of the social behaviour of *Presbytis* is probably a reflection of the genus' wide distribution and varied ecology. Langur groups tend to live in more or less fixed home ranges which partly overlap those of other groups. Core areas within the ranges include sleeping trees, food trees, and sometimes a source of water. However in areas of high population density (e.g. Polonnaruwa in Sri Lanka), purple-faced leaf monkeys exhibit territorial behaviour, the male adult actively defending boundaries against trespassers (Rudran, 1973). Group size varies from 3-120 individuals: bisexual groups have either several adult males (e.g. *P. entellus* at Lucknow, N. India: Jay, 1965; *P. cristata* in SW. Malaya: Furuya, 1963), or a single adult male (e.g. *P. entellus* at Dharwar: Sugiyama, 1964; *P. melalophos* in Sumatra: Wilson & Wilson, 1977); or, exceptionally, they form monogamous 'family' groups (e.g. *P. potenziani* in Siberut, Mentawai Islands: Tilson & Tenaza, 1976; Watanabe, 1981). Unisexual male groups and solitary males have also been observed. The adult sex ratio in a bisexual group may be as much as one male to 15 females. The most permanent members of a *P. entellus* group are the adult females which may spend the whole of their lives in the same group. The adult male component of the group, on the other hand, alters frequently, and these changes always involve some inter-male aggression (Boggess, 1979).

A notable feature of *Presbytis* behaviour is the care of the newborn young by females other than the mother. Almost from the moment of birth they handle and groom the infant, and even carry it some distance away from the mother. The distinctively different coat colour of the newborn (see below) may be adaptive, enabling the mother easily to monitor its whereabouts.

Reproduction The menstrual cycle in *P. entellus entellus* in captivity lasts on average 26.8 days (n=76), with a mode of 22 days (David & Ramaswami, 1969). The duration of gestation was reported by Jay (1965) to be approximately 6 months in *P. entellus,* but Hill (1937) considered that it must last for at least 196 days. Rudran (1973a) estimated it at 6½-7¼ months (198-220 days) in *P. vetulus.* Seasonality of births appears to be dependant upon ecological variables: births of *P. vetulus* occur throughout the year in the central mountains of Sri Lanka (2133m, 7000ft), but there is a distinct birth 'peak' between May and August in Polonnaruwa (*c.* 100m, 300ft) (Rudran, 1973*a*). Vogel (1971) notes similar differences between populations of *P. entellus.* In Sariska, Rajasthan (400m, 1312ft), births occur throughout the year, but in the Kumaon Hills in the Himalayas (900-1500m, 3000-5000ft), all births take place in April and May. Single births are usual. Lactation lasts 10-12 months. Sexual maturity is reached at 3-4 years in females, 4 years in males, though in the latter full growth is not complete until 6 or 7 years of age.

Species Species follow the arrangement of Thorington & Groves (1970) with the exception of *P. potenziani* which is placed in the *P. melalophos* group. Chasen (1940:78) noted its resemblance to *P. comata* [= *aygula* see p. 47] and more recently Tilson (1976) observed that the neonate is white but soon develops the 'cruciger' pattern like the *P. melalophos* group. Pocock's statement (1935:955) that the young are golden (thus linking the species with the *P. cristata* group) probably came about through the fortuitous association of a golden infant with the holotype of *Semnopithecus chrysogaster* (a junior objective synonym of *P. potenziani*) in a coloured illustration (Peters, 1879, pl. 4B).

The species groups cannot be distinguished solely on the basis of pelage colour, except in the case of the newborn young, but the following descriptive notes, together with the skull characters listed above, should assist in identification.

P. melalophos group (7 species: *P. melalophos, P. comata* [= *aygula*], *P. frontata, P. hosei, P. potenziani, P. rubicunda, P. thomasi*): all species of the *P. melalophos* group have an elongated occipital crest in the midline of the crown. The newborn are white, generally with a dark dorsal stripe which extends from crown to tailtip, spreading on to the shoulders to form a cross (the 'cruciger' pattern). During development of adult colouration, the dark pigmentation gradually encroaches upon the light parts. The final extent of the dark coloration varies from a few black-tipped hairs on the upper parts (*P.m.fluviatilis*) to the entire body with the exception of a small white patch on the abdomen and a white band down the inside of the leg to the ankle (hence the name 'banded leaf monkey' (*P.m.robinsoni*). *P. comata* [=*aygula*], *P. thomasi* and *P. hosei* are iron grey above and white or cream below with varied head patterns: *P. comata* has a completely black head with a prominent crest on the vertex; *P. thomasi* three broad black longitudinal stripes separated by two white stripes (the hairs of the median stripe are elongated, forming a sagittal crest); *P. hosei hosei* is sexually dimorphic, the males having the whole head white except for the black crown cap and nape, while in the female the white is reduced to a patch on the forehead and white whiskers, *P.h.canicrus* and *P.h.sabana* tending to resemble the female of *P.h.hosei*. *P. rubicunda* is entirely maroon red while *P. frontata* is grey brown with darker limbs; a whorl in front of the occipital crest parts the hairs, revealing a patch of naked white skin on the forehead. *P. potenziani* is entirely black except for the white forehead, cheeks, throat and pubic patch and some reddish colouration on chest and abdomen.

P. cristata group (6 species: *P. cristata, P. francoisi, P. geei, P. obscura, P. phayrei, P. pileata*): the *P. cristata* group is just as varied as the *P. melalophos* group. *P. geei* is an overall golden or creamy white; *P. pileata* is grey with a cap of erect hairs on the crown and long cheek whiskers; both *P. phayrei* and *P. obscura* have striking white spectacles around the eyes and a white patch on the lips. *P. obscura* is dark mushroom brown to almost black with pale nuchal patch, legs and tail, while *P. phayrei* is lighter and greyer, lacking contrastingly paler areas. *P. cristata* is dark grey brown to black with silver-tipped hairs which give a spangled appearance (*P. barbei* may belong here, see p. 53). The crown hairs are erect, forming an untidy peaked crest, or else curled forward, framing the face which lacks the striking white spectacles of *P. obscura* and *P. phayrei*. *P. francoisi* is a distinctive black leaf monkey with sharply defined patches of white or golden-yellow on the head and rump. The newborn young are orange in colour.

P. entellus group (1 species): *P. entellus* is an earthy or greyish brown with a slight purplish tinge, varying from very pale grey to dark grey brown. The head is lighter and the rump darker than the upper back. A radiating whorl on the forehead projects the brow hairs forward in a stiff fringe over the completely black face. A longitudinal crest on the vertex is present in some races. Newborn young are blackish brown.

P. vetulus group (2 species: *P. vetulus* [= *senex*], *P. johnii*): sometimes called the purple-faced leaf monkeys, the *P. vetulus* group are black or greyish brown with a well-defined short-haired rump patch, usually distinctly paler than the upper back. White wing-like whiskers contrast with the dark purplish-black face, except in *P. johnii* where the entire head is covered in long brown hairs. Newborn young of *P. vetulus* are pale grey with a brownish tinge (Rudran, 1973*a*) whereas the young of *P. johnii* are reddish-brown up to the age of 10 weeks (Poirier, 1969:24).

Presbytis melalophos group

PRESBYTIS MELALOPHOS (Raffles, 1821)

Banded Leaf Monkey, Black-crested Leaf Monkey
(17 subspecies)

Taxonomic notes Chasen's (1940) arrangement of the species has been followed with four exceptions: 1. the specific name *melalophos* is used instead of *femoralis* which it antedates by a considerable number of years; 2. *P.m.cruciger* Thomas, 1892 is considered to be an erythristic phase of the black *P.m.chrysomelas* (see under Subspecies, below); 3. *P.m.australis* is considered a synonym of *P.m.femoralis*; and *P.m.mitrata* is considered a subspecies of *P.m.melalophos*, and is kept separate pending a revision of the species.

Wilson & Wilson (1975) divide this species into two (*P. melalophos* and *P. femoralis*). This division, based largely upon field data, needs further investigation, incorporating the entire geographical range of the species.

Range The Malay Peninsula from about 13°30′N, including Batam, Bintan and Kundur Islands of the Riau Archipelago, Sumatra, Bangka, Belitung, Borneo and Bunguran Island in the Natuna Islands.

Dimensions Field measurements of specimens in the museum collections are as follows:

h & b	24 ♂♂	492 (423-565)	:	22 ♀♀	494 (420-555)
tail	24 ♂♂	711 (618-824)	:	22 ♀♀	711 (636-820)
weight	1 ♂	7031	:	1 ♀	6464

Key to the subspecies of *Presbytis melalophos* (excluding *P.m.mitrata*)

Name	Upperparts (inc. tail)	Underparts (inc. tail)	Outer thighs	Tibial streak	Ventral stripe on prox. ½ of tail	Other distinguishing characteristics/colour phases
robinsoni* (Peninsular Thailand, NW. Malaya)	Dark brown to blackish	Dark brown; white patch on lower abdomen extends to thighs	Dark like back	Occasionally absent	Absent	Alternative colour phase: yellowish white with grey-brown flanks, underparts and whiskers (= *robinsoni*)
femoralis (Johor and Singapore)	"	Greyish; white median line on chest; white patch on lower abdomen extends to thighs	As back with some white grizzling	Dark brown, slightly grizzled	"	Bases of crown hairs grizzled
percura (EC. Sumatra) (from Lyon, 1908; Miller, 1934)	"	Blackish; white median line on chest; whitish patch on lower adomen	"	"	Whitish, grizzled	Bases of crown hairs grey
batuana (Batu Is, W. Sumatra) (from Miller, 1903, 1934)	"	Black; greyish white median line on chest; white patch on lower abdomen	Dark like back	Faint	White (tail base distinctly bicoloured)	Smaller and with shorter tail than *P.m.sumatrana*
chrysomelas† (NW. Borneo)	"	Black; some white on chest; white patch on lower abdomen	"	White, narrow	White or whitish; (tail base distinctly bicoloured)	Alternative colour phases: 1. White areas replaced by sepia, with broader tibial streak (= *arwasca*); 2. Red with variable areas of black; white underparts (=*cruciger*)
sumatrana* (WC. Sumatra) (from Schlegel, 1876)	"	Buffy white from chin to pubis	"	White, narrowing distally	White (tail base distinctly bicoloured)	Alternative colour phase: yellowish red with paler underparts: blackish hands and feet
natunae (Natuna Is)	"	"	Buffy white, contrasting with back and tail	Whitish, narrowing and fading distally	Absent	Short white hairs on nose and at side of eye; bases of crown hairs black
paenulata (EC. Sumatra)	"	"	"	"	Light brown	Some white hairs at side of eye; bases of crown hairs grey
siamensis (NW. Malaya)	Pale grey-brown	"	"	Buffy white, reaching heel	"	—
rhionis (Bintan I., Riau Arch.)	"	"	Buffy or greyish white	Buffy white, fading distally	"	White patch at side of eye: white hairs sweep forward in front of ear; bases of crown hairs brown
catemana (EC. Sumatra) (from Lyon, 1908)	"	"	Grey, less contrast with back and tail	White, narrowing distally	"	Bases of crown hairs dirty white
nubigena (Melaka, Malaya)	"	"	"	Buffy white, broad, fading distally	"	Bases of crown hairs grey
cana (Kundur I., Riau Arch.)	"	"	Front dark grey, back whitish	Buffy white, reaching heel	"	White hairs at side of eye; bases of crown hairs grey
nobilis (WC. Sumatra)	Bright red, overlaid with long black or buff hairs in some	Buff, suffused with orange	Like back, no contrast with back or tail	Orange, fading distally	Orange	Black occipital crest; buff patch on nape extends down back in some
melalophos (WC. Sumatra)	Reddish fawn, overlaid with long black hairs	White	Golden or orange yellow, contrasting with back and tail	Yellowish, broad, reaching heel	Yellowish	Black occipital crest; buffy patch on nape
fluviatilis (S. Sumatra)	Buffy white, overlaid with long dark grey hairs; tail suffused with rufous	Buffy white	Entire limb buffy white, contrasting with back and tail	Imperceptible	Light orange	Occipital crest pale smoky grey
fuscomurina (S. Sumatra) (from Elliot, 1906)	Dark grey brown	White	Entire limb white, contrasting with back and tail	"	Ventral surface of entire tail white	Dark line runs from fore-head above ear to vertex where it joins occipital crest

* Two colour phases
† Three colour phases

Subspecies This species is characterized by much individual variation, by the presence of two or three different colour phases at a single locality, and by the occurrence of both red and white mutants.

There appear to be three phenotypes of *P.m.chrysomelas* in northwestern Borneo: *chrysomelas* is black with a sprinkling of white on the chest, whitish on abdomen and proximal ventral surface of tail, with a narrow whitish stripe on the inside of arm, and a conspicuous narrow white stripe on the inside of leg; *arwasca* is similar but with the white replaced by sepia brown, the stripes on the arms and legs being broader but less conspicuous; *cruciger* is bright red with variable amounts of black on the occipital crest, nape, shoulders, back, upper side of tail and entire upper limb; the black feet contrast sharply with the red legs. The *chrysomelas* phenotype is found at Pontianak, at Samarahan (SW. Sarawak), and at Baram (NE. Sarawak); the *arwasca* phenotype at Cape Datu (extreme NW. Sarawak) and at Baram; and *cruciger* at Batang Lupar and Betong (SW. Sarawak) and again at Baram. It is thus impossible to separate any of these taxa as a geographic race. The red *cruciger* phase has been observed at the same localities, in the same groups, and apparently interbreeding with the dark *chrysomelas* phenotype (Medway, 1970). Similarly with the most northwesterly subspecies: both the dark brown *keatii* and the yellowish white *robinsoni* were described from the same locality, Trang, Peninsular Thailand. It was not until 1915 that Robinson and Kloss noted that *robinsoni* was an aberrant whitish form of *keatii* and almost always found in association with it. The earlier name, *P.m.robinsoni*, of course takes precedence.

The taxon *australis* for specimens from S. Johore was separated from *P.m.femoralis* from Singapore by Miller (1913) on the grounds that the skulls were larger. This was shown to be inconclusive by Robinson & Kloss (1914). As the skins from S. Johore and Singapore appear to be identical, *P.m.australis* is considered a synonym of *P.m.femoralis*. The specimens from Segamat cited by Hill (1960:32) are very different from each other, and show intermediate characters between *P.m.femoralis* and *P.m.nubigena*.

Chasen (1940) and Hooijer (1962) have expressed different views on the red mutants of west central Sumatra. It is impossible with the material at hand to decide which races should be recognised. The holotype of *P.m.nobilis* is very different from that of *P.m.melalophos* (see table of characters), and it was relatively easy to allocate specimens to either one or the other. From the original description, *P.m.ferruginea* appears to be a synonym of *P.m.nobilis*. However, a specimen from Inderapura, the type locality of *P.m.nobilis* (restricted by Chasen, 1940:76) has been identified as *P.m.melalophos*, indicating that further study of this area is necessary.

Presbytis melalophos melalophos (Raffles, 1821)

Principal synonym *Semnopithecus flavimanus* I. Geoffroy, 1831:39.

Type localities Of *melalophos*, Bengkulu, SW. Sumatra; of *flavimanus*, Sumatra, restricted to Bengkulu by Chasen, 1940:76.

Presbytis melalophos batuana Miller, 1903

Type locality Pini Is, Batu Is, W. Sumatra.

Presbytis melalophos cana Miller, 1906

Type locality Kundur I., Riau Archipelago.

Presbytis melalophos catemana Lyon, 1908

Type locality Kateman R., E. Sumatra.

Presbytis melalophos chrysomelas (Müller, 1838)

Principal synonyms *Semnopithecus cruciger* Thomas, 1892:476; *Presbytis arwasca* Miller, 1934:126.

Type localities Of *chrysomelas*, Pontianak, W. Kalimatan; of *cruciger* and *arwasca*, Miri, NE. Sarawak, Borneo.

Presbytis melalophos femoralis (Martin, 1838)

Principal synonyms *Semnopithecus neglectus* Schlegel, 1876:47; *Presbytis australis* Miller, 1913.

Type localities Of *femoralis*, Sumatra, corrected to Singapore by Miller, 1913:29; of *neglectus*, Singapore; of *australis*, Jambu Luang, E. coast of Johor, Malaya.

Presbytis melalophos fluviatilis (Chasen, 1940)

Type locality Muaradua, Palembang, Sumatra, at 100m.

Presbytis melalophos fuscomurina Elliot, 1906

Type locality Telukbetung, Lampung, S. Sumatra.

Presbytis melalophos mitrata Eschscholtz, 1821

Type locality Sumatran coast near Zutphen I.

Presbytis melalophos natunae (Thomas & Hartert, 1894)

Type locality Bunguran I., N. Natuna Is.

Presbytis melalophos nobilis Gray, 1842

Principal synonym *Semnopithecus ferrugineus* Schegel, 1876:42.

Type localities Of *nobilis*, India, corrected to Inderapura, W. Sumatra by Chasen, 1940:76; of *ferrugineus*, Padang, W. Sumatra.

Presbytis melalophos nubigena Elliot, 1909

Type locality Malacca, Malaya.

Presbytis melalophos paenulata (Chasen, 1940)

Type locality Teluk Pandji, Barumun R., S. of Labuhanbilik, NE. Sumatra.

Presbytis melalophos percura Lyon, 1908

Type locality Kompei, Siak, E. Sumatra.

Presbytis melalophos rhionis Miller, 1903

Type locality Bintan I., Riau Archipelago.

Presbytis melalophos robinsoni Thomas, 1910

Principal synonym *Presbytis neglecta keatii* Robinson & Kloss, 1911:174.

Type locality Of *robinsoni* and *keatii*, Ko Khau, Trang, Peninsular Thailand.

Presbytis melalophos siamensis (Müller & Schlegel, 1841)

Principal synonyms *Semnopithecus nigrimanus* I. Geoffroy, 1843:546; *Presbytis dilecta* Elliot, 1909:270.

Type localities Of *siamensis*, Siam, corrected by Schlegel, 1876:38, to 'Malacca' (= Malay States, according to Chasen, 1940:74); of *nigrimanus*, Java (= Malay States, according to Chasen, 1940:74); of *dilecta*, Dusun Tua, Selangor, Malaya.

Presbytis melalophos sumatrana (Müller & Schlegel, 1841)

Type locality Sumatra, restricted by Schlegel (1876:46) to Mt Ophir, N. of Padang, W. Sumatra.

Locality	Date of Collection	Sex	Age	Nature of Specimen	Accession no.	References, Remarks, etc.
Presbytis melalophos melalophos (Raffles 1821)						
SUMATRA						
Inderapura, between Bengkulu and Padang, W.Sumatra, [2°02′S 100°56′E]	—	F	[Ad]	Skin	ZD.1842.1.19.94	Cited Gray,J.E. 1843: 2, Listed Gray, J.E. (unpubl.): 76, Cited Pocock,R.I. 1935: 914
—	—	F	Ad	Skin, Skull	ZD.1838.8.13.13	(*Presbytis melalophos*), Cited Gray,J.E. 1843: 2, Listed Gray,J.E. (unpubl.): 76
—	—	M	[Ad]	Skin	ZD.1852.5.9.1	
[SUMATRA]						
	—	M	Ad	Skin, Skull	ZD.1879.11.21.590	Syntype *Simia melalophos* Raffles 1821, (*Semnopithecus melalophos flavimanus*), Cited Raffles,T.S. 1821: 245, Cited Pocock,R.I. 1935: 914
Presbytis melalophos batuana Miller 1903						
NO SPECIMENS						
—	—	—	—	—	—	
Presbytis melalophos cana Miller 1906						
[MALAYA]						
Bliah [=Selatbliat], Kundur I., Riau Archipelago, [0°53′N 103°24′E]	17 Aug 1908	M	Ad	Skin, Skull, Radiograph of penis	ZD.1909.4.1.14	(*Presbytis siamensis rhionis*), Cited Pocock,R.I. 1935: 907, Cited Colyer,J.F. 1936: 38
,, ,, ,,	18 Aug 1908	M	Inf	Skin, Skull	ZD.1909.4.1.15	(*Presbytis siamensis rhionis*), Cited Pocock,R.I. 1935: 907
,, ,, ,,	21 Aug 1908	M	Juv	Skin, Skull	ZD.1909.4.1.16	,, ,, ,, ,,
,, ,, ,,	18 Aug 1908	F	Ad	Skin, Skull	ZD.1909.4.1.17	,, ,, ,, ,,
Presbytis melalophos catemana Lyon 1908						
NO SPECIMENS						
—	—	—	—	—	—	
Presbytis melalophos chrysomelas (Müller 1838)						
BORNEO						
Balingean, Sarawak, N.W.Borneo, [2°59′N 112°34′E]	18 Apr 1917	F	Ad	Skin, Skull	ZD.1955.721	Cited Pocock,R.I. 1935: 910, Cited Hill,J.E. 1960: 32
,, ,, ,,	26 Apr 1917	F	Ad	Skin, Skull	ZD.1955.722	,, ,, ,, ,,
Baram District, Sarawak, [c.4°30′N 114°12′E]	—	F	Ad	Skin, Skull	ZD.1956.213	(*Presbytis femoralis*)
,, ,, ,,	—	F	Juv	Skin, Skull	ZD.1956.214	,, ,, ,, ,,
Baram [Dist.], Sarawak, [c.4°30′N 114°12′E]	—	[M]	[Ad]	Skin	ZD.1939.3477	
Baram [Dist.], [Sarawak], N.W.Borneo, [c.4°30′N 114°12′E]	—	F	Ad	Skin, Skull	ZD.1889.1.8.2	
,, ,, ,,	—	M	Juv	Skin, Skull	ZD.1889.1.8.3	
[R.] Batang Lupar, S.W.Sarawak, [1°30′N 111°E]	Sep 1893	M	Ad	Skin, Skull	ZD.1894.2.1.1	(*Presbytis cruciger*)
,, ,, ,,	Sep 1893	M	[Inf]	Skin, Skull in skin	ZD.1894.2.1.2	,, ,, ,, ,,
Betong, Saribas, Sarawak, N.W.Borneo, [1°26′N 111°30′E]	2 Aug 1916	M	Ad	Skin, Skull	ZD.1955.719	Cited Pocock,R.I. 1935: 908, Cited Hill,J.E. 1960: 32
,, ,, ,,	8 Aug 1916	M	Ad	Skin, Skull	ZD.1955.720	,, ,, ,, ,,
Mora Barue, Samarahan, S.W.Sarawak	17 Nov 1919	F	Ad	Skin, Skull	ZD.1955.723	Cited Hill,J.E. 1960: 32
,, ,, ,,	17 Nov 1919	F	Inf	Skin, Skull	ZD.1955.724	,, ,, ,, ,,
R.Baram, N.W.Borneo, [c.4°30′N 114°12′E]	—	M	Ad	Skin, Skull, Radiograph of penis	ZD.1891.8.28.9	Figured Colyer,J.F. 1936: 39, Fig.49
R.Miri, R.Baram, N.E.Sarawak, [4°23′N 113°58′E]	Jan 1895	F	Ad	Skin, Skull	ZD.1895.5.7.2	(*Presbytis cruciger*), Correspondence Thomas,O. (unpubl.)
[R.Miri], [Sarawak], [4°23′N 113°58′E]	—	M	Ad	Skin, Skull	ZD.1908.7.17.4	Correspondence Thomas,O. (unpubl.)
R.Sadong, Sarawak, [c.1°33′N 110°45′E], Sea level	4 Apr 1947	F	[Ad]	Skin	ZD.1980.796	
Samarahan, S.W.Sarawak, [1°17′N 110°29′E]	—	F	Juv	Skin, Skull	ZD.1955.725	Cited Hill,J.E. 1960: 32
—	—	F	[Inf]	Skin	ZD.1844.10.17.6	
—	8 Feb 1843	F	[Inf]	Skin, Skull in skin	ZD.1857.12.14.6	
—	—	M	Ad	Skin, Skull, Skeleton	ZD.1879.8.30.6	
—	—	F	Ad	Skin, Skull, Skeleton	ZD.1879.8.30.7	
[BORNEO]						
Balingean, Sarawak, [2°59′N 112°34′E]	Apr 1903	F	Ad	Skin, Skull	ZD.1905.3.1.1	
,, ,, ,,	Apr 1903	M	Ad	Skin, Skull	ZD.1905.3.1.2	
Baram District, Sarawak, [c.4°30′N 114°12′E]	15 Jun 1898	F	Ad	Skin, Skull	ZD.1900.2.2.1	(*Semnopithecus cruciger*), Figured James,W.Warwick 1960: 228-229, figs.56a-c
,, ,, ,,	15 Jun 1898	M	Inf	Skin, Skull	ZD.1900.2.2.2	(*Semnopithecus cruciger*)
Baram [Dist.], Sarawak, [c.4°30′N 114°12′E]	1899	F	[Ad]	Skin	ZD.1899.12.9.1	(*Pithecus cruciger*)
,, ,, ,,	[1899]	I	[Inf]	Skin	ZD.1899.12.9.2	,, ,, ,, ,,
,, ,, ,,	—	M	Ad	Skull, Skeleton	ZD.1900.3.30.2	

Locality	Date of Collection	Sex	Age	Nature of Specimen	Accession no.	References, Remarks, etc.
Presbytis melalophos chrysomelas *continued*						
Betong Saribas, S.W.Sarawak, [1°26′N 111°30′E]	5 Aug 1916	M	Ad	Skin, Skull	ZD.1920.12.4.1	(*Pithecus cruciger*)
,, ,, ,,	7 Aug 1916	F	Ad	Skin, Skull	ZD.1920.12.4.2	,, ,, ,, ,,
Miri, Sea coast, 16 miles S.W. of Baram, N.E.Sarawak, [c.4°23′N 113°59′E]	1887	[F]	[Juv]	Skin, Bones of L foot	ZD.1892.11.15.1	Holotype *Semnopithecus cruciger* Thomas 1892, Cited Thomas,O. 1892: 476
R.Batang Lupar, S.W.Sarawak, [1°30′N 111°E]	Aug 1892	M	Ad	Skin, Skull	ZD.1893.1.30.1	(*Semnopithecus cruciger*)
R.Miri, Baram, Sarawak, [4°23′N 113°58′E]	Jan 1895	F	Ad	Skin, Skull	ZD.1895.5.7.1	
—	—	F	Ad	Skull	ZD.1858.5.4.191	Listed Anon. (unpubl.)a, Listed Gray,J.E. (unpubl.): 8
Presbytis melalophos chrysomelas? (Müller 1838)						
BORNEO						
Baram District, [Sarawak], [c.4°30′N 114°12′E]	—	F	Juv	Skull	CMZ E7226B	
,, ,, ,,	—	M	Ad	Skull, Skeleton	CMZ E7226C	
,, ,, ,,	—	[M]	Ad	Skull, Skeleton	CMZ E7226D	
,, ,, ,,	—	[F]	Ad	Skull, Skeleton	CMZ E7226E	
,, ,, ,,	—	[F]	Ad	Skull, Skeleton	CMZ E7226F	
Miri, Baram District, [Sarawak], [4°23′N 113°59′E]	4 Mar 1895	M	Ad	Skull	CMZ E7226A	(*Semnopithecus femoralis*)
[BORNEO]						
Baram [Dist.], Sarawak, [c.4°30′N 114°12′E]	—	M	Ad	Skull, Skeleton	CDA M49	,, ,, ,, ,,
,, ,, ,,	—	M	Ad	Skull, Skeleton	CDA M50	,, ,, ,, ,,
Presbytis melalophos femoralis (Martin 1838)						
MALAYA						
Changi, Singapore I., [1°23′N 103°59′E]	23 Jul 1908	F	Ad	Skin, Skull	ZD.1955.1564	Cited Hill,J.E. 1960: 32
[MALAYA]						
Gunong Pulai, Johor, [1°35′N 103°32′E], c.200ft	12 Sep 1897	F	Ad	Skin, Skull	ZD.1898.1.7.2	Cited Flower,S.S. 1900: 319, Cited Pocock,R.I. 1935: 902-904
,, ,, ,,	—	F	Ad	Skull	ZD.1898.1.7.3	,, ,, ,, ,,
[Gunong Pulai], [Johor], Malay Peninsula, [1°35′N 103°32′E]	25 Jul 1879	M	Ad	Skin, Skull	ZD.1885.8.1.9	Cited Thomas,O. 1886: 72, Cited Pocock,R.I. 1935: 901
Singapore	—	F	Ad	Skin	ZD.1857.4.17.2	Cited Pocock,R.I. 1935: 902
[Singapore]	—	M	Juv	Skin, Skull	ZD.1879.11.21.594	Syntype *Semnopithecus femoralis* Martin 1838, Cited Martin,W.C.Linnaeus 1838: 436, Cited Miller,G.S. 1913: 29, Cited Pocock,R.I. 1935: 901-902
Presbytis melalophos femoralis? (Martin 1838)						
MALAYA						
Padang Tuan, Segamat, N.W.Johor, [c.2°30′N 104°05′E]	24 Feb 1911	M	Ad	Skin, Skull	ZD.1955.1565	(*P.melalophos australis*), Cited Hill,J.E. 1960: 32
,, ,, ,,	24 Feb 1911	M	Ad	Skin	ZD.1955.1566	(*P.melalophos australis*), Figured Pocock,R.I. 1935: 902, Pl.1, fig.A. Cited Hill,J.E. 1960: 32
Presbytis melalophos fluviatilis (Chasen 1940)						
SUMATRA						
Highlands inland of Palembang, [c.4°S 104°E]	22 Mar 1930	F	Juv	Skin, Skull	ZD.1930.6.3.1	(*Pithecus robinsoni*), Cited Pocock,R.I. 1935: 916
Presbytis melalophos fuscomurina Elliot 1906						
NO SPECIMENS						
—	—	—	—	—	—	
Presbytis melalophos mitrata Eschscholtz 1821						
NO SPECIMENS						
—	—	—	—	—	—	
Presbytis melalophos natunae (Thomas and Hartert 1894)						
[NATUNA IS.]						
Bunguran I. [Gt.Natuna I.], [c.4°N 108°15′E]	19 Oct 1893	M	Ad	Skin, Skull	ZD.1894.9.28.1	Holotype *Semnopithecus natunae* Thomas and Hartert 1894, Cited Thomas,O. and Hartert,E. 1894: 652, Figured Pocock,R.I. 1935: 899, 908, Fig.1B
,, ,, ,,	10 Oct 1893	F	Ad	Skin, Skull	ZD.1894.9.28.2	Paratype *Semnopithecus natunae* Thomas and Hartert 1894, Cited Thomas,O. and Hartert,E. 1894: 652, Figured Pocock,R.I. 1935: 899, 908, Fig.1A

Locality	Date of Collection	Sex	Age	Nature of Specimen	Accession no.	References, Remarks, etc.
Presbytis melalophos natunae *continued*						
" " "	Oct 1893	[F]	Inf	Skin, Skull	ZD.1894.9.28.3	Paratype *Semnopithecus natunae* Thomas and Hartert 1894, Cited Pocock,R.I. 1935: 897
" " "	—	F	Foetus	Wet, cadaver	ZD.1894.9.28.22	Paratype *Semnopithecus natunae* Thomas and Hartert 1894, (*Semnopithecus natunae*)
" " "	4 Oct 1893	M	[Ad]	Skin, Skull in skin	ZD.1939.1141	Paratype *Semnopithecus natunae* Thomas and Hartert 1894, Cited Thomas,O. and Hartert,E. 1894: 652
" " "	20 Oct 1893	F	[Ad]	Skin, Skull in skin	ZD.1939.1142	" " " "
Gt. Natuna I., [c.4°N 108°15′E]	—	[F]	Ad	Skull, Skeleton	CMZ E7228A	(*Semnopithecus natunae*)
Mt.Ranai, Gt.Natuna I., [3°57′N 108°21′E], 2000ft	Aug 1894	M	Inf	Skin, Skull	ZD.1895.11.8.8	Cited Pocock,R.I. 1935: 897
Presbytis melalophos nobilis Gray 1842						
SUMATRA						
Kion Tanem [=?Kajutanam], W.Sumatra, [?0°33′S 100°20′E], 400ft	5 Jun 1938	[F]	Juv	Skin, Skull	ZD.1938.11.30.13	(*Presbytis melalophos*)
" " "	5 Jun 1938	[F]	Juv	Skin, Skull	ZD.1938.11.30.14	
Pajo, [on L.Singkarak], [Balipor Dist.], [c.0°37′S 100°33′E]	—	F	Ad	Skin, Skull	ZD.1879.6.28.3	Cited Pocock,R.I. 1935: 916
Siolakderas, Kerintji, [c.1°54′S 101°16′E], [3100ft]	16 Mar 1914	M	Ad	Skin, Skull	ZD.1919.11.12.4	Cited Robinson,H.C. and Kloss,C.Boden 1918: 4, Cited Pocock,R.I. 1935: 914-915
Sungei Kumbang, Kerintji, [c.1°48′S 101°16′E], 4500ft	5 Jun 1914	[M]	Ad	Skin, Skull	ZD.1919.11.12.5	" " " "
" " "	Apr 1914	F	Ad	Skin, Skull	ZD.1919.11.12.6	" " " "
" " "	May 1914	F	Ad	Skin, Skull	ZD.1919.11.12.7	" " " "
—	—	—	Ad	Skin	ZD.1848.10.31.1	(*Presbytis melalophos*)
—	Died 10 Nov 1901	M	Juv	Skin, Skull	ZD.1939.1144	(*Semnopithecus melalophos*)
—	—	—	[Ad]	Skin	ZD.1939.3481	(*Pithecus melanolophus*)
[SUMATRA]						
"India"	—	M	Ad	Skin, Skull	ZD.77a	Holotype *Presbytis nobilis* Gray 1842, Cited Gray,J.E. 1842: 256, Cited Pocock,R.I. 1935: 916, Listed Anon. (unpubl.)b: 77
—	—	M	Ad	Skin, Skull	ZD.1939.1145	
[West Central Sumatra]	—	M	Juv	Skin, Skull	ZD.1939.1146	
—	—	F	[Ad]	Skin	ZD.1939.1149	
Presbytis melalophos nobilis? Gray 1842						
[SUMATRA]						
—	—	M	Ad	Skull	ZD.1938.3.14.1	
—	—	M	[Ad]	Frozen cadaver	ZD.1975.1983	Correspondence Hill,C.A. (unpubl.)
—	Died 12 Mar 1975	F	[Ad]	Frozen cadaver	ZD.1975.1984	" " " "
—	—	F	[Ad]	Frozen cadaver	ZD.1975.1985	" " " "
—	—	M	[Juv]	Frozen cadaver	ZD.1975.1986	" " " "
—	—	F	[Juv]	Frozen cadaver	ZD.1975.1987	" " " "
—	—	F	[Juv]	Frozen cadaver	ZD.1975.1988	" " " "
—	Died Jan 1978	M	[Ad]	Wet, cadaver	ZD.1979.1661	
Presbytis melalophos nubigena Elliot 1909						
MALAYA						
Nyalas, Melaka, [2°26′N 102°28′E]	—	M	Ad	Skull	ZD.1955.1579	(*Presbytis melalophos siamensis*), Cited Hill,J.E. 1960: 32
[MALAYA]						
Melaka	—	M	Ad	Skin, Skull	ZD.1843.3.10.1	(*Presbytis albocinereus*)(*Presbytis cinerea*), Cited Gray,J.E. 1843: 193, Cited Pocock,R.I. 1935: 906
" " "	—	[M]	Ad	Skin, Skull	ZD.1854.10.5.2	(*Semnopithecus albocinereus*), Cited Pocock,R.I. 1935: 906
" " "	[May 1842 - Sep 1845]	F	Ad	Skin, Skull	ZD.1860.5.4.50	(*Semnopithecus albocinereus siamensis*)
" " "	[May 1842 - Sep 1845]	F	[Inf]	Skin, Skull in skin	ZD.1879.11.21.465	(*Semnopithecus albocinereus*), Listed Cantor,T. (unpubl.) Cited Pocock,R.I. 1935: 987
" " "	—	M	Ad	Skin, Skull	ZD.1879.11.21.595	Holotype *Presbytis nubigena* Elliot 1909, (*Presbytis femoralis siamensis*), Cited Elliot,D.G. 1909: 268
Presbytis melalophos nubigena? Elliot 1909						
NO DATA						
"Java"	—	M	Ad	Skin	ZD.1846.12.13.1	(*Presbytis cinerea*), Associated with next specimen (1850.2.27.25)
" " "	—	M	Ad	Skull	ZD.1850.2.27.25	(*Semnopithecus nigrimanus*), Listed Gray,J.E. (unpubl.): 939

Locality	Date of Collection	Sex	Age	Nature of Specimen	Accession no.	References, Remarks, etc.
Presbytis melalophos paenulata (Chasen 1940)						
SUMATRA						
Teluk Pandji, R.Baharumun, S. of Labuhanbilik, N.E.Sumatra, [2°02′N 100°14′E]	14 Aug 1937	M	Ad	Skin, Skull	ZD.1947.1499	Holotype *Pithecus femoralis paenulatus* Chasen 1940, Cited Chasen,F.N. 1940: 75
Presbytis melalophos percura Lyon 1908						
NO SPECIMENS						
—	—	—	—	—	—	
Presbytis melalophos rhionis Miller 1903						
[MALAYA]						
Sungei Biru, N.E.Bintan I., Riau Archipelago, [c.1°N 104°30′E]	13 Jun 1908	M	Ad	Skin, Skull	ZD.1909.4.1.18	Figured Pocock,R.I. 1935: 906, Pl.1, fig.B
,, ,, ,,	12 Jun 1908	F	Ad	Skin, Skull	ZD.1909.4.1.19	Figured Pocock,R.I. 1935: 906, Pl.1, fig.B, Figured Colyer,J.F. 1936: 37, Fig.47
Presbytis melalophos robinsoni Thomas 1910						
BURMA						
Bankachon, Tenasserim, [10°09′N 98°36′E]	15 Apr 1877	M	Ad	Skin, Skull	ZD.1885.8.1.8	(*P.femoralis keatii*) Cited Thomas,O. 1886: 66
[BURMA]						
Bankachon, V.P., S.Tenasserim, [10°09′N 98°36′E]	24 Dec 1913	F	Ad	Skin, Skull	ZD.1914.12.8.30	(*Presbytis femoralis keatii*), Cited Wroughton,R.C. 1915a: 702
Maliwun, V.Point, S.Tenasserim, [10°16′N 98°32′E]	9 Feb 1914	M	Ad	Skin, Skull	ZD.1914.12.8.29	,, ,, ,, ,,
Siam, 13°35′N 99°E, 1130m	[28] Feb 1913	M	Ad	Skin, Skull	ZD.1913.11.2.1	(*Presbytis femoralis keatii*), Cited Gairdner,K.G. (unpubl.): 1, Cited Pocock,R.I. 1939: 902
MALAYA						
Me Kuning, Taiping, Perak	14 Oct 1910	M	Ad	Skin, Skull	ZD.1955.1587	Cited Hill,J.E. 1960: 32
[MALAYA]						
Larut Hills, Perak, [4°47′N 100°45′E]	—	[M]	Juv	Skin, Skull	ZD.1902.5.2.2	(*Presbytis femoralis keatii*)
Pelarit, N.Malay Peninsula, Perlis, [6°39′N 100°11′E]	14 Nov 1911	F	Ad	Skin, Skull	ZD.1955.1586	Cited Hill,J.E. 1960: 32
Malay Peninsula, Perlis	—	F	Inf	Skin, Skull	ZD.1879.11.21.614	(*Presbytis femoralis*), Cited Cantor,T. 1846: 318
Ulu Ijok, Perak, [5°11′N 100°44′E], 400ft	26 Dec 1931	M	Ad	Skin, Skull	ZD.1934.7.18.7	(*Pithecus femoralis keatii*)
,, ,, ,,	29 Dec 1931	F	Ad	Skin, Skull	ZD.1934.7.18.8	(*Pithecus femoralis*)
[THAILAND]						
Chong, Trang, N.Malay Peninsula, [c.7°33′N 99°47′E]	11 Dec 1909	F	Juv	Skin, Skull	ZD.1910.10.1.2	(*Presbytis femoralis*)
,, ,, ,,	16 Dec 1909	F	Ad	Skin, Skull	ZD.1955.1585	Cited Hill,J.E. 1960: 32
Gap, Phatthalung, Trang, N.Malay Peninsula	24 Dec 1909	M	Ad	Skin, Skull	ZD.1910.10.1.3	(*Presbytis femoralis*)
Kao Nong, Bandon, N.E.Malay Peninsula, [8°49′N 99°30′E], 1200-1500ft	13 Jun 1913	M	Ad	Skin, Skull	ZD.1955.1582	Cited Hill,J.E. 1960: 32
,, ,, ,,	23 Jun 1915	M	Ad	Skin, Skull	ZD.1955.1588	Cited Pocock,R.I. 1935: 903, Cited Hill,J.E. 1960: 32
,, ,, ,,	13 Jun 1913	F	Ad	Skin, Skull	ZD.1955.1589	Cited Hill,J.E. 1960: 32
Khao Luang, Muang, Nakhon Si Thammarat, Siam, [c.8°30′N 99°44′E], 2000ft	19 Mar 1922	M	Ad	Skin, Skull	ZD.1955.1580	,, ,, ,, ,,
Khao Wang Hip, Peninsular Siam, [8°12′N 99°43′E]	24 Sep 1915	F	Ad	Skin, Skull	ZD.1920.7.3.4	(*Presbytis neglecta keatii*), Cited Pocock,R.I. 1935: 903
Klong Tung Sai, Tongka, [Phuket], Siam, [7°58′N 98°22′E]	26 Dec 1917	M	Ad	Skin, Skull	ZD.1955.1583	Cited Hill,J.E. 1960: 32
,, ,, ,,	19 Dec 1917	F	Ad	Skin, Skull	ZD.1955.1584	,, ,, ,, ,,
Ko Khau, Trang, N.Malay Peninsula	10 Jan 1910	M	Ad	Skin, Skull	ZD.1910.10.1.1	Holotype *Presbytis robinsoni* Thomas 1910, (*Presbytis femoralis keatii*), Cited Thomas,O. 1910: 25, Cited Pocock,R.I. 1935: 903
,, ,, ,,	10 Jan 1910	M	Ad	Skin, Skull	ZD.1949.425	Holotype *Presbytis neglecta keatii* Robinson and Kloss 1911, Cited Robinson,H.C. and Kloss,C.Boden 1911: 174
Tasan [=Ban Tha San], Chumphon, Peninsular Siam, [10°29′N 98°55′E], 220ft	19 Mar 1919	M	Ad	Skin, Skull	ZD.1955.1581	Cited Hill,J.E. 1960: 32
Presbytis melalophos siamensis (Müller and Schlegel 1841)						
MALAYA						
Bukit Cheraka, Jeram, Selangor coast, [3°12′N 101°23′E]	16 Nov 1910	F	Ad	Skin, Skull	ZD.1955.1576	(*Presbytis siamensis dilecta*), Cited Hill,J.E. 1960: 32
Changkat Menteri, Perak, [3°43′N 101°15′E]	11 Sep 1918	F	Ad	Skin, Skull	ZD.1955.1572	Cited Pocock,R.I. 1935: 905, Cited Hill,J.E. 1960: 32

Locality	Date of Collection	Sex	Age	Nature of Specimen	Accession no.	References, Remarks, etc.
Presbytis melalophos siamensis *continued*						
" " "	Sep 1918	M	Ad	Skin, Skull	ZD.1955.1573	" " " "
" " "	22 Sep 1918	F	Ad	Skin, Skull	ZD.1955.1574	Cited Hill,J.E. 1960: 32
Dusun Tua, Selangor, [3°08′N 101°50′E]	3 Oct 1906	M	Inf	Skin, Skull	ZD.1955.1577	
Genting Bidai, Selangor, [3°18′N 101°49′E]	23 Sep 1914	M	Ad	Skin, Skull	ZD.1955.1569	(*Presbytis siamensis dilecta*), Cited Pocock,R.I. 1935: 906, Cited Hill,J.E. 1960: 32
" " "	17 Sep 1914	M	Ad	Skin, Skull	ZD.1955.1570	Cited Pocock,R.I. 1935: 906, Cited Hill,J.E. 1960: 32
Gunong Mengkuang Lebah, Selangor, [3°24′N 101°48′E], 4800ft	24 Mar 1907	M	Ad	Skin, Skull	ZD.1955.1575	Cited Pocock,R.I. 1935: 905, Cited Hill,J.E. 1960: 32
Kuala Tembeling, Pahang River, [4°03′N 102°19′E], c.200ft	20 Aug 1905	F	Ad	Skin, Skull	ZD.1955.1571	(*Presbytis cinereus*), Cited Pocock,R.I. 1935: 906, Cited Hill,J.E. 1960: 32
Piah Valley, Upper Perak, [5°14′N 101°07′E]	—	M	Ad	Skin, Skull	ZD.1902.5.2.1	(*Semnopithecus siamensis*), Background Flower,S.S. 1900: 317
Ulu Gombak, Selangor, [c.3°20′N 101°50′E]	3 Nov 1910	M	Ad	Skin, Skull	ZD.1955.1568	Cited Pocock,R.I. 1935: 906, Cited Hill,J.E. 1960: 32
—	—	F	Ad	Skin	ZD.1844.1.18.31	(*Semnopithecus mitratus*), Associated with next specimen (1845.4.2.2)
—	—	F	Ad	Skull	ZD.1845.4.2.2	(*Presbytis cinereus*)
[MALAYA]						
Cheras, Selangor, [3°03′N 101°46′E]	16 Mar 1908	M	Ad	Skin, Skull	ZD.1908.7.20.2	(*Presbytis dilecta*), Cited Pocock,R.I. 1935: 906
" " "	24 Mar 1908	F	Juv	Skin, Skull	ZD.1908.7.20.4	(*Presbytis dilecta*), Cited Pocock,R.I. 1935: 906, Cited Colyer,J.F. 1936: 481
Dusun Tua, Selangor, [3°08′N 101°50′E]	3 Oct 1906	M	Ad	Skin, Skull	ZD.1909.4.1.12	(*Presbytis dilecta*), Cited Pocock,R.I. 1935: 906
" " "	3 Oct 1906	F	[Ad]	Skin	ZD.1955.1578	Cited Pocock,R.I. 1935: 906, Cited Hill,J.E. 1960: 32
Gunong Mengkuang Lebah, Selangor, [3°24′N 101°48′E], 4800ft	27 Mar 1907	F	Ad	Skin, Skull	ZD.1909.4.1.13	(*Presbytis dilecta/ P.albocinereus*), Cited Pocock,R.I. 1935: 905
Kuala Tembeling, Pahang River, [4°03′N 102°19′E], c.200ft	25 Aug 1905	M	Ad	Skin, Skull	ZD.1906.10.4.6	(*Presbytis femoralis/ Presbytis dilecta*), Cited Pocock,R.I. 1935: 906
Selangor, Malay Peninsula	—	F	Juv	Skin, Skull	ZD.1889.5.14.2	(*Presbytis dilecta*), Cited Pocock,R.I. 1935: 905
Mountains of Malay Peninsula beyond Kelang, Selangor	4 Apr 1879	M	Ad	Skin, Skull	ZD.1885.8.1.14	(*Presbytis dilecta*), Cited Thomas,O. 1886: 72, Cited Pocock,R.I. 1935: 905
Selangor	—	F	Ad	Skin, Skull	ZD.1889.5.14.1	Holotype *Presbytis dilecta* Elliot 1909, Cited Elliot,D.G. 1909: 270, Cited Pocock,R.I. 1935: 905
Ulu Gombak, [Selangor], [c.3°20′N 101°50′E]	3 Nov 1910	M	[Ad]	Skin	ZD.1955.1567	Cited Pocock,R.I. 1935: 906, Cited Hill,J.E. 1960: 32
Presbytis melalophos siamensis? (Müller and Schlegel 1841)						
MALAYA						
Gunong Besout, Perak, [3°49′N 101°12′E]	1972	F	[Ad]	Wet, cadaver	CDA	
" " "	1972	F	[Juv]	Wet, cadaver	CDA	
[MALAYA]						
Tapah, Perak, [4°07′N 101°16′E]	—	M	Ad	Skull	ZD.1913.3.6.6	(*Presbytis femoralis?*)
[THAILAND]						
Mabek, Malay Peninsula, [c.6°22′N 101°08′E]	27 Jul 1901	F	Juv	Skin, Skull	ZD.1903.2.6.10	(*Presbytis femoralis*), Cited Bonhote,J.L. 1903: 7, Cited Pocock,R.I. 1935: 902,904, Cited Colyer,J.F. 1936: 482
Presbytis melalophos sumatrana (Müller and Schlegel 1841)						
NO SPECIMENS						
—	—	—	—	—	—	
Presbytis melalophos (Raffles 1821)						
[MALAYA]						
Aring, [confluence of Rs.Aring and Lebeh], Kelantan, [c.5°01′N 102°23′E]	23 Aug 1899	M	Ad	Skull	CMZ E7231B	(*Semnopithecus obscurus*), Cited Bonhote,J.L. 1901: 872
SUMATRA						
—	—	F	Juv	Skull	ZD.1164c	Listed Gray,J.E. (unpubl.): 1164
—	—	F	Ad	Skull, Skeleton	ZD.1850.8.15.5	" " " "
—	Died 23 Feb 1904	F	Juv	Skull	ZD.1980.473	
NO DATA						
"Java"	—	M	Ad	Skull	ZD.1844.3.20.9	(*Presbytis cinereus? / Presbytis nigrimanus*)
—	—	M	Ad	Skull	ZD.1164b	Listed Gray,J.E. (unpubl.): 1164
—	—	M	Ad	Skull	ZD.1867.4.12.8	(*[Semnopithecus] siamensis*)

PRESBYTIS COMATA (Desmarest, 1822) [= *aygula* Linnaeus, 1758], see taxonomic note.

Javan Leaf Monkey
(2 subspecies)

Taxonomic note *Simia aygula* Linnaeus, 1758 has recently been re-identified as a Crab-eating Macaque. Linnaeus' description is based on (1) a young animal 'the size of a small cat' collected in Java by Osbeck and described by him in his *Ostindisk Resa* (1757); and (2) Osbeck's description of a live adult male 'markattor' (long-tailed monkey) belonging to Burgomaster Renhorn of Arboga in Sweden. At Linnaeus' request, Osbeck visited Renhorn and described his monkey in detail. The description is to be found in three letters from Osbeck to Linnaeus, dated 13th May, 3rd and 27th June, 1756 (Linnaean Society Archives). Crucial to this identification is a character not mentioned by Linnaeus but cited by Osbeck in his account of his voyage (1757): the longitudinal crest on the crown of the Javan monkey is described as the same colour as the rest of the upper parts (grey or grey yellow). This rules out the grey Javan *Presbytis* with its contrasting black-crested head, previously identified by Thomas & Wroughton (1909, *Proc. zool. Soc. Lond.* **1**:373) as *Simia aygula*. This latter species could in any case be excluded on the grounds that no leaf monkey could, at that time, have been brought to Europe and survived to adulthood. No type material is available in Sweden to confirm the identification. In order to preserve the name *Macaca fascicularis* Raffles, 1821 for the Crab-eating Macaque, a formal request has been made to the International Commission on Zoological Nomenclature for the suppression of the name *Simia aygula* Linnaeus, 1758 (Z.N.(S.) 2399) (Napier & Groves, 1983).

In Chasen's *Handlist of Malaysian Mammals* (1940:77), the next available name cited in the synonymy of *P. aygula* is *Presbytis mitrata* Eschscholtz, 1821, but as previously mentioned (p. 38) this has been re-identified as pertaining to *Presbytis melalophos*. The only other name cited by Chasen for the taxon is *Presbytis comata* (Desmarest, 1822), the description of which closely resembles the 11 western Javan specimens in the BM(NH) collections. The only difference is to be found in the colour of the hands and feet which are not reddish, as stated, but extremely variable, from whitish tinged with some yellow hairs to almost black. Desmarest (1822) gives the country of origin as Sumatra, but I. Geoffroy (1851) notes that this is erroneous and that the specimens upon which the name was based were sent to the Museum d'Histoire Naturelle by Diard in 1821 from Java.

Range Western and Central Java.

Dimensions Field measurements of specimens in the museum collections are as follows:

h & b	3 ♂♂ 505 (430-595)	: 3 ♀♀ 535 (475-570
tail	3 ♂♂ 655 (560-724)	: 3 ♀♀ 659 (590-720)

Subspecies *P. comata* has a black head with a prominent crest on the vertex. Two subspecies have been recognised: *P.c.comata* from western Java and *P.c.fredericae* from Mt Slamat in Central Java. They can be distinguished as follows:

P.c.comata	*P.c.fredericae* (from Sody, 1930*a*)
Back iron grey	Back black overlaid with long hairs (10cm) with ashy brown tips
Underparts white	Underparts black except for the following: grey throat, whitish pectoral band, white abdomen, and a white median line between the two latter areas

Presbytis comata comata (Desmarest, 1822)

Type locality Sumatra, corrected to Java by I. Geoffroy, 1851:16.

Presbytis comata fredericae (Sody, 1930)

Type locality Gunung Slamat, Central Java, at 1000m.

Locality	Date of Collection	Sex	Age	Nature of Specimen	Accession no.	References, Remarks, etc.
Presbytis comata comata (Desmarest 1822)						
JAVA						
Tjibodas, Gedeh, W.Java, [c.6°45′S 107°01′E], 5000ft	14 Feb 1916	M	Ad	Skin, Skull	ZD.1954.52	
Tjibodas, Gedeh, W.Java, [c.6°45′S 107°01′E], 4500ft	2 Mar 1916	F	Ad	Skin, Skull	ZD.1954.53	
Tjiwangi, Preanger, [7°04′S 107°02′E], 4000ft	27 Sep 1907	M	Inf	Skin, Skull	ZD.1909.1.5.20	(*Presbytis mitratus*/ *Presbytis pyrrhus*)
" " "	1 Oct 1907	M	[Ad]	Skin	ZD.1909.1.5.21	Cited Pocock,R.I. 1935: 919
" " "	25 Sep 1907	F	Ad	Skin, Skull	ZD.1909.1.5.22	" " " "
" " "	27 Sep 1907	F	Ad	Skin, Skull	ZD.1909.1.5.23	" " " "
West Java	—	F	Ad	Skin, Skull	ZD.1882.2.6.2	(*Semnopithecus mitratus*), Cited Pocock,R.I. 1935: 919
Wynkoops Bay, [= Teluk Pelabuanratu], W.Java, [c.7°S 106°32′E], 100ft	8 Mar 1920	M	Ad	Skin, Skull	ZD.1954.54	
" " "	18 Mar 1920	M	Ad	Skin, Skull	ZD.1954.55	
—	—	M	Ad	Skin	ZD.1843.12.7.4	(*P.mitratus*), Associated with next specimen(1845.4.2.4)
—	—	M	Ad	Skull	ZD.1845.4.2.4	Listed Gray,J.E. (unpubl.): 11
[JAVA]						
"Borneo"	—	[M]	[Ad]	Skin	ZD.1939.1139	(*Semnopithecus mitratus*)
"Malaysia"	—	I	Juv	Skin, Skull	ZD.1847.3.27.38	" " " "
NO DATA						
—	—	F	Ad	Skull	RCS(OM) A70.6	(*Presbytis aygula aygula*)
Presbytis comata fredericae (Sody 1930)						
NO SPECIMENS						
—	—	—	—	—	—	
Presbytis comata (Desmarest 1822)						
JAVA						
—	—	M	Ad	Skull	ZD.1844.3.20.10	(*Presbytis mitratus*), Listed Gray,J.E. (unpubl.): 11
—	—	I	Juv	Skull	ZD.1867.4.12.15	
—	—	M	Ad	Skull, Skeleton	ZD.1850.8.15.4	Listed Gray,J.E. (unpubl.): 11

PRESBYTIS FRONTATA (Müller, 1838)

White-fronted Leaf Monkey

Principal synonym *Presbytis nudifrons* Elliot, 1909:266.

Type localities Of *frontata*, SE. Borneo. According to Medway (1965: 82) the syntypes were collected at Pematang, Kuala Murung and 'Pulu Lampy', near Banjarmasin; of *nudifrons*, Sungai Bejalong, Mukah District, central Sarawak (*c.* 2°26′N 112°39′E).

Range Known only from central Sarawak and eastern Kalimantan. Said by Medway to be a lowland species not found above 1000ft, but in the original description Müller (1838:136) mentions that the Malays of the south coast of Borneo call it 'Djirangan Goenoeng' (= Mountain Djirangan) to distinguish it from *P. cristata* which they call simply 'Djirangan'.

Dimensions Adult specimens in the museum collections have the following field measurements (see p. 98 for comparative table):

h & b	1 ♂ 600	: 3 ♀♀ 545 (525-570)
tail	1 ♂ 740	: 3 ♀♀ 726 (720-730)

Subspecies *P. frontata* is medium to dark grey brown with black head and limbs, but paler tail. There is a prominent crest on the vertex and an intensely white bare patch on the forehead. The validity of *P.f.nudifrons*, which was recognised by Chasen (1940) as a separate race, cannot be confirmed from the material available.

Locality	Date of Collection	Sex	Age	Nature of Specimen	Accession no.	References, Remarks, etc.
Presbytis frontata (Müller 1838)						
BORNEO						
Poeroek Tjahoe [=Purukcahu], R.Barito, C.Borneo, [0°36′S 114°35′E], 115ft	5 Sep 1909	M	Ad	Skin, Skull	ZD.1910.4.5.12	Figured Elliot,D.G. 1913: Pl.5, Cited Pocock,R.I. 1935: 927
,, ,, ,,	2 Sep 1909	F	Ad	Skin, Skull	ZD.1910.4.5.13	Cited Pocock,R.I. 1935: 927
,, ,, ,,	2 Sep 1909	F	Ad	Skin, Skull	ZD.1910.4.5.14	,, ,, ,, ,,
,, ,, ,,	2 Sep 1909	F	Ad	Skin, Skull	ZD.1910.4.5.15	,, ,, ,, ,,
,, ,, ,,	5 [Sep] 1909	M	Juv	Skin, Skull	ZD.1910.4.5.16	,, ,, ,, ,,
"Probably Sarawak"	—	F	Juv	Skin, Skull	ZD.1955.714	(*Presbytis frontata nudifrons*), Cited Hill,J.E. 1960: 34
[Sungai] Bejalong, [Mukah Dist.], Sarawak, [2°26′N 112°39′E]	Jun 1903	M	Ad	Skin, Skull	ZD.1904.2.9.1	Holotype *Presbytis nudifrons* Elliot 1909, Cited Elliot,D.G. 1909: 266, Cited Pocock,R.I. 1935: 927
—	—	F	[Ad]	Skin	ZD.1845.7.16.4	Cited Pocock,R.I. 1935: 927
—	—	M	Ad	Skin, Skull in skin	ZD.1852.5.1.1	
[BORNEO]						
—	—	F	Ad	Skull, 4 limb bones	DLPA 71.0.5	

PRESBYTIS HOSEI (Thomas, 1889)

Hose's Leaf Monkey
(3 subspecies)

Taxonomic notes *P. hosei* from the Baram District of Sarawak and *P. everetti* from Mt Kinabalu in Sabah were originally thought to be separate species until Chasen & Kloss (1931:6) unequivocally stated that *P. everetti* was the adult female form of *P. hosei*. Pocock (1928:920) noted that the crown pattern of subadult females resembled that of adult males, indicating that the increasing pigmentation of the crown hairs was associated with approaching maturity. Of the 9 adult female skins (with skulls) in the museum collections, 7 have the typical *everetti* black crown with median white superciliary spot or band. Of the other two, one (BMNH1900. 8.4. 2) has a quantity of white hairs on either side of the vertex; and in the other (BMNH1951. 61) the white superciliary band is wider and deeper than usual, separating the black crown hairs from those of the brow. In both, the basicranial sutures are still open, indicating immaturity. One female of adult size (BMNH 1956. 211) has the typical male *hosei* crown pattern; the skull is missing, but since the nipples are not developed, she was probably nulliparous. Miller (1934:17) reserved judgment on the matter, treating *P. everetti* as a distinct species.

Range Northeast Borneo from Belaga on R. Rejang in Sarawak (2°42′N 113°47′E) to Kudat and Marudu Bay in Sabah (*c.* 6°45′N 116°55′E) and southwards to R. Karangan in Kalimantan (1°06′N 117°54′E).

Dimensions Adult specimens in the museum collections have the following field measurements: 2 ♂♂ h & b 520, 545; tail 670, 660 : 1 ♀ h & b 550, tail 650.

Subspecies *P. hosei* is iron grey with buffy-white underparts. Three subspecies are generally recognised. Sexual dichromatism is seen in *P.h.hosei* but neither in *P.h.sabana* (Davis, 1962, p.62) nor *P.h.canicrus* (Miller, 1943:17). The following table of distinguishing characters is based on fur and skin colour. Pigmentation of the facial skin in *P.h.sabana* is well illustrated in Davis, 1962 (pl. 12).

Distinguishing characters of the subspecies of *Presbytis hosei*

	Colour of face	Crown	Back and outside of limbs	Underparts and inside of limbs	Tail
hosei ♂♂ (NW. Sarawak)	Blackish mask; fleshpink nose; lips bluish; white hairs on cheeks and chin	Sagittal crest; crown, nape and brows black; rest of head white	Grizzled grey; lower leg blackish, blending with feet	Buffy white, fading on forearms and lower leg	As back, slightly paler below
hosei ♀♀ [= *everetti*]	"	Black; hairs slightly raised on vertex; white spot or band above brows	Grizzled grey; lower leg sometimes darkening towards ankle	As above but very slightly yellower in tone	"
canicrus (EC. Borneo) (from Miller, 1934)	Blackish mask includes nose and upper lip; lower lip, cheeks and chin whitish buff; dividing line runs from corner of mouth to ear	As above but brownish; no white spot or band on brow	Grizzled grey; arms and legs contrast sharply with black hands and feet	Buffy white, fading near wrists and ankles	Grizzled grey above and below
sabana (NE. Sabah)	Reddish fleshcolour mask and nose; lips bluish; large black patch between eye, nose and mouth, variable in extent	Narrow black crest on crown with grey patches on either side	"	Chin, sides of neck, throat and chest greyish; belly and insides of limbs whitish	"

Presbytis hosei hosei (Thomas, 1889)

Principal synonym *Semnopithecus everetti* Thomas, 1893:582, pl. 41.

Type localities Of *hosei*, Baram, Sarawak; of *everetti*, Mt Kinabalu, Sabah, at 3500ft.

Presbytis hosei canicrus Miller, 1934

Type locality R. Karangan (1°06′N 117°54′E), NE. Kalimantan.

Presbytis hosei sabana (Thomas, 1893)

Type locality Paitan (6°25′N 117°22′E), Sabah.

Locality	Date of Collection	Sex	Age	Nature of Specimen	Accession no.	References, Remarks, etc.
Presbytis hosei hosei (Thomas 1889)						
BORNEO						
Baram [Dist.], [Sarawak], [c.4°30′N 114°12′E]	—	[M]	[Ad]	Skin, Skull in skin	ZD.1939.3478	
Batu Blah, Sarawak, [4°02′N 114°29′E]	—	F	[Ad]	Skin	ZD.1956.211	Background Hose,C. 1893
[Gunung Mulu], Sarawak, [4°03′N 114°56′E]	—	[F]	[Juv]	Skin	ZD.1978.1548	
Long Lenin, R.Tinjar, Sarawak, N.W.Borneo, [c.3°20′N 114°16′E]	1 Sep 1919	M	Ad	Skin, Skull	ZD.1955.713	(*Presbytis aygula hosei*), Cited Hill,J.E. 1960: 33
Malingai Mt., N.Borneo	Jul 1892	F	Ad	Skin, Skull	ZD.1893.3.4.1	(*Semnopithecus everetti*), Background Hose,C. 1893, Figured Pocock,R.I. 1935: 899-900, fig.1F
Mt.Dulit, Baram Dist., Sarawak, [3°21′N 114°11′E], 2000ft	Dec 1898	M	Ad	Skull	CMZ E7234A	
Mt.Dulit, Sarawak, [3°21′N 114°11′E]	—	M	[Ad]	Skin	ZD.1956.209	Background Hose,C. 1893
,, ,, ,,	—	M	[Juv]	Skin	ZD.1956.210	,, ,, ,, ,,
,, ,, ,,	—	F	[Juv]	Skin	ZD.1956.212	,, ,, ,, ,,
Mt.Kalulong, E.Sarawak, [3°14′N 114°39E]	Feb [1893]	F	Ad	Skin, Skull	ZD.1893.6.2.1	(*Semnopithecus everetti*), Background Hose,C. 1893, Cited Pocock,R.I. 1935: 921, Cited Colyer,J.F. 1936: 37
[Mt.Kalulong], [Baram], Sarawak	Feb [1893]	M	Ad	Skin, Skull	ZD.1900.8.4.13	Background Hose,C. 1893
[Mt.Kalulong], [Tinjar, R.Baram], [Sarawak], [3°14′N 114°39′E]	Oct 1932	M	Ad	Skin, Skull	ZD.1951.60	Background Harrisson,T.H. 1933: 402
,, ,, ,,	Oct 1932	F	Ad	Skin, Skull	ZD.1951.61	,, ,, ,, ,,
,, ,, ,,	Oct 1932	F	Juv	Skin, Skull	ZD.1951.62	,, ,, ,, ,,
,, ,, ,,	—	M	Ad	Skull	ZD.1951.63	,, ,, ,, ,,
,, ,, ,,	Nov 1932	[M]	[Ad]	Skin, Radiograph of penis	ZD.1951.64	,, ,, ,, ,,
,, ,, ,,	Nov 1932	F	Ad	Skin, Skull	ZD.1951.65	,, ,, ,, ,,
Mt.Kinabalu, [Sabah], N.Borneo, [6°05′N 116°33′E], 3500ft	Mar 1892	F	Ad	Skin, Skull	ZD.1892.10.19.1	Holotype *Semnopithecus everetti* Thomas 1893, Figured Thomas,O. 1893: 582, pl.41, Cited Pocock,R.I. 1935: 920-922
Mt.Trekan, Baram Dist., [Sarawak], 2000ft	1891	I	Foetus	Wet, cadaver	CMZ E7234B	Background Hose,C. 1893
Niah, Baram [Dist.], [Sarawak], N.W.Borneo, [c.4°30′N 114°12′E]	—	M	Ad	Skin, Skull	ZD.1889.1.8.1	Holotype *Semnopithecus hosei* Thomas 1889, Figured Thomas,O. 1889: 159, pl.16, Cited Hose,C. 1893a: 11, Cited Pocock,R.I. 1935: 920-922, Listed Gray,J.E. (unpubl.): 1950
R.Baram, N.Borneo	—	M	Ad	Skin, Skull in skin	ZD.1891.8.28.6	Background Hose,C. 1893
R.Baram, N.W.Borneo	—	[?F]	[Inf]	Skin, Skull in skin	ZD.1891.8.28.7	,, ,, ,, ,,
,, ,, ,,	—	I	[Inf]	Skin, Skull in skin	ZD.1891.8.28.8	,, ,, ,, ,,
[Sarawak]	1932	M	[Ad]	Skin	ZD.1951.58	Background Harrisson,T.H. 1933: 402
,, ,, ,,	—	M	Ad	Skull	ZD.1951.59	(*Pithecus aygula hosei*), Background Harrisson,T.H. 1933: 402
Upper R.Berar, Mulu National Park, Sarawak	—	F	[Ad]	Photographs	ZD.1980.862	Photographs Medway,Lord (unpubl.)c
—	—	I	[Inf]	Skin	ZD.1908.10.23.4	
[BORNEO]						
Mt.Dulit, Baram Dist., Sarawak, [3°21′N 114°11′E]	5 Oct 1898	F	Ad	Skin, Skull	ZD.1900.8.4.1	(*Semnopithecus hosei everetti*)
,, ,, ,,	22 Dec 1896	F	Ad	Skin, Skull	ZD.1900.8.4.2	,, ,, ,, ,,
,, ,, ,,	9 Nov 1897	F	Ad	Skin, Skull	ZD.1900.8.4.3	,, ,, ,, ,,
Mt.Dulit, E.Sarawak, [3°21′N 114°11′E], 3000ft	Jun 1892	F	Ad	Skin, Skull	ZD.1892.10.20.1	Paratype *Semnopithecus everetti* Thomas 1893, Cited Thomas,O. 1893: 583
Mt.Dulit, Sarawak, [3°21′N 114°11′E], 3000ft	—	M	Ad	Skull, Skeleton	CDA M47	Background Hose,C. 1893
,, ,, ,,	—	M	Ad	Skull, Skeleton	CDA M48	,, ,, ,, ,,
Mt.Dulit, [Sarawak], [3°21′N 114°11′E], 2000ft	—	M	Ad	Skull, Skeleton	ZD.1892.6.1.1	,, ,, ,, ,,
Mt.Kinabalu, [Sabah], [6°05′N 116°33′E], 4000ft	—	M	Ad	Skull	ZD.1895.10.4.41	(*Semnopithecus everetti?*), Cited Thomas,O. 1889: 160, Cited Thomas,O. 1889a: 228
Presbytis hosei canicrus Miller 1934						
NO SPECIMENS						
—	—	—	—	—	—	

Locality	Date of Collection	Sex	Age	Nature of Specimen	Accession no.	References, Remarks, etc.
Presbytis hosei sabana (Thomas 1893)						
BORNEO						
Marudu Bay, [Sabah], N.Borneo, [6°45′N 116°55′E]	Feb 1894	M	Ad	Skin, Skull	ZD.1894.6.8.1	
" " "	—	[F]	[Inf]	Skin	ZD.1895.1.23.1	
N.Borneo	—	I	[Ad]	Wet, head only	ZD.1894.7.2.46	
Paitan, [Sabah], N.Borneo, [6°23′N 117°22′E]	Jul 1892	M	Ad	Skin, Skull	ZD.1893.3.4.2	Holotype *Semnopithecus sabanus* Thomas 1893, Figured Thomas,O. 1893a: 230, pl.7, Figured Colyer,J.F. 1936: 38, fig.48
" " "	Jul 1892	M	Ad	Skin, Skull	ZD.1893.3.4.3	Paratype *Semnopithecus sabanus* Thomas 1893, Cited Thomas,O. 1893a: 232
[BORNEO]						
Kudat, [Sabah], [6°53′N 116°50′E]	—	F	Ad	Skull, Skeleton	ZD.1894.6.8.6	

PRESBYTIS POTENZIANI (Bonaparte, 1856)

Mentawai Islands Leaf Monkey
(2 subspecies)

Taxonomic notes The position of *P. potenziani* within the genus has for many years been equivocal (Thorington & Groves, 1970:642). Recently however, Brandon-Jones (personal communication, 1980) has suggested that it is best referred to the *P. melalophos* group. The main reason for placing it with the *P. cristata* group was the supposed golden colour of the young. In 1879, Peters published a coloured illustration (pl. 4B) of the type of *Semnopithecus chrysogaster* Blyth, 1875 (a junior objective synonym of *P. potenziani*), showing the holotype together with an infant of a bright golden colour. There is nothing in the original descriptions of either *P. potenziani* or *S. chrysogaster* to connect the two specimens and their association could have been fortuitous (Mitchell & Weitzel, 1983). Tilson (1976) observed that the newborn *P. potenziani* is white but soon develops the 'cruciger' pattern as in *P. melalophos*. The association of *P. potenziani* with the *P. melalophos* group rather than the *P. cristata* group was recognised by both Thomas (1895:664) and Chasen (1940:78).

Range *P.p.potenziani* from Sipora and North and South Pagai Islands; *P.p.siberu* from Siberut.

Dimensions The holotype of *P.p.siberu*, an adult male, measures: h & b 500, tail 580. For further measurements, see Chasen & Kloss (1927:814).

Subspecies *P. potenziani* is mainly black, with white on forehead, crown, throat and pubis. The reddish underparts of *P.p.potenziani* are replaced in *P.p.siberu* by black or dull brown, the long black tips concealing the reddish hue of the hair bases. Thus the white pubic patch is a more striking feature in *P.p.siberu* than it is in *P.p.potenziani* where it blends with the light reddish ventrum. There are a few white hairs at the tip of the tail in both races.

Presbytis potenziani potenziani (Bonaparte, 1856)

Principal synonym *Presbytes chrysogaster* Blyth, 1875:10.

Type localities Of *potenziani* and *chrysogaster*, Tenasserim, corrected by Thomas (1895) to Sipora I., Mentawai Is, W. Sumatra.

Presbytis potenziani siberu (Chasen & Kloss, 1927)

Type locality Siberut I., Mentawai Is, W. Sumatra.

Locality	Date of Collection	Sex	Age	Nature of Specimen	Accession no.	References, Remarks, etc.
Presbytis potenziani potenziani (Bonaparte 1856)						
SUMATRA						
Sereinu (Sipora) I., Mentawai Is.	May 1894	F	Juv	Skin, Skull	ZD.1895.1.9.1	Cited Bonaparte,C. 1856: 412, Cited Thomas,O. 1895: 663
Presbytis potenziani siberu (Chasen and Kloss 1927)						
SUMATRA						
Siberut, Mentawai Is.	13 Sep 1924	M	Ad	Skin, Skull	ZD.1928.11.2.1	Holotype *Pithecus potenziani siberu* Chasen and Kloss 1927, Figured Chasen,F.N. and Kloss,C.Boden 1927: 811, pl.1

PRESBYTIS RUBICUNDA (Müller, 1838)

Maroon Leaf Monkey
(5 subspecies)

Taxonomic notes Chasen (1940:80) recognised 4 subspecies to which Davis (1962: 61) added *P.r. chrysea* from a relatively restricted area at the southeast end of Dewhurst Bay, Kinabatangan District, Sabah, N. Borneo.

Range Borneo and the Karimata Islands.

Dimensions Adult specimens in the museum collections have the following field measurements:

h & b	1 ♂ 520	: 3 ♀♀ 495 (470-508)
tail	1 ♂ 730	: 3 ♀♀ 702 (675-725)

Subspecies Lyon (1911:138) noted three regional colour types: *P.r.rubicunda* (SE. Borneo) is dark maroon with a blackish suffusion on hands and feet; *P.r.rubida* (SW. Borneo) is a lighter, brighter red with a few long black-tipped hairs on the back and short black hairs on the phalanges; and *P.r.ignita* from northern Borneo is similar to *rubida* but lacks the blackish suffusion of the hands and feet. More recently Davis (1962:63) has stressed the extent of local variability, which is exemplified by the museum's specimens.

All the localised specimens are from two areas: Baram District (including the type locality of *ignita*, Mt Mulu); and Paku, Saribas, near Kuching. Of those from the Baram District, the long hairs of the nape are generally of a paler hue, in accordance with Dollman's description of *ignita* (1909:204), though the fingers and toes are sometimes suffused with black. The six Saribas specimens lack the pale nape, and three have the hands and feet suffused with black. They were identified by Hill (1960:34) as *P.r.rubida*, but Medway (1970:552, fig. 2) has identified material from Saribas as *P.r.ignita*. The BM(NH) Saribas material has been catalogued as *P.r.rubida*?

P.r.carimatae resembles *rubida* in colour but the skull is distinguished by the long heavy post-glenoid process, the angle of the mandible being large and heavy (for further details see Lyon, 1911:139). *P.r.chrysea*, as its name implies, is golden-red in colour; it is restricted to a relatively small area in the extreme northeast of Borneo.

Presbytis rubicunda rubicunda (Müller, 1838)

Type locality Restricted by Müller & Schlegel (1841:71) to Mt Sakumbang on Tana-Lawut, SW. of Bandjarmasin, SE. Borneo.

Presbytis rubicunda carimatae Miller, 1907

Type locality Telok Edar, Karimata Is.

Presbytis rubicunda chrysea Davis, 1962

Type locality E. of R. Kretam Kechil, SE. end of Dewhurst Bay, Kinabatangan District, Sabah, Borneo.

Presbytis rubicunda ignita Dollman, 1909

Type locality Mt Mulu, N. Sarawak, Borneo, at 1000ft.

Presbytis rubicunda rubida (Lyon, 1911)

Type locality Batu Jurong (*c.* 2°37′S 110°09′E), south of the mouth of R. Kendawangan, SW. Borneo.

Locality	Date of Collection	Sex	Age	Nature of Specimen	Accession no.	References, Remarks, etc.
Presbytis rubicunda rubicunda (Müller 1838)						
BORNEO						
—	—	M	Ad	Skin, Skull	ZD.1842.1.19.93	[Syntype] *Semnopithecus rubicundus* Müller 1838, Cited Müller,S. 1838: 137, Cited Müller,S. and Schlegel,H. 1841: 71, Cited Gray,J.E. 1843: 2, Figured Elliot,D.G. 1913: Pl.4, Cited Pocock,R.I. 1935: 924-925
Presbytis rubicunda carimatae Miller 1907						
NO SPECIMENS						
—	—	—	—	—	—	
Presbytis rubicunda chrysea Davis 1962						
NO SPECIMENS						
—	—	—	—	—	—	
Presbytis rubicunda ignita Dollman 1909						
BORNEO						
Baram, [Sarawak]	—	F	Ad	Skin, Skull	ZD.1889.8.4.1	
,, ,, ,,	—	F	Juv	Skin, Skull	ZD.1889.8.4.2	
,, ,, ,,	—	F	[Juv]	Skin, Skull in skin	ZD.1889.8.4.3	
Baram Dist., Sarawak, [c.4°15′N 114°19′E]	Mar 1894	M	Ad	Skull, Skeleton	CMZ E7227A	(*Semnopithecus rubicundus*)
,, ,, ,,	—	[F]	Ad	Skull, Skeleton	CMZ E7227B	,, ,, ,, ,,
Baram Dist., Sarawak	—	F	[Ad]	Skin, Skull in skin	ZD.1956.215	(*Presbytis rubicundus*)
Baram [Dist.], Sarawak, [c.4°15′N 114°19′E]	—	[M]	[Ad]	Skin, Skull in skin	ZD.1939.3480	
Mt.Dulit, Baram [Dist.], Sarawak, 3400ft	6 Aug 1919	M	[Ad]	Skin	ZD.1955.730	
,, ,, ,,	—	F	Ad	Skull	ZD.1955.730a	
Mt.Mulu, Baram, Sarawak, 1000ft	Oct 1893	F	Ad	Skin, Skull	ZD.1894.6.2.31	Holotype *Presbytis ignita* Dollman 1909, Cited Dollman,G. 1909: 204, Cited Pocock,R.I. 1935: 924-925
Mt.Mulu, R.Baram, Sarawak, 1000ft	Nov 1893	[M]	Ad	Skull, Skeleton	ZD.1894.6.12.12	
Mt.Mulu	Nov	F	[Ad]	Skin	ZD.1908.7.17.2	Cited Pocock,R.I. 1935: 924
,, ,, ,,	Nov	M	Ad	Skull	ZD.1908.7.17.2a	,, ,, ,, ,,
R.Ridan, Baram Dist., [Sarawak], [4°12′N 114°19′E]	—		[Ad]	Skull in skin of head	CMZ E7227C	(*Semnopithecus rubicundus*)
[BORNEO]						
Baram, E.Sarawak	—	M	Ad	Skull	ZD.1892.11.28.1	
Presbytis rubicunda ignita? Dollman 1909						
BORNEO						
Sarawak, N.Borneo	—	M	[Ad]	Skin, Skull in skin	ZD.1906.8.26.1	
Presbytis rubicunda rubida? (Lyon 1911)						
BORNEO						
Paku, Saribas, Sarawak	Mar 1917	F	Ad	Skin, Skull	ZD.1920.12.4.3	Cited Pocock,R.I. 1935: 924
,, ,, ,,	Mar 1917	M	Juv	Skin, Skull	ZD.1955.726	(*Presbytis rubicunda rubida*), Cited Pocock,R.I. 1935: 924, Cited Hill,J.E. 1960: 34
,, ,, ,,	Mar 1917	M	Juv	Skin, Skull	ZD.1955.727	,, ,, ,, ,,
,, ,, ,,	Mar 1917	M	Ad	Skin, Skull	ZD.1955.728	,, ,, ,, ,,
,, ,, ,,	Mar 1917	F	Ad	Skin, Skull	ZD.1955.729	,, ,, ,, ,,

Locality	Date of Collection	Sex	Age	Nature of Specimen	Accession no.	References, Remarks, etc.
Presbytis rubicunda rubida? *continued*						
Paku[Sungai] Anyut, S.W.Sarawak, [1°33′N 111°33′E]	20 Jul 1914	F	Ad	Skin, Skull	ZD.1920.12.4.4	(*Presbytis rubicunda rubida*). Cited Pocock,R.I. 1935: 924
Presbytis rubicunda (Müller 1838)						
BORNEO						
[R.] Kidukarok, [Sabah], [5°36′N 116°31′E], 5500ft	9 Sep 1956	F	[Ad]	Skin, Skull	CMZ E7227D	
[?Sabah]	—	F	Juv	Skin, Skull	ZD.1908.7.17.3	(*Presbytis ignitus*)
—	—	[?F]	[Ad]	Skin	ZD.1878.7.4.4	
—	1899	F	Ad	Skull, Skeleton	CDA M51	
—	—	F	Ad	Skull, Skeleton	CDA M76	
—	—			7 skulls	RCS(OM) S70.7	

PRESBYTIS THOMASI (Collett, 1893)

Thomas' Leaf Monkey
(3 subspecies)

Range Atjeh, northern Sumatra but recorded only from near Mt Leuser at 3381m (*P.t.nubila*); from the east coast between Aru Bay (4°09′N) and the type locality of *P.t.thomasi* (3°57′N 98°20′E); and a little further south between 3°45′N and 2°57′N (*P.t.margae*).

Dimensions Field measurements of *P.t.thomasi* in the museum collections are as follows: 1 ♂ h & b 550, tail 690; 1 ♀ h & b 550, tail 760.

Subspecies Neither *P.t.nubila* nor *P.t.margae* are represented in the collections. *P.t.thomasi* is iron grey with creamy white underparts, insides of limbs and ventral aspect of tail; hands and feet are black. The head is strikingly marked with three black stripes which are separated by a white patch above each eye. *P.t.margae* is altogether darker, dark brownish grey with limbs and tail brownish black. The head markings are less striking, since they are dark brownish grey like the back. The throat, underparts, insides of limbs and ventral aspect of tail, are, as in *P.t.thomasi,* creamy white. Wilson & Wilson (1975) refer *P.t.margae* to *P. melalophos*. *P.t.nubila* is described as being like *P.t.thomasi* but darker and more slatey in colour with longer thicker fur as befits an animal living in the foothills of Mt Leuser up to 7000ft.

Presbytis thomasi thomasi (Collett, 1893)

Type locality Glen Bervie Estate (3°57′N 98°20′E), Langkat District, E. Sumatra.

Presbytis thomasi margae Hooijer, 1948

Type locality Serdang (3°42′N 98°52′E), Deli, E. Sumatra.

Presbytis thomasi nubila Miller, 1942

Type locality Blangnanga (20 miles W. of Blangkedjeren 3°59′N 97°20′E), Atjeh, N. Sumatra at 3600ft.

Locality	Date of Collection	Sex	Age	Nature of Specimen	Accession no.	References, Remarks, etc.
Presbytis thomasi thomasi (Collett 1893)						
SUMATRA						
Glen Bervie Estate, Langkat, N.E.Sumatra, [3°57′N 98°20′E]	6 Jul 1888	F	Ad	Skin, Skull	ZD.1890.1.20.1	Syntype *Semnopithecus thomasi* Collett 1893, Figured Collett,R. 1893: 613, pl.42, Cited Pocock,R.I. 1935: 918-919
Kuala Simpang, Atjeh, [4°17′N 98°03′E]	13 Jul 1938	M	Ad	Skin, Skull	ZD.1938.11.30.7	
" " "	13 Jul 1938	F	Ad	Skin, Skull	ZD.1938.11.30.8	
—	24 Feb 1928	F	Ad	Skin, Skull	ZD.1928.3.1.1	Cited Pocock,R.I. 1935: 918-919
—	—	I	[Ad]	Skin	ZD.1956.382	(*Presbytis aygula thomasi*)
—	—	I	[Ad]	Skin	ZD.1956.383	" " " "
—	—	I	[Ad]	Skin	ZD.1956.384	" " " "
—	—	I	[Ad]	Skin	ZD.1956.385	" " " "
—	—	[F]	Ad	Skull	RCS(OM) G68.871	
Presbytis thomasi margae Hooijer 1948						
NO SPECIMENS						
—	—	—	—	—	—	
Presbytis thomasi nubila Miller 1942						
NO SPECIMENS						
—	—	—	—	—	—	

Presbytis cristata group

PRESBYTIS CRISTATA (Raffles, 1821)

Silvered Leaf Monkey
(9 subspecies)

Taxonomic notes Two names often associated with this species are *Simia maura* Schreber, 1774 and *Cercopithecus auratus* E. Geoffrey, 1812. The former, depicted as a yellowish-brown long-tailed infant monkey, is generally regarded as indeterminable (Thomas & Wroughton, 1909; Elliot, 1913). The latter, based on a mounted female specimen in the Musée National d'Histoire Naturelle in Paris, was examined by Elliot. He identified it (1913:77) as the red phase of the black Javan leaf monkey, and therefore used the name *auratus*, which antedates both *Semnopithecus pyrrhus* Horsfield, 1823 and *Simia cristata* Raffles, 1821, for this taxon. Pocock (1928:667; 1935:920) found it impossible to agree with Elliot's identification, and as Chasen (1940:81) pointed out the problems will remain unsolved until there is a critical re-appraisal of the specimen, which has recently been made by Groves & Weitzel, who consider (personal communication) that Elliot was probably correct. The type locality was originally cited as: 'L'Inde: les Moluques, suivant M. Themminck'. Brandon-Jones (personal communication, 1980) has noted that Müller (1840:16) restricted it to Semarang, Java (6°58′S 110°29′E).

Subfossil material collected by A H Everett (1879:149-155) from Borneo cave sites has been catalogued under *P. cristata.*

Presbytis barbei This species has posed problems ever since it was tentatively proposed by Blyth in 1847. There have been four main areas of uncertainty:

1. The two syntypes, an adult male and female, were described as having pale markings on the face like *P. obscura* (i.e. on the lips and around the eyes). As Blyth was actually comparing specimens of *P. obscura* with *P. barbei* at the time of writing the description, it is difficult to see how he could have been in error on this point, but 16 years later in 1863, he redescribed the faces of *P. barbei* as black, without any comment on his previous statement to the contrary. Khajuria & Agrawal (1979) suggest that chemicals or paint used in the preparation of the specimens for exhibition may have altered the original colour of the facial skin.

2. In the pages preceding the description of *P. barbei*, Blyth described and figured the erect crest on the vertex of *P. phayrei*. He then described *P. barbei* as being unlike *P. phayrei* in this respect. However in a footnote he remarked that the taxidermist who had prepared the specimens stated categorically that they had had a thin raised crest upon the vertex when fresh.

3. The type locality, originally stated to be in the Province of Ye, Tenasserim, was altered in 1863 to Tipperah Hills (= Tripura) on information given to Blyth by 'M. Barbe', presumably a relative of the Rev. J. Barbe, the collector of the specimens, after whom the taxon was named. The Rev. J. Barbe was a missionary, and although his main collections were made in Tenasserim, he had also collected in Tripura. In an attempt to solve the problem, Agrawal (1974) has on three occasions collected leaf-monkeys in Tripura, but only *P. phayrei phayrei* has so far been found there.

4. The final difficulty, regarding the provenance of the syntypes, has recently been satisfactorily resolved, thanks to careful examination by Khajuria & Agrawal (1979). They observed in the female a unique feature mentioned by Blyth in both 1847 and 1863, a white stripe on the inside of the base of the thigh 'more developed on one side than the other'. Since the specimens have consecutive numbers — 36a and b in Blyth's Catalogue (1863) and 19a and b in Anderson's Catalogue (1881) — it is reasonable to assume, with Khajuria and Agrawal (1979), that these are indeed the original specimens.

Agrawal's description (1974) of the syntypes of *P. barbei* shows that, as originally surmised by Blyth, they are not readily assignable to either *P. phayrei* or *P. obscura*. The following table, adapted from Agrawal (1974), shows the differences between *P. phayrei phayrei* and *P. barbei*, and the resemblances between *P. barbei* and *P. cristata atrior.*

	P. phayrei phayrei	*Syntypes of P. barbei*	*P. cristata atrior* (n = 6)
	(from Tripura)	(Type locality: Ye, Tenasserim or ? Tripura)	(Type locality : Ye Tenasserim, *c.* 15°15′N)
1. Dorsal colour	Dusky grey	Sooty black	Dark greybrown
2.	Golden sheen on dorsum	Silvery lustre on shoulder and upper arm	Silvery or yellowish lustre on shoulder and upper arm in 4 specimens (*not* the type)
3. Lips	Perfectly white	Bluish black (but see para. 1)	Deeply pigmented (n = 4)
4. Lip hairs	Sparse white hairs on both upper and lower	Sparse white hairs on upper; blackish on lower	Sparse white hairs on both lips (n = 2); sparse black hairs on both lips (n = 2)
5. Tail	Tuft darker than proximal portion	Proximal portion black but paler distally	Proximal portion as back, tip slightly paler
6. Ventrum	Silvery white	Grey	Buff to dark brown

In view of the above-mentioned uncertainties, and without the opportunity of examining the syntypes, the status of *P. barbei* remains in doubt. If it should be found to fall within the rather variable taxon, *P.cristata atrior* (Pocock, 1928), it would of course antedate that name.

Range *P. cristata* is found south of the range of *P. phayrei* in Burma, Thailand and Indo-China. Both species have been collected at Lat Bua Kao in central Thailand, but Fooden (1976) has suggested that they may be allopatric, *P. phayrei* being found to the north and *P. cristata* to the south of the R. Mae Nam Mun. The range of *P. cristata* also includes Cambodia, S. Vietnam and S. Laos. The race *P.c.atrior* intrudes between the ranges of *P. phayrei* and *P. obscura* to the west of Bangkok in Thailand and in Tenasserim in Burma. Thence southwards, there supervenes a gap of some 550 miles (900 km) before further populations of *P. cristata* are found in the coastal mangrove forests of western Malaya between Province Wellesley (*c.* 5°30′N) and Kelang, Selangor (3°N)(Medway, 1969). The species is not found on Singapore Island, but is present on Bangka, Belitung and islands of the Riau and Lingga Archipelagos, as well as on Serasan Island of the South Natuna group. The range also includes Sumatra, Java and Borneo.

Dimensions As in *Macaca fascicularis*, the Javan representatives are the largest; *P.c.pullata* from the Riau Archipelago is the smallest race. Museum specimens are listed below in ascending order of size:

P.c.pullata (Riau Archipelago)

h & b	2 ♂♂ 516	(505, 526)	:	5 ♀♀ 439	(420-455)
tail	2 ♂♂ 692	(684, 700)	:	5 ♀♀ 651	(560-703)

P.c.ultima (S. Malaya, N. Sumatra, Borneo)

h & b	3 ♂♂ 544	(503-580)	:	3 ♀♀ 489	(460-514)
tail	3 ♂♂ 699	(671-750)	:	3 ♀♀ 715	(678-751)

P.c.germaini (S. Indo-China)

h & b	4 ♂♂ 553	(540-587)	:	2 ♀♀ 500	(490, 510)
tail	4 ♂♂ 778	(720-838)	:	2 ♀♀ 775	(755, 795)

P.c.pyrrhus (E. Java)

h & b	5 ♂♂ 566	(505-600)	:	1 ♀ 545	
tail	5 ♂♂ 741	(700-785)	:	1 ♀ 665	

P.c.sondaica (W. Java)

h & b	5 ♂♂ 571	(465-650)	:	4 ♀♀ 598	(550-630)
tail	5 ♂♂ 742	(680-810)	:	4 ♀♀ 725	(690-770)

The weight of an adult male *P.c.germaini* is recorded as 8626g.

Subspecies Two subspecies were recognised by Ellerman & Morrison-Scott (1951): *atrior* from Burma and *germaini* from Indo-China and Thailand. Chasen (1940) recognised six: *cristata* from S. Sumatra, *kohlbruggei* from Bali, *pyrrhus* from E. Java, *sondaica* from W. Java, *ultima* from N. Sumatra, W. Malaya and Borneo, and *vigilans* from Serasan Island, South Natuna Islands. Chasen included in *P.c.cristata* specimens from the Riau and Lingga Archipelagos (i.e. Batam, Bintan, Bulan, Sugi, Galang, Sebang, Bakong, Lingga and Chombok Islands) as well as Bangka Island. Later Hill (1960), after studying the museum collections, came to the conclusion that *P.c.pullata* from Bintan and Batam Islands should be given subspecific status, a decision reinforced by its considerably smaller size (see above).

With regard to Belitung Island, Chasen was unable to examine any specimens, but Hooijer (1960:580) recognised this population as a new subspecies, although he did not name it. In 1962 (p. 41) Hooijer proposed the name *Trachypithecus cristatus robustus* for a subspecies of Middle Pleistocene age from the Dubois Collections from Tegoean, Central Java. It has a more robust skull and larger teeth than the largest race living today, *P.c.pyrrhus*, which inhabits the same area.

Dào Van Tien (1977) has described a new subspecies of *P. cristata*, *P.cristata caudalis*, from an unknown locality in North Vietnam. It differs from *P.c.germaini* (and its synonym *P.c.margarita* Elliot, 1909) in having the whiskers and outside of the legs pale yellowish and the tail bicoloured, blackish above, pale yellowish below.

In 1935, Pocock described in detail many of the museum's specimens of *P. cristata*, showing that individuals vary both in basic coloration and in the quantity and distribution of the silver tipped hairs, as well as in the arrangement of the hairs on the crown. The provision of an adequate key was adjudged to be impossible, but the following table may be of some help in distinguishing the recognised subspecies.

Table of characters of the subspecies of *Presbytis cristata*

Subspecies	Crown hairs	General colour	Silvered effect of pale-tipped hairs	Whiskers	Size (see above)
atrior (Tenasserim)	Swept smoothly back; no crest	Dark greybrown	Absent, or minimal sprinkling on upper arms and legs	Dark greybrown	None recorded
sondaica (W. Java)	Long, erect, curling forward (with the whiskers) to frame the face	Black	"	Black, or with buff tips	Large
pyrrhus (E. Java)	"	Black (some reddish mutants)	Buff-tipped hairs on crown, nape, shoulders, arms, legs and rump	"	"
cristata (S. Sumatra)	Long, erect, forming an untidy peaked crest on vertex	Dark greybrown to black	White-tipped hairs on crown, nape, shoulders, upper arms and legs	Black	None recorded
pullata (Batam I.)	"	"	As above, plus some silvering on back in majority of specimens	Black with buff tips	Small
ultima (S. Malaya, N. Sumatra, Borneo)	"	Dark greybrown (some reddish mutants)	White-tipped hairs on all upper parts except hands, feet and distal portion of tail	Dark greybrown with white tips	Medium
germaini (S. Indo-China)	"	Silver grey to dark grey	Buff or white-tipped hairs on crown, nape, shoulders, arms, legs and rump; less conspicuous on paler specimens	Whiskers entirely whitish or buff	"

P.c.kohlbruggei (Sody, 1931) from Bali was described as being similar to *P.c.pyrrhus* but with a smaller skull; *P.c.vigilans* Miller, 1913 from Serasan Island was separated on characters of the facial skeleton (no skins were collected), and *P.c.caudalis* Dào, 1977 from North Vietnam is similar to *P.c.germaini* but with yellowish whiskers and legs, and a bicoloured tail.

Presbytis cristata cristata (Raffles, 1821)

Principal synonym *Semnopithecus pruinosus* Desmarest, 1822:533.

Type localities Of *cristata*, Bencoolen (= Bengkulu), SW. Sumatra: of *pruinosus*, Island of Sumatra.

Presbytis cristata atrior (Pocock, 1928)

Type locality 'Ye forest, 500ft, some 100 miles S. of Moulmein', Tenasserim, Burma (Pocock, 1935:953).

Presbytis cristata caudalis Dào Van Tien, 1977

Type locality North Vietnam, exact locality unknown.

Presbytis cristata germaini (Milne Edwards, 1876)

Principal synonyms *Presbytis margarita* Elliot, 1909:271; *Presbytis germaini mandibularis* Kloss, 1916:32; *Presbytis cristata koratensis* Kloss, 1919:340.

Type localities Of *germaini*, forests of Cochin China and Cambodia; of *margarita*, Langbian, Annam (= S. Vietnam); of *mandibularis*, Ko Chang Island, Chantabun (= Chanthaburi), SE. Thailand; of *koratensis*, Lat Bua Kao, 30 miles W. of Khorat (= Nakhon Ratchasima), Thailand.

Presbytis cristata kohlbruggei (Sody, 1931)

Principal synonym *Trachypithecus pyrrhus stresemanni* Pocock, 1935:931.

Type localities Of *kohlbruggei*, Bali; of *stresemanni*, Danau (= lake) Bratan, Bali.

Presbytis cristata pullata Thomas & Wroughton, 1909

Type locality Batam Island, Riau Archipelago.

Presbytis cristata pyrrhus (Horsfield, 1823)

Type locality Java, restricted by Robinson & Kloss, 1919, to the Province of Pasuruan, E. Java.

Presbytis cristata sondaica (Robinson & Kloss, 1919)

Type locality Tjibodas (= Cibodas), Preanger (= Priangan) Regencies, W. Java.

Presbytis cristata ultima (Elliot, 1910)

Type locality Mount Dulit, Sarawak, Borneo.

Presbytis cristata vigilans Miller, 1913

Type locality Sirhassen Island (= Serasan), S. Natuna Islands.

Locality	Date of Collection	Sex	Age	Nature of Specimen	Accession no.	References, Remarks, etc.
Presbytis cristata cristata (Raffles 1821)						
SUMATRA						
Lampar, R.Moesie, Palembang Residency, [3°40′S 102°58′E], 700ft	27 Jun 1881	F	Juv	Skin, Skull	ZD.1882.7.28.3	(*[S.] pruinosus*), Cited Pocock,R.I. 1935: 933-935
Telukbetung, S.Sumatra, [5°28′S 105°16′E]	[c.Nov 1880]	F	Inf	Skin, Skull	ZD.1881.3.15.2	Cited Pocock,R.I. 1935: 933
—	—	F	Ad	Skin, Skull	ZD.1838.3.13.14	(*Presbytis pruinosus*), Listed Anon. (unpubl.)b: 48
—	—	[M]	[Inf]	Skin, Skull in skin	ZD.1838.3.13.15	,, ,, ,, ,,
—	—	F	Inf	Skin, Skull	ZD.1838.3.13.16	,, ,, ,, ,,
—	—	M	[Inf]	Skin, Skull in skin	ZD.1838.3.13.17	,, ,, ,, ,,
—	—	M	Ad	Skull	ZD.1855.12.26.68	Syntype *Simia cristata* Raffles 1821, (*Trachypithecus cristatus cristatus*), Cited Raffles,T.S. 1821: 244, Listed Gray,J.E. (unpubl.): 7, Listed Anon. (unpubl.)a
[SUMATRA]						
[Inderapura], [between Bengkulu and Padang], [2°02′S 100°56′E]	—	F	Ad	Skull	ZD.1842.1.19.94	Listed Anon. (unpubl.)b: 76
Presbytis cristata atrior (Pocock 1928)						
[BURMA]						
Foot of Mt.Nwalaboo [=Nwalabo Taung], Tavoy Dist., Tenasserim, [c.14°01′N 98°25′E]	10 Apr 1878	F	Ad	Skin, Skull	ZD.1885.8.1.12	Paratype *Pithecus pyrrhus atrior* Pocock 1928, (*Semnopithecus barbei* Blyth), Cited Thomas,O. 1886: 66, Cited Pocock,R.I. 1928: 673, Figured Pocock,R.I. 1935: 928, fig.3E, 953
Ye Forest, S. of Moulmein, Ataran Division, Tenasserim, [c.15°15′N 98°E]	—	M	Ad	Skin, Skull, Skeleton	ZD.1910.7.11.1	Holotype *Pithecus pyrrhus atrior* Pocock 1928, Figured Pocock,R.I. 1928: 671, 673-674, Cited Pocock,R.I. 1935: 953, Cited Colyer,J.F. 1936: 37
[THAILAND]						
H. Maw Tee Maw [=Huai Mothimo], [Menam Noi], Sai Yoke, S.W.Siam, [14°31′N 98°38′E], 500m	24 Apr 1915	M	Ad	Skin, Skull	ZD.1915.12.1.2	Paratype *Pithecus pyrrhus atrior* Pocock 1928, (*Presbytis barbei*), Correspondence Gairdner,K.G. (unpubl.), Cited Pocock,R.I. 1928: 673, Cited Pocock,R.I. 1935: 953
Menam Noi, [Sai Yoke area], S.W.Siam, [c.14°25′N 98°51′E]	12 Apr 1915	F	[Ad]	Skin	ZD.1915.12.1.3	(*Presbytis barbei*), Correspondence Gairdner,K.G. (unpubl.), Cited Pocock,R.I. 1935: 953
[Menam Noi], [Sai Yoke area], S.W.Siam, [c.14°25′N 98°51′E]	[16 Feb 1915]	M	Inf	Skin, Skull	ZD.1915.12.1.7	(*Presbytis barbei*), Correspondence Gairdner,K.G. (unpubl.)
S.W.Siam, 14°25′N 98°45′E, 350m	18 Apr 1914	M	Ad	Skin, Skull	ZD.1914.8.22.7	Paratype *Pithecus pyrrhus atrior* Pocock 1928, (*Trachypithecus barbei barbei*), Cited Pocock,R.I. 1928: 673, Cited Pocock,R.I. 1935: 953
Presbytis cristata caudalis Dào Van Tien 1977						
NO SPECIMENS						
—	—	—	—	—	—	
Presbytis cristata germaini (Milne Edwards 1876)						
CAMBODIA						
[Phnom] Sambor, [12°53′N 105°04′E], 200ft	30 Dec 1927	M	Ad	Skin, Skull	ZD.1928.7.1.4	(*Pithecus pyrrhus germaini*), Cited Pocock,R.I. 1935: 936, 939, Listed Lowe,W.P. (unpubl.)a
,, ,, ,,	30 Dec 1927	F	Ad	Skin, Skull	ZD.1928.7.1.5	,, ,, ,, ,,
[?CAMBODIA]						
[Chaudoc Province], Camboja or Cochin China, [c.10°37′N 105°11′E]	[Dec 1867]	[F]	[Juv]	Skin	ZD.1878.6.17.1	(*Semnopithecus germaini*), Cited Pocock,R.I. 1935: 937
Camboja or Cochin China	—	[M]	Juv	Skin, Skull	ZD.1878.6.17.2	,, ,, ,, ,,
[S.VIETNAM]						
Cochin China	—	F	Ad	Skin, Skull	ZD.1881.6.30.1	Cited Pocock,R.I. 1935: 936, 939
,, ,, ,,	23 Apr 1926	[F]	[Juv]	Skin	ZD.1927.12.1.13	(*Pithecus germaini*), (Semnopithèque), Background Lowe,W.P. (unpubl.)
Lang Bian, Annam, [12°03′N 108°27′E]	—	M	Juv	Skin, Skull	ZD.1908.11.1.5	Holotype *Presbytis margarita* Elliot 1909, Cited Elliot,D.G. 1909: 271, Cited Pocock,R.I. 1935: 937, Figured Elliot,D.G. 1913: Pl.2
Lào-Bao[?], Annam, [?16°37′N 106°36′E]	17 Jan 1929	[?F]	[Ad]	Skin	ZD.1933.4.1.12	Cited Osgood,W.H. 1932: 202, Cited Pocock,R.I. 1935: 937
,, ,, ,,	17 Jan 1929	[?M]	[Juv]	Skin	ZD.1933.4.1.13	,, ,, ,, ,,
[THAILAND]						
Klong Yai, S.E.Siam, [11°46′N 102°54′E]	5 Jan 1915	M	Ad	Skin, Skull	ZD.1915.11.4.12	Cited Kloss,C.Boden 1916a: 32, Cited Pocock,R.I. 1935: 937, 939
Ko Chang I., S.E.Siam, [12°N 102°23′E]	7 Dec 1914	M	Ad	Skin, Skull	ZD.1915.11.4.5	Holotype *Presbytis germaini mandibularis* Kloss 1916, Cited Kloss,C.Boden 1916a: 32, 67, Cited Pocock,R.I. 1935: 937, 939

Locality	Date of Collection	Sex	Age	Nature of Specimen	Accession no.	References, Remarks, etc.
Presbytis cristata germaini *continued*						
Ko Chang I., Chanthaburi, S.E.Siam, [12°N 102°23′E]	7 Dec 1914	F	Ad	Skin, Skull	ZD.1915.11.4.6	Paratype *Presbytis germaini mandibularis* Kloss 1916, Cited Kloss,C.Boden 1916a: 33, 67, Cited Pocock,R.I. 1935: 937
,, ,, ,,	8 Dec 1914	M	Juv	Skin, Skull	ZD.1915.11.4.13	Paratype *Pithecus germaini mandibularis* Kloss 1916, Cited Kloss,C.Boden 1916a: 33
Siam, 13°40′N 99°25′E, 70m	[14] Jul [1913]	M	Ad	Skin, Skull	ZD.1913.11.2.3	Cited Gairdner,K.G. (unpubl.): 10, Cited Pocock,R.I. 1935: 937
,, ,, ,,	—	F	Ad	Skin, Skull	ZD.1913.11.2.4	,, ,, ,, ,,
Siam, 13°45′N 99°25′E, 80m	Jan 1913	M	Ad	Skin, Skull	ZD.1914.8.22.8	Cited Gairdner,K.G. (unpubl.): 3, Cited Pocock,R.I. 1935: 937
,, ,, ,,	Feb 1914	[?F]	[Ad]	Skin	ZD.1914.8.22.9	,, ,, ,, ,,
Siam, 13°40′N [99°25′E]	7 Jul 1913	F	Ad	Skull	ZD.1914.8.22.10	Cited Gairdner,K.G. (unpubl.): 10
Tahkamen, on R.Bangpakong between Pachim and Kabin, [Bang Pakong Dist.], Siam, [c.13°33′N 100°58′E]	[20 Mar 1897]	M	Ad	Skin, Skull, Radiograph of penis	ZD.1897.8.3.1	Cited Flower,S.S. 1900: 319, Cited Pocock,R.I. 1935: 936, 939
Presbytis cristata kohlbruggei (Sody 1931)						
[JAVA]						
Danau [=lake] Bratan, Bali I., [8°15′S 115°10′E], 2500ft	28 Mar 1911	M	Ad	Skin, Skull	ZD.1913.3.6.1	Holotype *Trachypithecus pyrrhus stresemanni* Pocock 1935, Cited Pocock,R.I. 1935: 931
Danau [=lake] Bratan, Bali I., [8°15′S 115°10′E], [2500ft]	—	F	Ad	Skull	ZD.1913.3.6.2	Paratype *Trachypithecus pyrrhus stresemanni* Pocock 1935, Cited Pocock,R.I. 1935: 931
,, ,, ,,	—	F	Ad	Skull	ZD.1913.3.6.3	Paratype *Trachypithecus pyrrhus stresemanni* Pocock 1935, Cited Pocock,R.I. 1935: 931, Figured Colyer,J.F. 1936: 217, Fig.260
,, ,, ,,	—	M	Ad	Skull	ZD.1913.3.6.4	Paratype *Trachypithecus pyrrhus stresemanni* Pocock 1935, Cited Pocock,R.I. 1935: 931
[LESSER SUNDA IS.]						
Mangsit Hill, N.Lombok, Lombok I., [8°29′S 116°03′E]	Jun 1896	M	Ad	Skin, Skull	ZD.1939.1143	(*Pygathrix cristata pullata*)
Presbytis cristata pullata Thomas and Wroughton 1909						
[MALAYA]						
Pasir Panjang, Bintang I., Riau Archipelago, [c.1°N 104°30′E]	9 Jun 1908	M	Ad	Skin, Skull	ZD.1909.4.1.3	Cited Thomas,O. and Wroughton,R.C. 1909c: 106
,, ,, ,,	9 Jun 1908	F	Juv	Skin, Skull	ZD.1909.4.1.4	,, ,, ,, ,,
Tanjong Sau, Batam I., Riau Archipelago, [c.1°N 104°E]	11 Jul 1908	F	Ad	Skin, Skull	ZD.1909.4.1.11	,, ,, ,, ,,
,, ,, ,,	10 Jul 1908	[M]	Inf	Skin, Skull	ZD.1980.175	,, ,, ,, ,,
Tanjong Turut, Batam I., Riau Archipelago, [c.1°N 104°E]	14 Jul 1908	M	Ad	Skin, Skull	ZD.1909.4.1.5	Cited Thomas,O. and Wroughton,R.C. 1909c: 106, Figured Colyer,J.F. 1936: 668, Fig.909
,, ,, ,,	12 Jul 1908	F	Ad	Skin, Skull	ZD.1909.4.1.6	Cited Thomas,O. and Wroughton,R.C. 1909c: 106
,, ,, ,,	12 Jul 1908	F	Ad	Skin, Skull	ZD.1909.4.1.7	,, ,, ,, ,,
,, ,, ,,	12 Jul 1908	F	Inf	Skin, Skull	ZD.1909.4.1.8	,, ,, ,, ,,
,, ,, ,,	12 Jul 1908	F	Ad	Skin, Skull	ZD.1909.4.1.9	Holotype *Presbytis cristata pullata* Thomas and Wroughton 1909, (*Trachypithecus pyrrhus cristatus*), Cited Thomas,O. and Wroughton,R.C. 1909b: 439, Cited Thomas,O. and Wroughton,R.C. 1909c: 106, Cited Pocock,R.I. 1935: 934
,, ,, ,,	14 Jul 1908	F	Ad	Skin, Skull	ZD.1909.4.1.10	Cited Thomas,O. and Wroughton,R.C. 1909c: 106
Presbytis cristata pyrrhus (Horsfield 1823)						
JAVA						
Badjoelmati, E.Java, [7°55′S 114°23′E], 100ft	27 Jan 1920	M	Ad	Skin, Skull	ZD.1954.61	
E.Java	Died 2 May 1938	[M]	Juv	Skin, Skull	ZD.1938.11.30.10	(*T.pyrrhus sondaicus*)
,, ,, ,,	Died 3 Apr 1938	[M]	Inf	Skin, Skull	ZD.1938.11.30.11	,, ,, ,, ,,
Madiun, E.Java, [7°37′S 111°33′E], 300ft	9 Aug 1938	F	[Juv]	Skin, Skull	ZD.1938.11.30.9	,, ,, ,, ,,
Mojokerto, E.Java, [7°25′S 112°31′E], 700ft	26 Apr 1938	M	Juv	Skin, Skull	ZD.1938.11.30.12	
Ongop ongop, Idjen Massif, E.Java, [c.8°05′S 114°14′E], 5700ft	5 Apr 1916	M	Ad	Skin, Skull	ZD.1954.56	
Rembang, E.Java, [6°45′S 111°22′E]	Died Nov 1937	F	[Ad]	Skin	ZD.1938.3.14.1	
,, ,, ,,	—	F	Inf	Skin, Skull	ZD.1938.3.14.2	
Sodong Jerok, Idjen Massif, nr. Banjuwangie, E.Java, 3900ft	8 Apr 1916	M	Ad	Skin, Skull	ZD.1954.57	
,, ,, ,,	10 Apr 1916	M	Ad	Skin, Skull, Radiograph of penis	ZD.1954.58	
,, ,, ,,	9 Apr 1916	F	Ad	Skin, Skull	ZD.1954.59	

Locality	Date of Collection	Sex	Age	Nature of Specimen	Accession no.	References, Remarks, etc.
Presbytis cristata pyrrhus *continued*						
Tamansari, E.Java, [7°58′S 111°31′E], 1600ft	17 Jan 1920	M	Ad	Skin, Skull	ZD.1954.60	
—	—	F	Ad	Skin, Skull	ZD.22a	Syntype *Semnopithecus pyrrhus* Horsfield 1823, Listed Anon. (unpubl.)b: 22, Figured Horsfield,T. 1823, Cited Horsfield,T. 1851: 10, Cited Thomas,O. and Wroughton,R.C. 1909: 372
—	—	F	Ad	Skin, Skull	ZD.1855.12.24.11	[Syntype] *Semnopithecus pyrrhus* Horsfield 1823
[JAVA]						
[E.Java]	—	M	Juv	Skin, Skull	ZD.1846.3.23.18	(*Presbytis auratus*)
—	—		Ad	3 skulls	RCS(OM) S70.4(A)	
NO DATA						
—	—	F	Ad	Skull	RCS(OM) G68.87	
Presbytis cristata sondaica (Robinson and Kloss 1919)						
JAVA						
Cheribon [=Cirebon], [6°46′S 108°33′E]	—			10 skulls	RCS(OM) S70.43(A)	
" " "	—	M	Ad	2 skulls	RCS(OM) S70.43(G)	
Gunung Pantjar, W.Java, [6°35′S 106°54′E]	—	[M]	[Juv]	Skin, Skull	ZD.1907.6.18.1	(*[T.] pyrrhus sondaicus*)
Kalipoetjang, Tji Tandoei R., S.Java, [7°39′S 108°44′E]	Mar 1908	F	Inf	Skin, Skull	ZD.1909.1.5.18	(*P.pyrrhus sondaicus*), Cited Thomas,O. and Wroughton,R.C. 1909: 372
Kosala, Bantam, W.Java, [c.6°40′S 106°22′E], 2500ft	—	M	Ad	Skin, Skull	ZD.1882.2.6.1	(*[T.] pyrrhus sondaicus*)
Pangandaran, Dirk de Vries Bay, S.Java, [7°41′S 108°40′E]	15 Apr 1908	M	Ad	Skin, Skull	ZD.1909.1.5.4	(*P.pyrrhus sondaicus*)(*Presbytis maurus*), Cited Thomas,O. and Wroughton,R.C. 1909: 372
" " "	2 Apr 1908	F	Inf	Skin, Skull	ZD.1909.1.5.5	" " " "
" " "	—	F	Ad	Skin, Skull	ZD.1909.1.5.6	" " " "
" " "	—	M	Inf	Skin, Skull	ZD.1909.1.5.6a	" " " "
" " "	8 Apr 1908	F	Inf	Skin, Skull	ZD.1909.1.5.7	" " " "
Tjibodas, Preanger Regencies, W.Java, [c.6°45′S 107°E], 5500ft	12 Feb 1916	M	Ad	Skin, Skull	ZD.1949.423	Holotype *Pithecus pyrrhus sondaicus* Robinson and Kloss 1919, Cited Robinson,H.C. and Kloss,C.Boden 1919: 374
Tjibodas, W.Java, [c.6°45′S 107°E], 5000ft	24 Feb 1916	M	Ad	Skin, Skull	ZD.1954.62	Paratype *Pithecus pyrrhus sondaicus* Robinson and Kloss 1919, Cited Robinson,H.C. and Kloss,C.Boden 1919: 374
Tjilatjap [=Cilacap], [7°44′S 109°E], Sea level	19 Oct 1907	M	Ad	Skin, Skull	ZD.1909.1.5.8	(*Presbytis maurus*), Cited Thomas,O. and Wroughton,R.C. 1909: 372
" " "	19 Oct 1907	F	Ad	Skin, Skull	ZD.1909.1.5.9	(*P.pyrrhus sondaicus*), Cited Thomas,O. and Wroughton,R.C. 1909: 372
" " "	19 Oct 1907	F	[Inf]	Skin	ZD.1909.1.5.10	" " " "
" " "	7 Nov 1907	F	Ad	Skin, Skull	ZD.1909.1.5.11	" " " "
Tjilatjap [=Cilacap], [7°44′S 109°E]	17 Nov 1907	F	Ad	Skin, Skull	ZD.1909.1.5.12	" " " "
" " "	17 Nov 1907	F	Inf	Skin, Skull	ZD.1909.1.5.13	" " " "
Tji Wangie, Preanger, [7°04′S 107°02′E], 4000ft	4 Oct 1907	M	Ad	Skin, Skull	ZD.1909.1.5.15	(*Presbytis pyrrhus*), Cited Thomas,O. and Wroughton,R.C. 1909: 372
" " "	28 Sep 1907	F	Ad	Skin, Skull	ZD.1909.1.5.16	" " " "
" " "	28 Sep 1907	F	Ad	Skin, Skull	ZD.1909.1.5.17	" " " "
[Tji Wangie], Preanger, [7°04′S 107°02′E], [4000ft]	[c.26 Sep 1907]	M	Ad	Skull	ZD.1909.1.5.14	" " " "
—	—	F	Ad	Skin	ZD.1843.12.7.3	(*[T.] pyrrhus sondaicus*) Associated with next specimen (1845.4.2.6)
—	—	F	Ad	Skull	ZD.1845.4.2.6	(*Semnopithecus maurus*)
—	Died 3 Apr 1903	F	[Juv]	Skin	ZD.1939.1140	" " " "
[JAVA]						
Borneo [?]	—	M	Ad	Skin, Skull	ZD.1846.1.9.4	(*Semnopithecus maurus*)(*Presbytis pyrrhus*)
Pinang [?]	—	F	[Ad]	Skin	ZD.1879.11.21.301	(*[T.] pyrrhus sondaicus*)
Presbytis cristata ultima (Elliot 1910)						
BORNEO						
Baram, N.W.Borneo	—	[F]	Inf	Skin, Skull	ZD.1889.8.4.4	(*Semnopithecus cristatus*)
" " "	—	F	Ad	Skin, Skull	ZD.1890.1.28.2	(*cristatus*), Cited Pocock,R.I. 1935: 935
" " "	—	[M]	Inf	Skin, Skull	ZD.1890.1.28.3	" " " "
Baram Dist., Sarawak	—	M	[Ad]	Skin	ZD.1956.216	
Karanginton, R.Martapoera, S.Borneo, [3°25′S 114°47′E], 30ft	19 Aug 1909	M	Ad	Skin, Skull	ZD.1910.4.5.17	Cited Pocock,R.I. 1935: 935, 938-939
—	—	[F]	[Inf]	Skin, Skull in skin	ZD.1845.10.2.4	(*Presbytis pyrrhus cristatus*)
—	—	M	Ad	Skin, Skull	ZD.1879.8.30.2	(*Trachypithecus pyrrhus cristatus*), Cited Pocock,R.I. 1935: 935
—	—	F	Ad	Skin, Skull	ZD.1879.8.30.3	" " " "
—	—	M	Juv	Skin, Skull	ZD.1879.8.30.4	(*Semnopithecus maurus*)
—	—	I	[Juv]	Skin	ZD.1879.8.30.5	(*Trachypithecus pyrrhus cristatus*)

Locality	Date of Collection	Sex	Age	Nature of Specimen	Accession no.	References, Remarks, etc.
Presbytis cristata ultima *continued*						
[BORNEO]						
[Balingian], Sarawak, [2°59′N 112°34′E]	[c.14 May 1917]	[F]	[Ad]	Skin, Skull	ZD.1955.715	Cited Hill,J.E. 1960: 34
Entawa, Samarahan, Sarawak, [1°17′N 110°29′E]	21 Nov 1919	F	[Ad]	Skin, Skull	ZD.1955.717	Cited Pocock,R.I. 1935: 935, Cited Hill,J.E. 1960: 34
[Sungai] Abai, [Sabah], [6°23′N 116°21′E]	Jul 1895	F	Ad	Skin, Skull	ZD.1896.5.1.1	Cited Pocock,R.I. 1935: 935, 939
Sungei Pelandok, Paku, Sarebas, S.W.Sarawak, [1°31′N 111°31′E]	22 Aug 1916	F	Ad	Skin, Skull	ZD.1955.718	Cited Pocock,R.I. 1935: 935, Cited Hill,J.E. 1960: 34
—	—	[M]	[Ad]	Skin, Skull	ZD.1955.716	Cited Hill,J.E. 1960: 34
MALAYA						
Kuala Selangor, Selangor Coast, [3°21′N 101°15′E]	11 Oct 1909	M	Ad	Skin, Skull	ZD.1955.1562	Cited Pocock,R.I. 1935: 933
Kuala Selangor, Selangor coast, F.M.S., [3°21′N 101°15′E]	9 Feb 1907	F	Ad	Skin, Skull	ZD.1955.1563	,, ,, ,, ,,
Selangor	Died 7 Dec 1956	F	Juv	Skin, Skull	ZD.1971.2612	
[MALAYA]						
Kuala Selangor, Selangor, F.M.S., [3°21′N 101°15′E]	9 Dec 1904	M	[Ad]	Skin	ZD.1909.4.1.1	(*Trachypithecus pyrrhus cristatus*), Cited Thomas,O. and Wroughton,R.C. 1909c: 106, Cited Pocock,R.I. 1935: 934
,, ,, ,,	9 Feb 1907	F	[Ad]	Skin	ZD.1909.4.1.2	(*Trachypithecus pyrrhus cristatus*), Cited Thomas,O. and Wroughton,R.C. 1909c: 106, Cited Pocock,R.I. 1935: 933
Pinang I., [5°24′N 100°14′E]	—	F	Inf	Skin, Skull	ZD.1879.11.21.300	(*Trachypithecus pyrrhus cristatus*), Listed Cantor,T. (unpubl.)
Near sea coast, Wellesley Province, [Pinang], [c.5°25′N 100°25′E]	—	F	Ad	Skin, Skull	ZD.1900.2.4.1	(*Trachypithecus pyrrhus cristatus*), Cited Pocock,R.I. 1935: 934, 939
SUMATRA						
Deli, [c.3°45′N 98°41′E]	—	M	Ad	Skin, Skull	ZD.1890.3.3.1	(*Trachypithecus pyrrhus cristatus*), Cited Pocock,R.I. 1935: 933, 939
—	—	[F]	[Ad]	Skin	ZD.1956.380	
—	—	[M]	[Juv]	Skin	ZD.1956.381	
Presbytis cristata vigilans Miller 1913						
NO SPECIMENS						
—	—	—	—	—	—	
Presbytis cristata (Raffles 1821)						
BORNEO						
Baram Dist., [Sarawak], [4°35N 113°58′E]	18 Jan 1896	M	Ad	Skull, Skeleton	CMZ E7246A	(*Semnopithecus cristatus*)
[MALAYA]						
[Selangor]	—	F	Ad	Skull	ZD.1955.1578a	(*Presbytis dilecta*)(*Presbytis melalophos siamensis*)
SUMATRA						
[Atjeh], [N.W.Sumatra]	—	M	Ad	Skull	RCS(OM) A70.424	(*Presbytis cristata cristata*)
NO DATA						
—	[Died] 21 Sep 1839	I	Juv	Skull	ZD.1850.11.22.24	(*Trachypithecus cristatus cristatus*), Listed Anon. (unpubl.)a, Listed Gray,J.E. (unpubl.): 7, 952
—	—	I	Juv	Skull	ZD.1853.12.16.1	(*Presbytis maurus*)
—	—	F	Ad	Skull, Skeleton	ZD.1855.12.17.22	(*Presbytis maurus*)(*Presbytis pyrrhus*), Listed Gray,J.E. (unpubl.): 9, Figured Colyer,J.F. 1936: 216, 219, Fig.259, 262
—	—	I	Juv	Skull	ZD.1855.12.26.55	(*Presbytis maura*), Listed Gray,J.E. (unpubl.): 9
—	—	M	Ad	Skull, Skeleton	ZD.1866.4.25.11	(*Presbytis maurus*), Listed Gray,J.E. (unpubl.): 9
—	—	F	Ad	Skull	ZD.1867.4.12.9	(*Trachypithecus cristatus?*), Listed Gray,J.E. (unpubl.): 1164
Presbytis cristata? (Raffles 1821)						
BORNEO						
Jambusan, (Cave 5, Excavation D), [Sarawak], [1°25′N 110°11′E]	[1879]	I	[Ad]	Bone fragments (21)	BM(NH)A M20565	(Subfossil) (*Macaca cf. nemestrina*), Cited Everett,A.H. 1879: 152
,, ,, ,,	[1879]	I	[Ad]	L femur	BM(NH)A M20566	,, ,, ,, ,,
,, ,, ,,	[1879]	I	[Ad]	L tibia	BM(NH)A M20567	,, ,, ,, ,,
,, ,, ,,	[1879]	I	[Ad]	R tibia	BM(NH)A M20568	,, ,, ,, ,,
,, ,, ,,	[1879]	I	[Ad]	L humerus	BM(NH)A M20569	,, ,, ,, ,,
,, ,, ,,	[1879]	I	[Ad]	L ulna	BM(NH)A M20570	,, ,, ,, ,,
,, ,, ,,	[1879]	I	[Ad]	R radius	BM(NH)A M20571	,, ,, ,, ,,
,, ,, ,,	[1879]	I	[Ad]	L fibula	BM(NH)A M20572	,, ,, ,, ,,
,, ,, ,,	[1879]	I	[Ad]	L fibula	BM(NH)A M20573	,, ,, ,, ,,
,, ,, ,,	[1879]	I	[Ad]	R innominate bone	BM(NH)A M20574	,, ,, ,, ,,

Locality	Date of Collection	Sex	Age	Nature of Specimen	Accession no.	References, Remarks, etc.
Presbytis cristata? *continued*						
Jibong Hills [= Gunong Jebong], (Cave), [Sarawak], [1°24′N 110°09′E]	[1879]	F	Ad	Skull	BM(NH)A M20539	(Subfossil)(*Macaca cf. nemestrina*), Cited Everett,A.H. 1879: 153
„ „ „	[1879]	I	[Ad]	L and R femoral heads	BM(NH)A M20540	(Subfossil)(*Macaca cf. nemestrina*), Cited Everett,A.H. 1879: 153-154
„ „ „	[1879]	I	[Ad]	R humerus	BM(NH)A M20541	„ „ „ „
„ „ „	[1879]	I	[Ad]	L humerus	BM(NH)A M20542	„ „ „ „
„ „ „	[1879]	I	[Ad]	R humerus	BM(NH)A M20543	(Subfossil)(*Macaca cf. nemestrina*) Background Everett, A.H. 1879: 153-154
„ „ „	[1879]	I	[Ad]	R ulna	BM(NH)A M20544	(Subfossil)(*Macaca cf. nemestrina*), Cited Everett A.H. 1879: 153-154
„ „ „	[1879]	I	[Ad]	L ulna	BM(NH)A M20545	„ „ „ „
„ „ „	[1879]	I	[Ad]	L radius	BM(NH)A M20546	„ „ „ „
„ „ „	[1879]	I	[Ad]	L radius	BM(NH)A M20547	„ „ „ „
„ „ „	[1879]	I	[Ad]	R radius	BM(NH)A M20548	„ „ „ „
„ „ „	[1879]	I	[Ad]	8 vertebrae	BM(NH)A M20550	„ „ „ „
„ „ „	[1879]	I	[Ad]	R innominate	BM(NH)A M20551a	„ „ „ „
„ „ „	[1879]	I	[Ad]	Foot and tail bones (10), bone fragments	BM(NH)A M20552	(Subfossil)(*Macaca cf. nemestrina*), Cited Everett,A.H. 1879: 153
„ „ „	[1879]	I	[Ad]	Sacrum	BM(NH)A M20553	(Subfossil)(*Macaca cf. nemestrina*), Cited Everett,A.H. 1879: 153-154
Jibong Hills [=Gunong Jebong], (Cave), Jambusan Valley, [Sarawak], [1°24′N 110°09′E]	[1879]	I	[Ad]	Sacrum	BM(NH)A M20556	(Subfossil)(Primate), Cited Everett,A.H. 1879: 153
Cave in hill near Jibong [=Gunung Jebong], Jambusan Valley, [Sarawak], [1°24′N 110°09′E]	[1879]	I	[Ad]	9 vertebrae, L. calcaneum, bone fragments	BM(NH)A M20555	„ „ „ „
[BORNEO]						
[Samarahan], [S.W.Sarawak], [1°17′N 110°29′E]	—	F	Ad	Skull	ZD.1955.723	(*P.femoralis chrysomelas*), Cited Hill,J.E. 1960: 32
[JAVA]						
—	—	M	Ad	Skull	ZD.1867.4.12.10	(*Presbytis*)(?typical *pyrrhus*)
NO DATA						
—	—	F	Ad	Skull	ZD.[48a]	Listed Anon. (unpubl.)b: 48
—	—	F	Ad	Skull	ZD.1846.5.11.4	(*Presbytes maura*), Listed Gray,J.E. (unpubl.): 9
—	—	I	Juv	Skull	ZD.1867.4.12.13	(*?Trachypithecus cristatus*)

PRESBYTIS FRANCOISI (Pousargues, 1898)

François' Leaf Monkey
(?6 subspecies)

Taxonomic notes Since Ellerman & Morrison-Scott (1951) provisionally recognised four subspecies, two further forms have been described: *Trachypithecus* (*Presbytis*) *leucocephalus* T'an Pang-Chieh, 1957 and *Presbytis francoisi hatinhensis* Dào Van Tien, 1970. Osgood (1932:206) also cited a completely black skin, with skull, collected in French Indo-China, which may represent yet another taxon. Whether these taxa should be considered as species or subspecies is at present uncertain, but Fooden, (1976:109), unites them in a single species, an arrangement which is followed here. Li & Ma (1981) have recently reported intergradation on the boundary between *P.f.francoisi* and *P.f.leucocephala*.

Range Central Laos, N. Vietnam and southwestern Kwangsi, China. *P. francoisi* has been found mainly in rocky habitats (Fooden, 1976).

Dimensions Adult specimens in the museum collections have the following field measurements:

h & b	4 ♂♂ 548 (485-635)	: 3 ♀♀ 570 (550-590)
tail	4 ♂♂ 849 (820-872)	: 3 ♀♀ 852 (830-887)

Subspecies Six subspecies are provisionally recognised. All are crested and mainly black, relieved by varying amounts of white or gold on the head, body and tail.

Key to the subspecies of *P. francoisi* (adapted from Osgood, 1932, and Dào Van Tien, 1970)

1.	Head mainly black with white stripes on cheeks and behind ears	2
1a.	Head mainly white or golden yellow	4
2.	Narrow white stripe behind ear not passing beyond base of pinna on to nape; (rump and thighs black like rest of body)	**P.f.francoisi**
2a.	Broad white stripe behind ear passes beyond base of pinna on to nape	3
3.	Rump and thighs black like rest of body	**P.f.hatinhensis**
3a.	Rump and outer side of thighs pure creamy white	**P.f.delacouri**
4.	Head white with black crest and brow band; (rump and thighs black like rest of body)	**P.f.laotum**
4a.	Head completely white or golden	5
5.	Rump and thighs black like rest of body; some white hairs on hands and feet; tailtip white	**P.f.leucocephala***
5a.	Rump and outer side of thighs frosted, due to white-tipped hairs; hands, feet and entire tail black; (slight blackening of crest in some)	**P.f.poliocephala**

*From T'an Pang-chieh (1957), Li & Ma (1981), and photograph by Dr M R Brambell (1974).

Presbytis francoisi francoisi (Pousargues, 1898)

Type locality Lung-chou (22°24′N 106°59′E), S. Kwangsi, China.

Presbytis francoisi delacouri (Osgood, 1932)

Type locality Hoi Xuan (20°22′N 105°07′E), N. Vietnam.

Presbytis francoisi hatinhensis Dào Van Tien, 1970

Type locality Xom-cuc (*c.* 17°55′N 105°48′E), Ha Tinh, N. Vietnam.

Presbytis francoisi laotum (Thomas, 1921)

Type locality Ban Na Sao (= Ban Naxao), R. Mekong (left bank) at *c.* 17°30′N, Laos.

Presbytis francoisi leucocephala T'an Pang-Chieh, 1957

Type locality Fu-sui (22°40′N 107°56′E), S. Kwangsi, China.

Presbytis francoisi poliocephala (Trouessart, 1911).

Type locality Caï-Khin (= Cai kinh, *c.* 20°22′N 106°25′E), N. Vietnam.

Locality	Date of Collection	Sex	Age	Nature of Specimen	Accession no.	References, Remarks, etc.
Presbytis francoisi francoisi (Pousargues 1898)						
[N.VIETNAM]						
Lang Son, Tonkin, [20°50′N 106°45′E], 500ft	23 Jan 1927	M	Ad	Skin, Skull	ZD.1927.12.1.16	Listed Lowe,W.P. (unpubl.)
„ „ „	23 Jan 1927	F	Ad	Skin, Skull	ZD.1927.12.1.17	„ „ „ „
Tonkin	18 Jan 1928	F	Ad	Skin, Skull	ZD.1928.7.1.8	„ „ „ „
Presbytis francoisi delacouri (Osgood 1932)						
N. VIETNAM						
Hoi Xuan, Annam, [20°22′N 105°07′E]	15 Feb 1930	M	Ad	Skin, Skull, Baculum	ZD.1932.4.19.2	Holotype *Pithecus delacouri* Osgood 1932, Figured Osgood, W.H. 1932: 205, Pl.10, Listed Lowe, W.P. (unpubl.)a
Hoi Xuan, Annam, [20°22′N 105°07′E], [c.500ft]	25 Jan 1930	F	[Ad]	Skin	ZD.1933.4.1.13a	Listed Lowe,W.P. (unpubl.)a
Presbytis francoisi hatinhensis Dào Van Tien 1970						
NO SPECIMENS						
—	—	—	—	—	—	
Presbytis francoisi laotum (Thomas 1921)						
LAOS						
Ban Na Sao [=Ban Naxao], French side of R.Mekong, S.W.Laos, c.17°30′N [104°45′E]	24 Feb 1920	M	Ad	Skin, Skull	ZD.1920.12.10.1	Holotype *Pithecus laotum* Thomas 1921, Cited Thomas,O. 1921: 181
Presbytis francoisi leucocephala T'an Pang-Chieh 1957						
NO SPECIMENS						
—	—	—	—	—	—	
Presbytis francoisi poliocephala (Trouessart 1911)						
[N.VIETNAM]						
Cac Ba Island, Baie d'Along, Tonkin, [20°47′N 107°E]	7 Jan 1930	M	Ad	Skin, Skull	ZD.1933.4.1.9	Listed Lowe,W.P. (unpubl.)a
Cac Ba Island, Bay of Along, Tonkin, [20°47′N 107°E]	12 Dec 1930	F	Juv	Skin, Skull	ZD.1933.4.1.10	„ „ „ „
Cac Ba Island, Baie d'Along, Tonkin, [20°47′N 107°E]	12 Jan 1930	F	Ad	Skin, Skull	ZD.1933.4.1.11	„ „ „ „
Tonkin	28 Jan 1927	M	[Ad]	Skin	ZD.1927.12.1.12	Listed Lowe,W.P. (unpubl.) Cited Osgood,W.H. 1932: 204

PRESBYTIS GEEI Gee, 1956

Gee's Langur, Golden Langur

Type locality Forest around Jamduar Forest Rest House, E. bank of R. Sankosh, Goalpara District, Assam, India.

Taxonomic note The authorship of *Presbytis geei* has been in question ever since it was inadvertently named by Gee himself on January 20th, 1956, a few weeks before a formal description by Khajuria appeared in the *Annals and Magazine of Natural History* in February of the same year. However, as the criteria of availability cited in Articles 11-15 of the International Code of Zoological Nomenclature are satisfied in Gee's publication, as well as the provisions of Article 50 regarding authorship, Gee must be cited as the author and his article accepted as the original description. Biswas (1967), in arguing for the citation of the name from Khajuria (1956), appears to have misinterpreted Article 50 of the Code, and his view that there is no indication in Gee's publication that the name *Presbytis geei* on the sketch map refers to the Golden Langur is untenable.

Range India: district of Goalpara north of the Brahmaputra between the Rs. Sankosh and Manas, and adjacent areas of Bhutan. Reports of its occurrence in Meghalaya, south of the R. Brahmaputra were noted by Gee (1961) but have not been confirmed (Mukherjee & Saha, 1974; Mukherjee, 1978).

Dimensions An adult female in the museum collections has a h & b length of 490mm, and a tail length of 713mm.

Locality	Date of Collection	Sex	Age	Nature of Specimen	Accession no.	References, Remarks, etc.
Presbytis geei Gee 1956						
[INDIA]						
Jamduar Forest Rest House, E. bank of R.Sankosh [near Bhutan border], Goalpara Dist., Assam, [c.26°43′N 89°52′E]	7-9 Apr 1955	F	Juv	Skin, Skull	ZD.1956.379	Paratype *Presbytis geei* Gee 1956, Cited Gee,E.P. 1956: 252-254, Cited Gee,E.P. 1961: 3, Cited Khajuria,H. 1956: 86-88

PRESBYTIS OBSCURA (Reid, 1837)

Dusky Leaf monkey
(7 subspecies)

Taxonomic notes In 1928, Pocock united *P. obscura* with *P. cristata* on the grounds that *P. phayrei* probably intergraded with both in the area where the Malay Peninsula joins the landmass of southeast Asia at about 13-15°N. Later in 1935 he separated them since evidence of such intergradation was lacking. He noted that on the contrary *P.o.flavicauda*, the subspecies nearest to *P. phayrei*, was much more distinct from it in pelage characters than was the more distant southern form, *P.o.obscura*. He also noted that this latter form did not intergrade with the sympatric Malayan *P. cristata* in Perak and Selangor.

Range From Tavoy in Burma (*c.* 14°N) and Phet Buri in Thailand (*c.* 13°N) southwards to the tip of the Malay Peninsula in Johore at *c.* 1°30′N. *P. obscura* is not found on Singapore Island (Chasen, 1935). The most northerly specimen comes from the mouth of R. Tavoy, only about 20 miles (32km) from Nwalabo Taung where *P. cristata atrior* is found. Fooden's map (1976, fig.4) shows clearly the intrusion of *P. cristata atrior* between *P. obscura* and *P. phayrei*.

Dimensions Adult specimens in the museum collections have the following field measurements:

h & b	27 ♂♂	534	(420-675)	:	38 ♀♀	519	(425-595)
tail	27 ♂♂	734	(570-790)	:	38 ♀♀	723	(635-813)
weight	12 ♂♂	7319	(6129-9080)	:	22 ♀♀	6602	(4994-8626)

Subspecies Seven are recognised, following Chasen (1940) and Ellerman & Morrison-Scott (1951), two from the peninsula itself and five from adjacent islands. Subspecific identification follows Hill (1960) who drew a dividing line between the northern *P.o.flavicauda* and the southern *P.o.obscura* at about 7°N, between Trang in Thailand (7°38′N) and Perlis in Malaya (6°39′N). The main difference between these two races lies in the strongly contrasting colours of the former (head and body dark brown, almost black, with nuchal patch, legs and tail whitish or silvery), compared with the more uniform colours of the latter (head and body medium brown with fawn nuchal patch and greybrown legs and tail). This cline from south to north of the peninsula, from medium mushroom brown to almost black, was noted by Chasen (1935) and Hill (1960). Two specimens from either side of Latitude 7°N are remarkably similar and show intermediate characters between the two races: BMNH1903. 2.6.7 from Laem Pho (6°55′N 101°19′E) and BMNH1955. 1548 from Chong, Trang, the type locality of *P.o.flavicauda* (*c.* 7°32′N 99°47′E).

Both Chasen's and Pocock's (1935) accounts of the species indicate that there is so much individual variation as to make the provision of a satisfactory key virtually impossible. The following notes are adapted from Chasen (1935):

1. Back medium mushroom ('purplish') brown; legs and tail greybrown, nuchal patch fawn or greyish (little contrast with back) .... **P.o.obscura** (Malay Peninsula, S. of 7°N)
2. Back dark mushroom brown; legs and tail dark greybrown (little contrast with back); nuchal patch pale fawn or greyish (greater contrast with back) .... **P.o.halonifer** (Pinang I., Langkawi I.)
3. Back dark purplish brown to almost black;
   (a) legs and tail grey or whitish, nuchal patch buffy or whitish (maximum contrast with back) .... **P.o.flavicauda** (Malay peninsula, N. of 7°N)
   (b) legs and tail dark greybrown; nuchal patch dirty white:
      (i) frontal band black .... **P.o.carbo** (Ko Tarutao)
      (ii) frontal band browner .... **P.o.seimundi** (Ko Phangan)
   (c) legs dark greybrown;
      (i) tail silvery ashy grey (slightly paler than legs; nuchal patch buffy grey .... **P.o.styx** (E. Perhentian I.)
      (ii) tail pale buffy grey (much paler than legs); nuchal patch creamy white .... **P.o.sanctorum** *(Zadetkyi Kyun)

*Holotype not examined; colours stated to be like *P.o.carbo* with the exception of the paler tail and nuchal patch (Elliot, 1910:352).

Presbytis obscura obscura (Reid, 1837)

Principal synonym *Semnopithecus leucomystax* Müller & Schegel, 1841:59.

Type localities Of *obscura*, unknown, restricted to Melaka (Chasen, 1940); of *leucomystax*, Melaka.

Presbytis obscura carbo Thomas & Wroughton, 1909

Principal synonym *Presbytis corvus* Miller, 1913:27.

Type localities Of *carbo* and *corvus*, Terutao Island (= Ko Tarutao), Peninsular Thailand.

Presbytis obscura flavicauda (Elliot, 1910)

Principal synonyms *Presbytis obscura smithi* Kloss, 1916:5; *Trachypithecus obscurus corax* Pocock, 1935:944.

Type localities Of *flavicauda*, Trong (= Trang), Peninsular Thailand; of *smithi*, Khlong Bang Lai, Pathiu, Peninsular Thailand; of *corax*, Tenasserim Town, Burma.

Presbytis obscura halonifer (Cantor, 1845)

Type locality Prince of Wales Island (= Pinang), Malaya.

Presbytis obscura sanctorum (Elliot, 1910)

Type locality St Matthew's Island (= Zadetkyi Kyun), Mergui Archipelago, Burma.

Presbytis obscura seimundi (Chasen, 1940)

Type locality Koh Pennan Island (= Ko Phangan), NE. Malay Peninsula, Thailand.

Presbytis obscura styx Kloss, 1911

Type locality East Perhentian Island, off Terengganu, east coast of Malay Peninsula, Malaya.

Presbytis obscura obscura (Reid 1837)

MALAYA

Locality	Date of Collection	Sex	Age	Nature of Specimen	Accession no.	References, Remarks, etc.
Ayer Kroh, Melaka, [2°16′N 102°17′E]	1972	F	Ad	Wet, cadaver	CDA	,, ,, ,, ,,
,, ,, ,,	1972	F	Juv	Wet, cadaver	CDA	,, ,, ,, ,,
Bukit Merah, Perak, [5°02′N 100°39′E]	4 Oct 1967	M	Ad	Skull, Skeleton	ZD.1971.728	Listed Medway,Lord (unpubl.)b
,, ,, ,,	4 Oct 1967	M	Ad	Skull, Skeleton	ZD.1971.729	,, ,, ,, ,,
,, ,, ,,	4 Oct 1967	F	Juv	Skull, Skeleton	ZD.1971.730	,, ,, ,, ,,
,, ,, ,,	4 Oct 1967	F	Ad	Skull, Skeleton	ZD.1971.731	,, ,, ,, ,,
,, ,, ,,	4 Oct 1967	M	Ad	Skull, Skeleton	ZD.1971.732	,, ,, ,, ,,
3 miles from Bukit Merah, Perak, [c.5°02′N 100°39′E]	31 Oct 1967	M	Ad	Skull, Skeleton	ZD.1971.733	,, ,, ,, ,,
,, ,, ,,	31 Oct 1967	M	Ad	Skull, Skeleton	ZD.1971.734	,, ,, ,, ,,
,, ,, ,,	31 Oct 1967	M	Ad	Skull, Skeleton	ZD.1971.735	,, ,, ,, ,,
,, ,, ,,	31 Oct 1967	F	Ad	Skull, Skeleton	ZD.1971.736	,, ,, ,, ,,
4.5 miles from Bukit Merah, Perak, [c.5°02′N 100°39′E]	1 Nov 1967	F	Ad	Skull, Skeleton	ZD.1971.737	,, ,, ,, ,,
,, ,, ,,	1 Nov 1967	M	Juv	Skull, Skeleton	ZD.1971.738	,, ,, ,, ,,
Changkat Cheko, Parit, [Perak], [c.4°29′N 100°55′E]	9 Jan 1968	F	Ad	Skull, Skeleton	ZD.1971.719	,, ,, ,, ,,
,, ,, ,,	9 Jan 1968	F	Ad	Skull, Skeleton	ZD.1971.720	,, ,, ,, ,,
,, ,, ,,	9 Jan 1968	F	Ad	Skull, Skeleton	ZD.1971.721	,, ,, ,, ,,
,, ,, ,,	9 Jan 1968	M	Ad	Skull, Skeleton	ZD.1971.722	,, ,, ,, ,,
,, ,, ,,	9 Jan 1968	F	Ad	Skull, Skeleton	ZD.1971.723	,, ,, ,, ,,
Changkat Rambai, Jalan (=road to) Kampong Gajah, Perak, [c.4°10′N 100°58′E]	1 Jun 1968	F	Ad	Skull, Skeleton	ZD.1971.724	,, ,, ,, ,,
,, ,, ,,	1 Jun 1968	M	Juv	Skull, Skeleton	ZD.1971.725	,, ,, ,, ,,
,, ,, ,,	1 Jun 1968	M	Juv	Skull, Skeleton	ZD.1971.726	,, ,, ,, ,,
,, ,, ,,	1 Jun 1968	F	Juv	Skull, Skeleton	ZD.1971.727	,, ,, ,, ,,
Changkat Tamboh, Kampong Gajah Hilir, Perak, [c.4°11′N 100°57′E]	5 Feb 1967	M	Juv	Skull	ZD.1971.680	Correspondence Medway, Lord (unpubl.)a
,, ,, ,,	5 Feb 1967	F	Ad	Skull	ZD.1971.681	,, ,, ,, ,,
,, ,, ,,	5 Feb 1967	F	Ad	Skull	ZD.1971.682	,, ,, ,, ,,
,, ,, ,,	5 Feb 1967	F	Ad	Skull	ZD.1971.683	,, ,, ,, ,,
,, ,, ,,	5 Feb 1967	M	Ad	Skull	ZD.1971.684	,, ,, ,, ,,
,, ,, ,,	5 Feb 1967	M	Juv	Skull	ZD.1971.685	,, ,, ,, ,,
,, ,, ,,	2 Mar 1967	M	Juv	Skull	ZD.1971.686	,, ,, ,, ,,
,, ,, ,,	2 Mar 1967	F	Ad	Skull	ZD.1971.687	,, ,, ,, ,,
,, ,, ,,	2 Mar 1967	F	Ad	Skull	ZD.1971.688	,, ,, ,, ,,
,, ,, ,,	2 Mar 1967	F	Juv	Skull	ZD.1971.689	,, ,, ,, ,,
,, ,, ,,	8 Jul 1967	F	Ad	Skull	ZD.1971.690	,, ,, ,, ,,
,, ,, ,,	8 Jul 1967	F	Ad	Skull	ZD.1971.691	,, ,, ,, ,,
,, ,, ,,	8 Jul 1967	F	Ad	Skull	ZD.1971.692	,, ,, ,, ,,
,, ,, ,,	8 Jul 1967	F	Ad	Skull	ZD.1971.693	,, ,, ,, ,,
,, ,, ,,	8 Jul 1967	F	Juv	Skull	ZD.1971.694	,, ,, ,, ,,
,, ,, ,,	8 Jul 1967	F	Juv	Skull	ZD.1971.695	,, ,, ,, ,,
,, ,, ,,	5 Sep 1967	F	Ad	Skull	ZD.1971.696	,, ,, ,, ,,
,, ,, ,,	5 Sep 1967	F	Ad	Skull	ZD.1971.697	,, ,, ,, ,,
,, ,, ,,	5 Sep 1967	F	Ad	Skull	ZD.1971.698	,, ,, ,, ,,
,, ,, ,,	5 Sep 1967	M	Ad	Skull	ZD.1971.699	,, ,, ,, ,,
,, ,, ,,	5 Sep 1967	M	Ad	Skull	ZD.1971.700	,, ,, ,, ,,
,, ,, ,,	3 Aug 1967	M	Ad	Skull	ZD.1971.701	,, ,, ,, ,,
,, ,, ,,	3 Aug 1967	M	Ad	Skull	ZD.1971.702	,, ,, ,, ,,
,, ,, ,,	2 Jan 1967	F	Ad	Skull, Skeleton	ZD.1971.703	,, ,, ,, ,,
,, ,, ,,	2 Jan 1967	[M]	Ad	Skull, Skeleton	ZD.1971.704	,, ,, ,, ,,
,, ,, ,,	2 Jan 1967	F	Ad	Skull, Skeleton	ZD.1971.705	,, ,, ,, ,,
,, ,, ,,	29 Nov 1967	F	Ad	Skull, Skeleton	ZD.1971.706	,, ,, ,, ,,
,, ,, ,,	29 Nov 1967	F	Ad	Skull, Skeleton	ZD.1971.707	,, ,, ,, ,,
,, ,, ,,	29 Nov 1967	F	Ad	Skull, Skeleton	ZD.1971.708	,, ,, ,, ,,
,, ,, ,,	29 Nov 1967	M	Ad	Skull, Skeleton	ZD.1971.709	,, ,, ,, ,,
,, ,, ,,	29 Nov 1967	F	Ad	Skull, Skeleton	ZD.1971.710	,, ,, ,, ,,
,, ,, ,,	29 Nov 1967	F	Ad	Skull, Skeleton	ZD.1971.711	,, ,, ,, ,,
Gunong Benom, Base Camp, Pahang, [3°51′30″N 102°11′25″E], [097/667], 700ft	3 Mar 1967	F	Ad	Skull, Skeleton	ZD.1971.751	Listed Medway,Lord (unpubl.)
Gunong Benom, [Camp II], Pahang, [c.3°51′30″N 102°11′25″E], [081/667], 1800ft	23 Mar 1967	M	Ad	Skull, Skeleton	ZD.1971.749	,, ,, ,, ,,
,, ,, ,,	23 Mar 1967	M	Juv	Skull, Skeleton	ZD.1971.750	,, ,, ,, ,,
Gunong Benom, [Camp II], Pahang, [c.3°51′30″N 102°11′25″E], [081/667], 2800ft	23 Mar 1967	M	Juv	Skull, Skeleton	ZD.1971.752	,, ,, ,, ,,
Kampong Kedah, Bagan Serai, [Perak], [4°58′N 100°33′E]	3 Oct 1967	M	Ad	Skull, Skeleton	ZD.1971.718	Listed Medway,Lord (unpubl.)b
Palau Tiga, Kampong Gajah, Perak, [4°13′N 100°56′E]	7 Apr 1967	F	Ad	Skull	ZD.1971.713	Correspondence Medway,Lord (unpubl.)a, Listed Medway,Lord (unpubl.)b
,, ,, ,,	7 Apr 1967	F	Ad	Skull	ZD.1971.714	,, ,, ,, ,,
,, ,, ,,	7 Apr 1967	F	Ad	Skull	ZD.1971.715	,, ,, ,, ,,
,, ,, ,,	7 Apr 1967	F	Juv	Skull	ZD.1971.716	,, ,, ,, ,,
[Palau Tiga], [Kampong Gajah], Perak, [4°13′N 100°56′E]	7 Apr 1967	F	Ad	Skull	ZD.1971.743	Listed Medway,Lord (unpubl.)b
Perak	3 Aug 1967	F	Ad	Skull	ZD.1971.712	Correspondence Medway,Lord (unpubl.)a,Listed Medway,Lord (unpubl.)b
,, ,, ,,	4 May 1967	M	Ad	Skull	ZD.1971.717	,, ,, ,, ,,

Locality	Date of Collection	Sex	Age	Nature of Specimen	Accession no.	References, Remarks, etc.
Presbytis obscura obscura *continued*						
,, ,, ,,	3 Aug 1967	F	Ad	Skull	ZD.1971.739	Listed Medway,Lord (unpubl.)b
,, ,, ,,	3 Aug 1967	F	Ad	Skull	ZD.1971.740	,, ,, ,, ,,
,, ,, ,,	3 Aug 1967	F	Ad	Skull	ZD.1971.741	,, ,, ,, ,,
,, ,, ,,	3 Aug 1967	F	Ad	Skull	ZD.1971.742	,, ,, ,, ,,
,, ,, ,,	4 May 1967	F	Ad	Skull	ZD.1971.744	,, ,, ,, ,,
,, ,, ,,	4 May 1967	F	Ad	Skull	ZD.1971.745	,, ,, ,, ,,
,, ,, ,,	4 May 1967	F	Ad	Skull	ZD.1971.746	,, ,, ,, ,,
,, ,, ,,	4 May 1967	M	Ad	Skull	ZD.1971.747	,, ,, ,, ,,
,, ,, ,,	4 May 1967	M	Ad	Skull	ZD.1971.748	,, ,, ,, ,,
—	—	F	Ad	Skull	RCS(OM) A70.444	
[MALAYA]						
Genting Bidai, Selangor, [3°18′N 101°49′E]	14 Sep 1914	M	Ad	Skin, Skull	ZD.1955.1535	Cited Pocock,R.I. 1935: 941-942, Cited Hill,J.E. 1960: 33
[Kampong] Gurun, Kedah, [5°49′N 100°29′E]	14 Dec 1915	F	Ad	Skin, Skull	ZD.1955.1543	Cited Pocock,R.I. 1935: 941, Cited Hill,J.E. 1960: 33
Kuala Teku, Tahan R., Pahang, [c.4°33′N 102°18′E], c.500ft	7 Aug 1905	F	Ad	Skin, Skull	ZD.1906.10.4.4	
,, ,, ,,	7 Aug 1905	M	Inf	Skin, Skull	ZD.1906.10.4.5	
Melaka	—	F	Ad	Skin	ZD.1843.3.10.2	Associated with next specimen (1845.9.20.3)
,, ,, ,,	—	F	Ad	Skull	ZD.1845.9.20.3	Listed Gray,J.E. (unpubl.): 8
,, ,, ,,	—	F	Ad	Skin, Skull	ZD.1879.11.21.596	
Nyalas, Melaka, [2°26′N 102°28′E]	27 Oct 1910	F	Ad	Skin, Skull	ZD.1955.1538	Cited Pocock,R.I. 1935: 941-942, Cited Hill,J.E. 1960: 33
,, ,, ,,	28 Oct 1910	F	Juv	Skin, Skull	ZD.1955.1539	,, ,, ,, ,,
Padang Tuan, Segamat, Johor, [c.2°30′N 102°49′E]	9 Feb 1911	M	Ad	Skin, Skull	ZD.1955.1536	,, ,, ,, ,,
Pahang R., Malay Peninsula	May 1898	[F]	Inf	Skin, Skull	ZD.1899.12.9.76	
Pelarit, N.Malay Peninsula, Perlis, [6°39′N 100°11′E]	1 Nov 1911	M	Ad	Skin, Skull	ZD.1955.1542	Cited Pocock,R.I. 1935: 941-942, Cited Hill,J.E. 1960: 33
Port Dickson, Negeri Sembilan, F.M.S., [2°31′N 101°48′E]	—	F	[Ad]	Skin	ZD.1939.300	
,, ,, ,,	—	[M]	[Ad]	Skin	ZD.1939.301	
,, ,, ,,	—	I	[Juv]	Skin	ZD.1939.302	
[Port Dickson], Negeri Sembilan, F.M.S., [2°31′N 101°48′E]	—	F	Ad	Skull	RCS(OM) A70.443	
,, ,, ,,	—	M	Ad	Skull	RCS(OM) A70.442	
,, ,, ,,	—	I	Juv	Skull	RCS(OM) A70.441	
Semangko Pass, Selangor-Pahang Boundary, [3°40′N 101°45′E], 2500-4500ft	27 Feb 1908	F	Ad	Skull	ZD.1908.7.20.5	(*Presbytis obscurus*), Cited Pocock,R.I. 1935: 941
Semangko Pass, Selangor-Pahang Boundary, [3°40′N 101°45′E], 2700ft	25 Oct 1909	M	Ad	Skin, Skull	ZD.1955.1533	Cited Pocock,R.I. 1935: 941, Cited Hill,J.E. 1960: 33
,, ,, ,,	26 Oct 1909	M	Ad	Skin, Skull	ZD.1955.1534	,, ,, ,, ,,
[Semangko Pass], [Selangor-Pahang Boundary], Selangor, [3°40′N 101°45′E], [2500-4500ft]	20 Jul 1908	M	Juv	Skin, Skull	ZD.1908.7.20.3	(*Presbytis obscurus*), Cited Pocock,R.I. 1935: 940, 942
Si Karang, S.E.Johor, Johor, [c.1°25′N 104°05′E]	2 Aug 1908	M	Juv	Skin, Skull	ZD.1955.1537	Cited Pocock,R.I. 1935: 941, Cited Hill,J.E. 1960: 33
Sungei Kedah, Sea level	14 Apr 1895	F	Ad	Skull	ZD.1896.1.13.3	Cited Flower,S.S. 1900: 318
Tanjong Hantu, Dindings, Perak, [4°19′N 100°34′E]	23 Jul 1918	M	Ad	Skin, Skull	ZD.1955.1540	Cited Pocock,R.I. 1935: 941, Cited Hill,J.E. 1960: 33
,, ,, ,,	21 Jul 1918	F	Ad	Skin, Skull	ZD.1955.1541	,, ,, ,, ,,
Tebrau, S.Johor, Johor, [1°32′N 103°35′E]	—	F	Ad	Skull	ZD.1955.1544	Cited Hill,J.E. 1960: 33
[THAILAND]						
[Ban Sai Kau], [Nawngchik], Malay Peninsula, [c.6°38′N 101°08′E]	18 Sep 1901	M	Inf	Skin, Skull	ZD.1903.2.6.8	Cited Bonhote,J.L. 1903: 6
C.Pattani [=Laem Pho], Malay Peninsula, [c.6°55′N 101°19′E]	10 Jun 1901	M	Ad	Skin, Skull	ZD.1903.2.6.5	Cited Bonhote,J.L. 1903: 6, Cited Pocock,R.I. 1935: 941-942
,, ,, ,,	1 Sep 1901	F	Ad	Skin, Skull	ZD.1903.2.6.7	Cited Bonhote,J.L. 1903: 6, Cited Pocock,R.I. 1935: 941
Pattani, Malay Peninsula, [6°50′N 101°20′E]	10 Jun 1901	F	Ad	Skull	ZD.1903.2.6.9	Cited Bonhote,J.L. 1903: 6
Tenelong, [Jalor], Malay Peninsula, [c.6°30′N 101°10′E]	11 Aug 1901	M	[Ad]	Skin	ZD.1903.2.6.6	,, ,, ,, ,,
NO DATA						
—	—	F	Ad	Skull	RCS(OM) A70.44	
Presbytis obscura obscura? (Reid 1837)						
[MALAYA]						
Melaka	—	[F]	[Inf]	Skin, Skull in skin	ZD.1879.11.21.598	
Presbytis obscura carbo Thomas and Wroughton 1909						
[THAILAND]						
Pulau Tarutao I., Straits of Malacca, [6°35′N 99°40′E]	26 Feb 1909	M	Ad	Skin, Skull	ZD.1909.11.1.4	Holotype *Presbytis obscura carbo* Thomas and Wroughton 1909, Cited Thomas,O. and Wroughton,R.C 1909a: 534, Cited Pocock,R.I. 1935: 947-948

Locality	Date of Collection	Sex	Age	Nature of Specimen	Accession no.	References, Remarks, etc.
Presbytis obscura carbo *continued*						
" " "	26 Feb 1909	F	Ad	Skin, Skull	ZD.1909.11.1.5	Paratype *Presbytis obscura carbo* Thomas and Wroughton 1909, Cited Pocock,R.I. 1935: 947-948
Pulau Tarutao I., Straits of Melaka, off W. coast Peninsular Siam, [6°35′N 99°40′E]	26 Feb 1909	M	Ad	Skin, Skull	ZD.1955.1555	Cited Pocock,R.I. 1935: 947-948, Cited Hill,J.E. 1960: 33
Presbytis obscura flavicauda (Elliot 1910)						
BURMA						
Mouth of Tavoy R., Tavoy Dist., [c.13°45′N 98°12′E]	21 Apr 1936	F	Ad	Skin, Skull	ZD.1936.9.10.2	Figured Pocock,R.I. 1939: 139-141, Fig.42
[BURMA]						
Bankachon, Nr. Victoria Point, S.Tenasserim, [10°09′N 98°36′E]	12 Jan 1914	M	Ad	Skin, Skull	ZD.1914.12.8.23	
Bankachon, S.Tenasserim, [10°09′N 98°36′E]	28 Dec 1913	M	Ad	Skin, Skull	ZD.1914.12.8.22	
" " "	10 Dec 1913	F	Ad	Skin, Skull	ZD.1914.12.8.24	
Bankachon, Tenasserim, [10°09′N 98°36′E]	25 May 1877	F	Ad	Skin, Skull	ZD.1885.8.1.13	(*Semnopithecus obscurus*), Cited Pocock,R.I. 1928: 673
" " "	9 Dec 1913	F	Ad	Skin, Skull	ZD.1937.9.10.1	
" " "	—	F	Ad	Skull	ZD.1980.160	
Bankachon, Victoria Point, S.Tenasserim, [10°09′N 98°36′E]	15 Dec 1913	F	Ad	Skin, Skull	ZD.1914.12.8.25	
Bankachon, V.P., S.Tenasserim, [10°09′N 98°36′E]	9 Dec 1913	F	Inf	Skin, Skull	ZD.1914.12.8.26	
" " "	—	M	[Ad]	Skin	ZD.1951.611	(*Trachypithecus obscurus*)
Banlaw, Great Tenasserim R., [Tenasserim], [12°07′N 99°03′E]	22 Apr 1914	M	Ad	Skin, Skull	ZD.1914.12.8.27	Paratype *Trachypithecus obscurus corax* Pocock 1935, Cited Pocock,R.I. 1928: 673, Cited Pocock,R.I. 1935: 945
King I. [=Kadan Kyun], Mergui Archipelago, [12°30′N 98°22′E], Sea level	29 Sep 1921	M	Ad	Skin, Skull	ZD.1937.9.10.2	Paratype *Trachypithecus obscurus corax* Pocock 1935, (*Trachypithecus obscurus flavicauda (corax)*), Cited Pocock,R.I 1928: 673, Cited Pocock,R.I. 1935: 945, Cited Lindsay,H.M. 1926b: 43
King I. [=Kadan Kyun], Mergui [Archipelago], [12°30′N 98°22′E]	—	M	Ad	Skull	ZD.1980.161	(*P.obscura corax*)
Malcolm I. [=Ale-man Kyun], [Mergui Archipelago], [11°17′N 98°33′E]	14 Feb 1922	F	Inf	Skull	ZD.1933.7.29.12	Cited Lindsay,H.M. 1926b: 43
Tenasserim Town, [Tenasserim], [12°05′N 99°E]	18 Mar 1914	M	Ad	Skin, Skull	ZD.1914.12.8.28	Holotype *Trachypithecus obscurus corax* Pocock 1935, Cited Pocock,R.I. 1928: 673-674, Cited Pocock,R.I. 1935: 944
[THAILAND]						
[Ban] Khlong Bang Lai, Pathiu, Peninsular Siam, [Chumphon], [10°45′N 99°10′E]	21 Jan 1916	M	Ad	Skin, Skull	ZD.1920.7.3.3	Holotype *Presbytis obscura smithi* Kloss 1916, Cited Kloss,C.Boden 1916: 5, Cited Pocock,R.I. 1935: 943-946
Ban Kok Klap, Bandon, N.E.Malay Peninsula, [8°53′N 99°17′E], 50ft	2 Jul 1913	M	Juv	Skin, Skull	ZD.1955.1946	Cited Robinson,H.C. and Kloss,C.Boden 1915: 113, Cited Pocock,R.I. 1935: 948, Cited Chasen,F.N. 1940: 79, Cited Hill,J.E. 1960: 33
Chong, Trang, Peninsular Siam, [c.7°32′N 99°47′E]	11 Dec 1909	F	Ad	Skin, Skull	ZD.1955.1548	Cited Hill,J.E. 1960: 33, Cited Pocock,R.I. 1935: 941-942
Khao Wang Hip, Peninsular Siam, [8°12′N 99°43′E]	16 Sep 1915	M	Ad	Skin, Skull	ZD.1920.7.3.1	Cited Kloss,C.Boden 1916: 4, Cited Pocock,R.I. 1935: 943, 945
" " "	16 Sep 1915	F	Ad	Skin, Skull	ZD.1920.7.3.2	" " " "
Nong Kok, Krabi, Siam, [8°06′N 98°52′E]	5 Jan 1918	M	Ad	Skin, Skull	ZD.1955.1547	Cited Pocock,R.I. 1935: 943-946, Cited Hill,J.E. 1960: 33
Siam, 12°50′N, 800m	27 Jan 1912	M	[Ad]	Skin	ZD.1913.11.2.2	Cited Gairdner,K.G. (unpubl.)a, Cited Pocock,R.I. 1935: 945
Tasan [=Ban Tha San], Peninsular Siam, Chumphon, [10°29′N 98°55′E], 220ft	15 Mar 1919	M	Ad	Skin, Skull	ZD.1955.1545	Cited Pocock,R.I. 1935: 943, 945, Cited Hill,J.E. 1960: 33
Presbytis obscura flavicauda? (Elliot 1910)						
[BURMA]						
[Bankachon], Tenasserim, [10°09′N 98°36′E]	—	F	Ad	Skull	ZD.1980.162	(*Trachypithecus obscurus*)
Tenasserim	—	M	Ad	Skull	ZD.1856.5.6.13	(*Presbytis obscura flavicauda* or *corax*), Listed Gray,J.E. (unpubl.): 8
" " "	—	M	Ad	Skull	ZD.1856.5.6.14	(*P.obscura*), Listed Gray,J.E. (unpubl.): 8
Presbytis obscura halonifer (Cantor 1845)						
[MALAYA]						
Langkawi I., Straits of Melaka, [6°22′N 99°48′E]	15 Feb 1909	M	Ad	Skin, Skull	ZD.1909.11.1.6	(*Presbytis obscura carbo*), Cited Pocock,R.I. 1935: 946, 948
Pantai Kerchut, Pinang I., [5°27′N 100°11′E]	23 Mar 1911	M	Ad	Skin, Skull	ZD.1955.1549	Cited Pocock,R.I. 1935: 946, Cited Hill,J.E. 1960: 33

Locality	Date of Collection	Sex	Age	Nature of Specimen	Accession no.	References, Remarks, etc.
Presbytis obscura halonifer *continued*						
Pinang I., [5°24′N 100°14′E]	—	F	Ad	Skull	ZD.1879.11.21.217	(*Presbytis obscurus*), Listed Gray,J.E. (unpubl.): 8, Cited Pocock,R.I. 1935: 948
Pinang [I.], [5°24′N 100°14′E]	—	M	Ad	Skin, Skull	ZD.1879.11.21.298	Cited Pocock,R.I. 1935: 946, 948
Pulau Langkawi I., Straits of Melaka, [6°22′N 99°48′E]	8 Feb 1909	F	Ad	Skin, Skull	ZD.1955.1553	Cited Pocock,R.I. 1935: 946, 948, Cited Hill,J.E. 1960: 33
,, ,, ,,	8 Feb 1909	F	Ad	Skin, Skull	ZD.1955.1554	,, ,, ,, ,,
Telok Bahang, Pinang I., [5°28′N 100°13′E]	28 Mar 1911	M	Ad	Skin, Skull	ZD.1955.1550	,, ,, ,, ,,
,, ,, ,,	5 Apr 1911	M	Ad	Skin, Skull	ZD.1955.1551	,, ,, ,, ,,
,, ,, ,,	9 Mar 1911	F	Ad	Skin, Skull	ZD.1955.1552	,, ,, ,, ,,
Presbytis obscura sanctorum (Elliot 1910)						
NO SPECIMENS						
—	—	—	—	—	—	
Presbytis obscura seimundi (Chasen 1940)						
[THAILAND]						
Koh Pennan [=Ko Phangan], N.E.Malay Peninsula, [9°45′N 100°E]	25 May 1913	M	Ad	Skin, Skull	ZD.1947.1500	Holotype *Pithecus obscurus seimundi* Chasen 1940, Cited Chasen,F.N. 1940: 80, Cited Robinson,H.C. and Kloss,C.Boden 1914: 131
S.W.Koh Pennan [=Ko Phangan], N.E.Malay Peninsula, [9°45′N 100°E]	27 May 1913	M	Ad	Skin, Skull	ZD.1955.1556	Paratype *Pithecus obscurus seimundi* Chasen 1940, Cited Robinson,H.C. and Kloss,C.Boden 1914: 131, Cited Hill,J.E. 1960: 33
,, ,, ,,	27 May 1913	M	Ad	Skin, Skull	ZD.1955.1557	,, ,, ,, ,,
,, ,, ,,	26 May 1913	M	Juv	Skin, Skull	ZD.1955.1558	,, ,, ,, ,,
,, ,, ,,	24 May 1913	F	Ad	Skin, Skull	ZD.1955.1559	,, ,, ,, ,,
Presbytis obscura styx Kloss 1911						
[MALAYA]						
E.Perhentian I., [c.5°55′N 102°45′E]	12 Sep 1910	M	Ad	Skin, Skull	ZD.1949.424	Holotype *Presbytis obscura styx* Kloss 1911, Cited Kloss,C.Boden 1911: 116, Cited Pocock,R.I. 1935: 947-948
,, ,, ,,	9 Sep 1910	M	Ad	Skin, Skull	ZD.1955.1560	Paratype *Presbytis obscura styx* Kloss 1911, Cited Pocock,R.I. 1935: 947-948, Cited Hill,J.E. 1960: 33
Presbytis obscura (Reid 1837)						
MALAYA						
Singapore[?]	—	M	Inf	Skin, Skull	ZD.1879.11.21.613	
—	Died 7 Feb 1923	F	[Ad]	Skin	ZD.1939.596	
[THAILAND]						
Biserat, [R.Pattani], Jalor, Malay Peninsula, [6°32′N 101°14′E]	[26 May 1899]	I	Inf	Skull	CMZ E7220D	(*Semnopithecus* sp.), Cited Bonhote,J.L. 1901: 872
Koh Nam Kam I., Thale Sap, [Songkhla], Malay Peninsula, [7°20′N 100°25′E]	12 Apr 1899	F	Juv	Skull	CMZ E7231A	(*Semnopithecus obscurus*), Listed Bonhote,J.L. 1901: 872
Telok Poh, Pulau Panjang [=Ko Yao Yai], [8°N 98°35′E]	22 Jan 1918	M	Ad	Skin, Skull	ZD.1955.1561	(*Presbytis obscura* subsp.), Cited Pocock,R.I. 1935: 947-948, Cited Hill,J.E. 1960: 34
NO DATA						
—	—	F	Juv	Skin, Skull	ZD.78c	(*Presbytis obscura obscura*), Listed Anon. (unpubl.)b: 78
—	—	[M]	Inf	Skin, Skull	ZD.[78d]	,, ,, ,, ,,
—	—	M	Inf	Skin, Skull	ZD.1852.8.27.3	
—	—	I	Inf	Skin	RSM 1904/5/3	
Presbytis obscura? (Reid 1837)						
[MALAYA]						
Melaka	—	[M]	[Inf]	Skin, Skull in skin	ZD.1879.11.21.299	
[MALAYA/THAILAND BORDER]						
Malay Peninsula	[1899]	F	Ad	Skull	CMZ E7220E	(*Semnopithecus* sp.)(*Trachypithecus [obscurus]*), Background Bonhote,J.L. 1901: 869-883
,, ,, ,,	[1899]	I	Juv	Skull, Skeleton	CMZ E7220C	,, ,, ,, ,,
NO DATA						
—	—	M	Inf	Skin, Skull	ZD.1852.8.27.4	(*Semnopithecus maurus*)
—	—	F	Ad	Skull	ZD.1844.11.8.1	(*Trachypithecus* sp.)

PRESBYTIS PHAYREI Blyth, 1847

Phayre's Leaf Monkey
(3 subspecies)

Taxonomic notes All species of the *P. cristata* group are considered by Fooden (1976:109, fig. 4) to be allopatric. He notes however that both *P. phayrei crepuscula* and *P. cristata koratensis* (= *P.c.germaini*) have been collected at Lat Bua Khao, South-Central Thailand, but possibly from opposite banks of R. Mae Nam Mun. *Semnopithecus holotephreus* Anderson, 1879, probably belongs to one of the darker subspecies of *P. phayrei*, and is tentatively allocated to the synonymy of *P.p.phayrei*.

Presbytis ruhei Knottnerus-Meyer, 1933, doubtfully included here by Ellerman & Morrison-Scott, 1951, cannot be identified with any confidence from the description alone. The white lips and (? white) 'mask' suggest *P. phayrei* or *P. obscura*; the ash-grey coloration and black tail resemble *P. cristata*; and in its whitish-grey thighs which are paler than the legs, it bears some resemblance to the *P. melalophos* group. The type locality is given as Singora (= Songkhla), SE. Peninsular Thailand.

For a discussion of the affinites of *Presbytis barbei*, see *P. cristata* taxonomic notes, p. 53.

Range Burma, SW. China, E. India and Bangladesh, NW. Laos, N. Vietnam and NE. Thailand, (see Fooden, 1976, fig. 4).

Dimensions Adult specimens in the museum collections have the following field measurements:

h & b	19 ♂♂	542	(420-600)	: 8 ♀♀	516	(442-570)
tail	19 ♂♂	770	(745-858)	: 8 ♀♀	764	(719-795)
weight	8 ♂♂	7931	(5675-9080)	: 5 ♀♀	6946	(6356-7491)

Subspecies The three races are greybrown in colour, darker in the northwestern (Burmese) forms and paler in the southeastern (Indo-Chinese) forms. There is often a paler sheen on the nape and upper back which sometimes extends to shoulders. The greybrown of the arm grades imperceptibly into the brownish black hands, whereas that of the leg contrasts sharply with the brownish black feet. The presence of a tuft or coconut-like crest is individually variable (Pocock, 1939), as is the extent of the white ring around the eyes (Fooden, 1976).

The presence of a white pubic patch and white stripe down the inside of the upper thigh is also variable. In BM(NH) specimens of *P.p.phayrei*, they are present in all females (n = 11) and in 10 out of 13 males. Of the other 3 males, two have only the white pubic patch, and one lacks any trace of white in the pubic region. In *P.p.shanica*, both characters are present in all females (n = 4) and absent in 4 out of 5 males. *P.p.crepuscula* shows complete sexual dichromatism in both characters which are present in all females (n = 8) and absent in all males (n = 12), as confirmed by Fooden (1976), resembling the syntypes of *P. barbei* in this respect (see p. 53).

The following key is adapted from Pocock (1939):

1.	A whorl or parting in the hairs just behind the brow	**shanica**
1a.	No whorl or parting in the hairs just behind the brow	2
2.	Colour of head and back darker on average (dark greybrown), contrasting more strongly with whitish or buff ventrum; white pubic patch generally present in both males and females	**phayrei**
2a.	Colour of head and back lighter on average (silvery grey-brown), scarcely contrasting with pale grey ventrum; white pubic patch present in females only	**crepuscula**

Presbytis phayrei phayrei Blyth, 1847

Principal synonyms ?*Semnopithecus holotephreus* Anderson, 1879:27; *Presbytis melamera* Elliot, 1909:267.

Type localities Of *phayrei*, Arakan, Burma; of *holotephreus*, unknown; of *melamera*, Cadu Ciaung (= Kadu Hka), near Bhamo, North Burma.

Presbytis phayrei crepuscula Elliot, 1909

Principal synonyms *Presbytis crepuscula wroughtoni* Elliot, 1909:272; *Presbytis argenteus* Kloss, 1919:338.

Type localities Of *crepuscula*, Mt Mooleyit (= Mulayit Taung), Tenasserim, Burma; of *wroughtoni*, Pachebon (= Phetchabun), Thailand; of *argenteus*, Lat Bua Khao, W. of Khorat (= Nakhon Ratchasima), Thailand.

Presbytis phayrei shanica (Wroughton, 1917)

Type locality Se'en, Hsipaw State, North Shan States, Burma.

Locality	Date of Collection	Sex	Age	Nature of Specimen	Accession no.	References, Remarks, etc.
Presbytis phayrei phayrei Blyth 1847						
[BANGLADESH]						
Chittagong Hill Tracts	23 Feb 1901	F	Ad	Skull	ZD.1926.10.8.2	(*Pithecus phayrei*)
BURMA						
Bhamo Division, Upper Burma, 500ft	12 Jan 1936	M	Ad	Skin, Skull	ZD.1936.12.26.1	Cited Pocock,R.I. 1939: 132
Cadu Ciaung, Near Bhamo, N.Burma, [24°20′N 97°15′E]	19 Apr 1886	F	Juv	Skin, Skull	ZD.1888.12.1.64	Holotype *Presbytis melamera* Elliot 1909, Cited Elliot,D.G. 1909: 267, Cited Pocock,R.I. 1928: 668-9, Cited Pocock,R.I. 1939: 131
Kadat Reserve, 60 miles N. of Pegu, 1000ft	15 Mar 1916	M	[Juv]	Skin	ZD.1937.9.10.16	Cited Wroughton,R.C. and Davison,W.M. 1918: 475
" " "	15 Mar 1916	F	[Ad]	Skin	ZD.1937.9.10.17	" " " "
Kin, Lower R.Chindwin, W. bank, [22°46′N 94°42′E]	7 Jun 1914	M	Ad	Skin, Skull	ZD.1937.9.10.12	Cited Wroughton,R.C. 1916: 297
" " "	7 Jun 1914	M	Ad	Skin, Skull	ZD.1915.5.5.9	" " " "
" " "	7 Jun 1914	M	[Ad]	Skin	ZD.1979.2346	" " " "
Letsegan, 20 miles W. of Kindat, [23°37′N 94°15′E], 3000ft	10 Feb 1915	F	Ad	Skin, Skull	ZD.1937.9.10.13	Cited Wroughton,R.C. 1916d: 763
[Letsegan], [20 miles W. of Kindat], Chin Hills, [23°37′N 94°15′E], [3000ft]	[10 Feb 1915]	F	Ad	Skull	ZD.1979.2347	Cited Wroughton,R.C. 1916d: 763, Cited Pocock,R.I. 1939: 137
Mt.Popa, Upper Burma, [Myingyan Dist.], [20°55′N 95°15′E], 4961ft	11 Sep 1913	M	Ad	Skin, Skull	ZD.1914.7.19.3	Cited Wroughton,R.C. 1915: 464, Cited Pocock,R.I. 1928: 670, Cited Pocock,R.I. 1939: 133
" " "	27 Sep 1913	M	Ad	Skin, Skull	ZD.1914.7.19.4	" " " "
" " "	3 Sep 1913	F	[Ad]	Skin	ZD.1914.7.19.5	" " " "
" " "	3 Sep 1913	F	Juv	Skin, Skull	ZD.1937.9.10.11	" " " "
Near Mt.Arakan Yoma, Upper Burma, Magwe Dist.	1914	[M]	[Juv]	Skin	ZD.1937.9.10.14	Cited Wroughton,R.C. 1915a: 720
30 miles S.E. of Prome, 300ft	24 Nov 1916	[M]	[Juv]	Skin	ZD.1979.2362	Cited Wroughton,R.C. 1921a: 553, Cited Pocock,R.I. 1928: 670, Cited Pocock,R.I. 1939: 133
" " "	24 Nov 1916	[F]	[Ad]	Skin	ZD.1979.2363	" " " "

Locality	Date of Collection	Sex	Age	Nature of Specimen	Accession no.	References, Remarks, etc.
Presbytis phayrei phayrei *continued*						
Ramree I., Arakan, [19°06′N 93°48′E]	—	[F]	[Inf]	Skin, Skull in skin	ZD.1879.11.21.599	Cited Pocock,R.I. 1928: 670, Cited Pocock,R.I. 1939: 133
S.Zamayi Reserve, 60 miles N. of Pegu, [18°21′N 95°59′E], 500ft	10 Mar 1916	M	Ad	Skin, Skull	ZD.1917.4.24.1	Cited Wroughton,R.C. and Davison,W.M. 1918: 475, Cited Pocock,R.I. 1928: 670, Cited Pocock,R.I. 1939: 133, 137
30 miles N.W. Toungoo, 500ft	22 Nov 1928	F	[Ad]	Skin	ZD.1937.9.10.9	Cited Pocock,R.I. 1939: 133
30 miles N.W. of Toungoo, 800ft	7 Jan 1928	M	Ad	Skin, Skull	ZD.1937.9.10.4	Cited Fry,T.B. 1929: 640, Cited Pocock,R.I. 1939: 133
" " "	8 Jan 1928	M	Ad	Skin, Skull	ZD.1937.9.10.5	" " " "
" " "	8 Jan 1928	M	Ad	Skin, Skull	ZD.1937.9.10.6	" " " "
" " "	9 Jan 1928	M	Ad	Skin, Skull	ZD.1937.9.10.7	" " " "
30 miles N.W. of Toungoo (W), 600ft	18 Jan 1928	F	Juv	Skin, Skull	ZD.1937.9.10.8	" " " "
30 miles N.W. of Toungoo, Lower Burma, 100ft	27 Apr 1927	F	Ad	Skin, Skull	ZD.1927.11.18.2	" " " "
35 miles N.W. of Toungoo, 500ft	12 Jan 1928	F	[Juv]	Skin	ZD.1937.9.10.10	" " " "
Yin Village, Lower R.Chindwin, [22°47′N 94°42′E]	Oct 1913	I	[Ad]	Skin	ZD.1937.9.10.15	Cited Wroughton,R.C. 1915: 464, Cited Pocock,R.I. 1928: 670
Presbytis phayrei crepuscula Elliot 1909						
BURMA						
Mulayit Taung, [Dawna Range], Tenasserim, [16°11′N 98°32′E], 5000ft	30 Jan 1877	M	Ad	Skin, Skull	ZD.1885.8.1.10	Paratype *Presbytis crepuscula* Elliot 1909, (*Semnopithecus obscurus*), Cited Elliot,D.G. 1909: 272, Cited Pocock,R.I. 1928: 672, Cited Pocock,R.I. 1935: 950, 952, Cited Pocock,R.I. 1939: 134-135, 137
" " "	30 Jan 1877	M	Ad	Skin, Skull	ZD.1885.8.1.11	Holotype *Presbytis crepuscula* Elliot 1909, Cited Elliot,D.G. 1909: 271, Cited Pocock,R.I. 1928: 672, Cited Pocock,R.I. 1935: 950, 952, Cited Pocock,R.I. 1939: 135, 137
[BURMA]						
Lampha, Tenasserim, [16°18′N 98°19′E], 1000ft	30 Dec 1923	F	Ad	Skin, Skull	ZD.1924.9.2.14	Cited Pocock,R.I. 1928: 672, Cited Pocock,R.I. 1935: 950-951, Cited Pocock,R.I. 1939: 135
LAOS						
Xieng Khouang, [c.19°20′N 103°22′E], [c.4000ft]	8 Jan 1926	F	Ad	Skin, Skull	ZD.1926.10.4.6	(*Pithecus phayrei*), Cited Thomas,O. 1927: 43, Cited Pocock,R.I. 1928: 672, Cited Pocock,R.I. 1935: 950-952, Listed Lowe,W.P. (unpubl.)
—	—	F	Ad	Skull, Skeleton	ZD.1882.12.15.2	(*Semnopithecus obscurus*)
[N.VIETNAM]						
Hoi Xuan, Annam, [20°22′N 105°07′E]	19 Jan 1930	M	Ad	Skin, Skull	ZD.1933.4.1.7	(*Pithecus pyrrhus argenteus*), Listed Lowe,W.P. (unpubl.)
" " "	14 Feb 1930	M	Ad	Skin, Skull	ZD.1933.4.1.8	" " " "
Phu Qui, Annam, [19°19′N 105°25′E], 100ft	29 Feb 1928	M	Ad	Skin, Skull	ZD.1928.7.1.6	(*Pithecus obscurus crepusculus*), Cited Thomas,O. 1929: 832, Cited Pocock,R.I. 1935: 950-952, Listed Lowe,W.P. (unpubl.)
[THAILAND]						
Headquarters Camp, [Sai Yoke area, S.W.Siam], [c.14°25′N 98°51′E]	—	F	Ad	Skin, Skull	ZD.1915.12.1.4	(*Trachypithecus phayrei atrior?*), Correspondence Gairdner,K.G. (unpubl.)
Khao Kamphaeng, Seeswad [=Si Sawat], Siam, [14°37′N 99°18′E], 980-1000m	18 May 1915	M	Ad	Skin, Skull	ZD.1915.12.1.5	Correspondence Gairdner,K.G. (unpubl.), Cited Gairdner,K.G. 1915a: 252, Cited Pocock,R.I. 1928: 672, Cited Pocock,R.I. 1935: 951-952
[Khao Kamphaeng], Seeswad [=Si Sawat], Siam, [c.14°37′N 99°18′E], 940m	19 May 1915	F	Ad	Skin, Skull	ZD.1915.12.1.6	Correspondence Gairdner,K.G. (unpubl.), Cited Gairdner,K.G. 1915a: 252, Cited Pocock,R.I. 1928: 672, Cited Pocock,R.I. 1935: 950, Cited Pocock,R.I. 1939: 137
Lat Bua Kao, W. of Korat, S.W.Siam, [14°52′N 101°36′E]	13 Sep 1916	F	Ad	Skin, Skull	ZD.1917.8.4.1	(Topotype of *argenteus* Kloss), Background Kloss,C.Boden 1919: 338, Cited Pocock,R.I. 1935: 950-952, Cited Pocock,R.I. 1939: 134
[Muang] Nan, Siam, [18°47′N 100°50′E]	1929	[F]	[Ad]	Skin	ZD.1934.11.14.1	
[Phetchabun], [16°25′N 101°08′E]	—	M	Ad	Skin, Skull	ZD.1861.10.8.1	Holotype *Presbytis crepuscula wroughtoni* Elliot 1909, (*Semnopithecus obscurus*), Cited Elliot,D.G. 1909: 272, Cited Pocock,R.I. 1928: 672, Cited Pocock,R.I. 1935: 950, Cited Pocock,R.I. 1939: 134
" " "	—	I	[Inf]	Skin, Skull in skin	ZD.1861.10.8.2	
[R.] Khlong Khlung, S.W. of Kempenpet [=Kamphaeng Phet], Siam, [c.16°05′N 99°15′E], c.600ft	27 Feb 1924	F	Ad	Skin, Skull	ZD.1924.9.2.15	(*Pithecus melamerus*), Cited Pocock,R.I. 1928: 672, Cited Pocock,R.I. 1935: 950, 952, Cited Pocock,R.I. 1939: 135

Locality	Date of Collection	Sex	Age	Nature of Specimen	Accession no.	References, Remarks, etc.
Presbytis phayrei crepuscula *continued*						
R.Mewong [Nam Mae Wong], 40 miles E. of Um Pang [=Umphang], Siam, 1000ft	9 Feb 1924	M	Ad	Skin, Skull	ZD.1924.9.2.9	(*Pithecus obscurus crepusculus*), Cited Pocock,R.I. 1928: 672, Cited Pocock,R.I. 1935: 950, Cited Pocock,R.I. 1939: 135
,, ,, ,,	10 Feb 1924	M	Ad	Skin, Skull	ZD.1924.9.2.10	,, ,, ,, ,,
R.Mewong [Nam Mae Wong], 53 miles E. of Umphang, Siam, 800ft	18 Feb 1924	M	Ad	Skin, Skull	ZD.1924.9.2.11	(*Pithecus crepusculus wroughtoni*), Cited Pocock,R.I. 1928: 672, Cited Pocock,R.I. 1935: 950, Cited Pocock,R.I. 1939: 135
R.Mewong [Nam Mae Wong], 40 miles E. of Umphang, Siam, 1000ft	9 Feb 1924	[M]	[Juv]	Skin, Skull	ZD.1924.9.2.12	(*Pithecus crepusculus wroughtoni*), Cited Pocock,R.I. 1935: 950, Cited Pocock,R.I. 1939: 135
R.Mewong [Nam Mae Wong], 53 miles E. of Umphang, Siam, 800ft	18 Feb 1924	M	Inf	Skin, Skull	ZD.1924.9.2.16	(*Pithecus melamerus*), Cited Pocock,R.I. 1935: 950, Cited Pocock,R.I. 1939: 135
28 miles S.E. of Umphang, Siam, 1750ft	3 Feb 1924	M	Juv	Skin, Skull	ZD.1924.9.2.13	(*Pithecus crepuscula wroughtoni*)(*Pithecus melamerus*), Cited Pocock,R.I. 1928: 672, Cited Pocock,R.I. 1935: 950
Presbytis phayrei shanica (Wroughton 1917)						
BURMA						
Byibon, Mandalay Dist., [22°20′N 96°15′E], 800ft	13 Feb 1937	M	Juv	Skin, Skull	ZD.1937.12.3.74	
Gokteik, [Hsipaw State], N.Shan States, [22°20′N 96°52′E], 2133ft	—	M	Ad	Skull	ZD.1914.7.8.1	Cited Ryley,K.V. 1914b: 715, Cited Pocock,R.I. 1928: 670,Figured Colyer,J.F. 1936: 220, Fig.263, Cited Pocock,R.I. 1939: 136
,, ,, ,,	20 Apr 1913	M	Ad	Skin, Skull	ZD.1914.7.8.2	Cited Ryley,K.V. 1914b: 715, Cited Pocock,R.I. 1928: 670, Cited Pocock,R.I. 1939: 136
,, ,, ,,	21 Apr 1913	F	Ad	Skin, Skull	ZD.1914.7.8.3	,, ,, ,, ,,
,, ,, ,,	24 Apr 1913	F	Ad	Skin, Skull	ZD.1914.7.8.4	,, ,, ,, ,,
Nattaung, Mandalay Dist., [22°20′N 96°32′E], 700ft	15 Mar 1937	F	Ad	Skin, Skull	ZD.1937.12.3.73	Cited Pocock,R.I. 1939: 138
Nattaung, Maymyo F.D., [22°20′N 96°32′E], 700ft	18 Dec 1935	F	Juv	Skin, Skull	ZD.1936.12.26.2	,, ,, ,, ,,
Ngapyinin, R.Irrawady, E. bank opposite Kyauk Myaung, W.Burma, [22°36′N 95°57′E], 250ft	31 Jul 1913	M	[Ad]	Skin	ZD.1914.7.19.6	Background Wroughton,R.C. 1915: 465, Cited Pocock,R.I. 1928: 670, Cited Pocock,R.I. 1939: 136, 138
Pyaunggaung, N.Shan States, [22°35′N 97°05′E], 2794ft	8 May 1913	F	Inf	Skin, Skull	ZD.1937.9.10.3	Cited Ryley,K.V. 1914b: 715, Cited Pocock,R.I. 1928: 670, Cited Pocock,R.I. 1939: 136
Sedaw, Mandalay Canal, Mandalay Dist., 500ft	14 Feb 1937	M	Ad	Skin, Skull	ZD.1937.12.3.72	
Se′en, Hsipaw State, N.Shan States, [c.22°43′N 97°31′E], 1411ft	26 May 1913	M	Ad	Skin, Skull	ZD.1914.7.8.5	Holotype *Pithecus shanicus* Wroughton 1917, Cited Ryley,K.V. 1914b: 715, Cited Wroughton,R.C. 1917: 47, Cited Pocock,R.I. 1928: 671, Figured Pocock,R.I. 1939: 136-137, Fig 41
,, ,, ,,	26 May 1913	F	[Ad]	Skin	ZD.1980.16	(*Pithecus melamerus*), Cited Ryley,K.V. 1914b: 715, Background Wroughton,R.C. 1917: 47, Cited Pocock,R.I. 1928: 670, Cited Pocock,R.I. 1939: 136
—	—	[F]	[Inf]	Skin	ZD.1980.17	(*Trachypithecus shanicus*)

PRESBYTIS PILEATA (Blyth, 1843)

Capped Langur
(5 subspecies)

Range Eastern India, both north and south of the R. Brahmaputra; eastern Bangladesh including the Chittagong Hills and Sylhet (Gittins & Akonda, 1982); and northern and western Burma.

Dimensions Adult specimens in the museum collections have the following field measurements:

h & b	12 ♂♂	618	(533-710)	: 6 ♀♀	543	(490-660)
tail	12 ♂♂	946	(859-1025)	: 6 ♀♀	867	(830-955)
weight	4 ♂♂	5774	(4310-6630)	: 2 ♀♀	5139	(4752, 5525)

Subspecies *P. pileata* is grey with a cap of erect hairs on the crown (hence the common name) and long cheek whiskers. Five subspecies are recognised, following Ellerman & Morrison-Scott (1951). They can be distinguished in the following key, adapted from Pocock (1939):

1.	Little contrast between crown and cheek whiskers: whiskers short, grey, scarcely paler than grey crown..........	**shortridgei** (N. Burma, E. of upper R. Chindwin)
1a.	Marked contrast between crown and cheek whiskers: whiskers long (white, yellow, or orange-red) sharply contrasting with grey crown	2
2.	No sharp contrast in colour between upper and undersides of body and between outer and inner sides of limbs; throat and chest paler than abdomen; foot uniformly dark grey-brown	**brahma** (Dafla Hills, Arunachal Pradesh, India)
2a.	Sharp contrast in colour between upper and undersides of body and between outer and inner sides of limbs; throat and chest not paler than abdomen; foot not uniformly coloured (grey-brown to black, darkest on proximal phalanges; orange to yellowish white hue of inner side of leg extends to tip of hallux)..	3

3. Whiskers, throat and entire underparts and inner sides of limbs yellowish white .... **pileata** (E. India: Assam, Meghalaya and Nagaland; NE. Bangladesh)
3a. Whiskers, throat and chest orange to orange-red 4
4. Abdomen orange-red, matching chest; crown, nape and back uniformly coloured (brownish grey) .... **durga** (India: E. Assam, Nagaland, Manipur, and NW. Burma as far as W. bank of upper R. Chindwin)
4a. Abdomen buff, contrasting with orange chest; crown, nape and back not uniformly coloured (buff collar separates dark grey crown from brownish grey back) .... **tenebrica** (India: Kamrup, E. of R. Manas and N. of R. Brahmaputra)

Where the two races are sympatric in Nagaland, the yellow-bellied *P.p.pileata* is found at 5000ft and the red-bellied *P.p.durga* at 2000ft, and below. This was noted by the collector, P R Mills, cited in Pocock (1939).

Presbytis pileata pileata (Blyth, 1843)

Principal synonym *Semnopithecus argentatus* Horsfield, 1851:7.

Type localities Of *pileata*, unknown: the type was received from the Vice-Regal Menagerie at Barrackpore and stated to be from Malaya; later in 1863 (p.12), Blyth gave its habitat as 'Nágas, and hill regions of Sylhet, Tipperá, Chittagong'. Of *argentatus*, not cited; in 1918, Wroughton traced the holotype, a specimen (BMNH1879. 11.21.597) from Sylhet, NE. Bangladesh.

Presbytis pileata brahma (Wroughton, 1916)

Type locality Seajuli, Dafla Hils, N. Lakhimpur, Arunachal Pradesh, India.

Presbytis pileata durga (Wroughton, 1916)

Principal synonym *Pithecus pileatus saturatus* Hinton, 1923:81.

Type localities Of *durga*, Lower Assam; of *saturatus*, Bara Hapjan, Lakhmipur, Upper Assam, India, at 200ft.

Presbytis pileata shortridgei (Wroughton, 1915)

Principal synonym *Pithecus belliger* Wroughton, 1915:57.

Type localities Of *shortridgei*, Homalin, E. bank upper R. Chindwin, Burma, at 400ft; of *belliger*, Singkaling Hkamti, E. bank of upper R. Chindwin, Burma, at 500ft.

Presbytis pileata tenebrica (Hinton, 1923)

Type locality R. Matunga, northern Kamrup, Assam, India, at 1200ft.

Locality	Date of Collection	Sex	Age	Nature of Specimen	Accession no.	References, Remarks, etc.
Presbytis pileata pileata (Blyth 1843)						
[BANGLADESH]						
Sylhet, Assam, [24°53′N 91°51′E]	—	I	Juv	Skin, Skull	ZD.1879.11.21.597	Holotype *Semnopithecus argentatus* Horsfield 1851, (*Trachypithecus pileatus pileatus*), Cited Horsfield,T. 1851: 7, Cited Hinton,M.A.C. 1923a: 79, Cited Pocock,R.I. 1928: 663
[INDIA]						
Konshnong, Jaintia Hills, Assam, [c.25°30′N 94°31′E], 3000ft	1 Aug 1920	M	Inf	Skin, Skull	ZD.1921.7.13.3	Cited Hinton,M.A.C. 1923a: 79, Cited Hinton,M.A.C. and Lindsay,H.M. 1926: 386, Cited Pocock,R.I. 1928: 662, Cited Pocock,R.I. 1939: 123
Laitkynsao, Khasi Hills, Assam, [c.25°12′N 91°40′E], 2000ft	30 Apr 1920	M	Juv	Skin, Skull	ZD.1921.7.13.2	,, ,, ,, ,,
Mokokchung, Naga Hills, Assam, [26°20′N 94°30′E], 5000ft	3 Jan 1921	M	Ad	Skin, Skull	ZD.1921.7.16.1	Cited Hinton,M.A.C. 1923a: 79, Cited Pocock,R.I. 1928: 661-662, Cited Pocock,R.I. 1939: 123, 127
,, ,, ,,	26 Jan 1921	F	[Juv]	Skin	ZD.1921.7.16.2	,, ,, ,, ,,
Tura, Garo Hills, [Meghalaya], Assam, [25°32′N 90°14′E], 1400ft	24 Feb 1920	M	Ad	Skin, Skull	ZD.1920.11.1.1	Cited Hinton,M.A.C. 1923a: 79, Cited Hinton,M.A.C. and Lindsay,H.M. 1926: 386, Cited Pocock,R.I. 1928: 662, Cited Pocock,R.I. 1939: 123, 127
,, ,, ,,	8 Mar 1920	I	[Juv]	Skin	ZD.1921.7.13.4	Cited Hinton,M.A.C. 1923a: 79, Cited Hinton,M.A.C. and Lindsay,H.M. 1926: 386
Presbytis pileata brahma (Wroughton 1916)						
INDIA						
Seajuli, N.Lakhimpur, Upper Assam, north side of Rajghur, [c.27°21′N 94°06′E]	23 Nov 1911	M	Ad	Skin, Skull	ZD.1913.2.21.1	Holotype *Pithecus brahma* Wroughton 1916, (*Trachypithecus pileatus brahma*), Cited Wroughton,R.C. 1916c: 654, Cited Hinton,M.A.C. 1923a: 83, Cited Pocock,R.I. 1928: 664,Cited Pocock,R.I. 1939: 127-128
Presbytis pileata durga (Wroughton 1916)						
[BANGLADESH]						
Chittagong Hill Tracts	18 Jan 1901	M	Ad	Skull	ZD.1926.10.8.1	
INDIA						
N.Bengal	—	M	Ad	Skull	RCS(OM) A70.5	
[INDIA]						
Assam	—	M	Ad	Skin, Skull	ZD.1875.1.8.1	Cited Colyer,J.F. 1936: 37

Locality	Date of Collection	Sex	Age	Nature of Specimen	Accession no.	References, Remarks, etc.
Presbytis pileata durga *continued*						
Bara Hapjan, Lakhimpur, Assam, [27°32′N 95°30′E], 200ft	3 Nov 1919	M	Ad	Skin, Skull	ZD.1921.7.13.5	Holotype *Pithecus pileatus saturatus* Hinton 1923, Cited Hinton,M.A.C. 1923a: 81-83, Cited Hinton,M.A.C. and Lindsay,H.M. 1926: 386, Cited Pocock,R.I. 1928: 663, Cited Pocock,R.I. 1939: 126-127
Golaghat, [Sibsagar], Assam, [26°30′N 93°59′E], 300ft	31 Dec 1919	F	Ad	Skull	ZD.1920.11.1.2	Cited Hinton,M.A.C. 1923a: 81-82, Cited Pocock,R.I. 1939: 125-127
Golaghat, Sibsagar, Assam, [26°30′N 93°59′E], 400ft	28 Jan 1920	M	Ad	Skin, Skull	ZD.1921.7.13.1	(*Pithecus pileatus saturatus*), Cited Hinton,M.A.C. 1923a: 81-82, Cited Hinton,M.A.C. and Lindsay,H.M. 1926: 386, Cited Pocock,R.I. 1928: 664, Cited Pocock,R.I. 1939: 125-127
Lakhuni, Naga Hills, Assam, [26°33′N 94°27′E], 2000ft	10 Sep 1919	F	Ad	Skin, Skull	ZD.1921.7.16.3	(*Pithecus pileatus saturatus*), Cited Mills,J.P. 1923: 222, Cited Pocock,R.I. 1928: 663, Cited Colyer,J.F. 1936: 36, Cited Pocock,R.I. 1939: 125, 127
Lamsakhang, Cachar Hills, Assam, [25°48′N 93°04′E], 800ft	7 Sep 1920	F	Ad	Skin, Skull	ZD.1921.7.13.7	Cited Pocock,R.I. 1928: 663-664, Cited Pocock,R.I. 1939: 125-127
Lanka, N.Cachar, Assam, [25°56′N 92°57′E], 400ft	22 Oct 1920	M	Ad	Skin, Skull	ZD.1921.7.13.6	Cited Hinton,M.A.C. 1923a: 81-83, Cited Pocock,R.I. 1928: 663-664, Cited Pocock,R.I. 1939: 125-127
[Lower Assam]	—	M	Ad	Skin, Skull, Skeleton	ZD.1886.10.18.1	Holotype *Pithecus durga* Wroughton 1916, (*Pithecus durga, Macacus pileatus*), Listed Gray,J.E. (unpubl.): 1103, Cited Wroughton,R.C. 1916c: 655, Cited Hinton,M.A.C. 1923a: 83, Cited Pocock,R.I. 1928: 663, Cited Pocock,R.I. 1939: 125-127
Presbytis pileata durga? (Wroughton 1916)						
BURMA						
Nr. Kabaw Valley, Upper R.Chindwin, W. bank, [c.24°20′N 94°25′E]	Apr 1914	[M]	Ad	Skull	ZD.1937.3.24.15	Cited Pocock,R.I. 1939: 125, Cited Fooden,J. 1976: 111
Tamanthi, Upper R.Chindwin [W. bank], [25°20′N 95°18′E]	—	M	Ad	Skull	ZD.1937.3.24.14	(*Trachypithecus pileatus shortridgei*)
Presbytis pileata shortridgei (Wroughton 1915)						
BURMA						
Chati Hka, Triangle, Upper Burma, 26°25′N 97°50′E, 1000ft	19 Apr 1939	M	Ad	Skin, Skull	ZD.1950.368	
,, ,, ,,	19 Apr 1939	F	Ad	Skin, Skull	ZD.1950.369	
,, ,, ,,	19 Apr 1939	F	Ad	Skin, Skull	ZD.1950.370	
Chati Hka, Triangle, Upper Burma, 26°25′N 97°50′E, 3000ft	19 Apr 1939	F	Ad	Skull	ZD.1950.371	
Homalin, Upper R.Chindwin, E.bank, [24°50′N 94°59′E], 400ft	17 Jul 1914	M	Ad	Skin, Skull	ZD.1915.5.5.10	Holotype *Pithecus shortridgei* Wroughton 1915, Cited Wroughton, R.C. 1915d: 56, Cited Wroughton,R.C. 1916: 297, Cited Pocock,R.I. 1928: 665-666, Cited Pocock,R.I. 1939: 129, Figured Colyer,J.F. 1936: 37,Fig.46
,, ,, ,,	18 Jul 1914	F	Ad	Skin, Skull	ZD.1915.5.5.11	Cited Wroughton,R.C. 1916: 297, Cited Pocock,R.I. 1928: 665, Cited Pocock,R.I. 1939: 129, Cited Colyer,J.F. 1936: 220
Htingnan, Upper Burma, 26°36′N 97°52′E, 4500ft	11 Jan 1939	M	Ad	Skin, Skull	ZD.1950.364	Cited Pocock,R.I. 1941: 468-469
Htingnan, Upper Burma, 26°36′N 97°52′E, 3500ft	7 Apr 1939	M	Ad	Skin, Skull	ZD.1950.365	
,, ,, ,,	16 Mar 1939	F	[Ad]	Skin, Skull	ZD.1950.366	Cited Pocock,R.I. 1941: 468-469
,, ,, ,,	16 Feb 1939	M	Ad	Skull	ZD.1950.367	Cited Pocock,R.I. 1941: 468
Janrawng Bum, Kumon Range, N.Burma, [26°42′N 97°12′E]	Mar 1961	I	[Ad]	Skin	ZD.1963.1492	Correspondence Keenan,J. (unpubl.)
Janrawng Bum, Kumon Range, N.Burma, [26°42′N 97°12′E], 8500ft	Mar 1961	[F]	[Ad]	Skin	ZD.1963.1495	,, ,, ,, ,,
Kachin Hills	—	M	Ad	Skin, Skull	ZD.1932.11.1.1	Cited Dollman,G. 1932: 477
Kasun Chaung, Pidaung Reserve, Myitkyina Dist., [c.25°25′N 97°15′E], 500ft	23 May 1936	M	Ad	Skin, Skull	ZD.1937.12.3.68	Cited Pocock,R.I. 1939: 129
,, ,, ,,	24 May 1936	F	Ad	Skin, Skull	ZD.1937.12.3.69	,, ,, ,, ,,
,, ,, ,,	23 May 1936	F	Ad	Skin, Skull	ZD.1937.12.3.70	,, ,, ,, ,,
,, ,, ,,	24 May 1936	F	Inf	Skin, Skull	ZD.1937.12.3.71	,, ,, ,, ,,
Minsin, Upper R.Chindwin, E. bank, [25°35′N 95°25′E], 450ft	15 Aug 1914	M	Ad	Skin, Skull	ZD.1915.5.5.12	Cited Wroughton,R.C. 1916: 297, Cited Pocock,R.I. 1939: 129
N′changyang, Upper Burma, 25°50′N 97°48′E, 1500ft	8 Jul 1939	M	Ad	Skin, Skull	ZD.1950.362	
Nogmung, Upper Burma, 27°30′N 97°50′E, 2500ft	[Jul 1938]	I	[Ad]	Skin	ZD.1950.363	Cited Pocock,R.I. 1941: 468-469
Singkaling Hkamti, Upper R.Chindwin, E. bank, [26°01′N 95°39′E], 500ft	27 Jul 1914	M	Juv	Skin, Skull	ZD.1915.5.5.13	(*Presbytis shortridgei belliger*), Cited Wroughton,R.C. 1916: 298, Cited Pocock,R.I. 1939: 129

Locality	Date of Collection	Sex	Age	Nature of Specimen	Accession no.	References, Remarks, etc.
Presbytis pileata shortridgei *continued*						
„ „ „	3 Aug 1914	M	Ad	Skin, Skull	ZD.1915.5.5.14	Holotype *Pithecus belliger* Wroughton 1915, Cited Wroughton,R.C. 1915d: 57, Cited Wroughton,R.C. 1916: 298, Cited Pocock,R.I. 1928: 666, Cited Pocock,R.I. 1939: 129
[BURMA]						
[Homalin], [Upper R.Chindwin, E. bank], [24°50′N 94°59′E], [400ft]	[17 Jul 1914]	F	Ad	Skull	ZD.1979.1630	(*Presbytis homalinus*)
Presbytis pileata? shortridgei? (Wroughton 1915)						
[BURMA]						
Nmai Hka Valley, Mekh Rame confluence, 26°30′N, 4-6000ft	17 Sep 1914	F	Inf	Skin, Skull	ZD.1915.2.1.1	(*Trachypithecus [?crepusculus]*)
Presbytis pileata tenebrica (Hinton 1923)						
[INDIA]						
Bogranadi, N.Kamrup, [Assam], 3000ft	4 Jan 1921	F	Ad	Skin, Skull	ZD.1921.7.13.9	Cited Pocock,R.I. 1928: 664, Cited Pocock,R.I. 1939: 126-127
Bogranadi, N.Kamrup, [Assam], 2000ft	14 Jan 1921	M	Inf	Skin, Skull	ZD.1921.7.13.10	Cited Pocock,R.I. 1928: 664, Cited Pocock,R.I. 1939: 126
R.Matunga, N.Kamrup, 1200ft	31 Dec 1920	M	Ad	Skin, Skull	ZD.1921.7.13.8	Holotype *Pithecus pileatus tenebricus* Hinton 1923, Cited Hinton,M.A.C. 1923a: 81, Cited Pocock,R.I. 1928: 664-665, Cited Pocock,R.I. 1939: 126-127
Presbytis pileata (Blyth 1843)						
NO DATA						
—	Died 7 Apr 1915	F	Inf	Skin, Skull	ZD.1979.1663	
—	—	F	Juv	Skin, Skull	ZD.1939.1151	(*Presbytis entellus x Presbytis pileatus hybrid*)

Presbytis vetulus group

PRESBYTIS JOHNII (Fischer, 1829)

John's Leaf Monkey, Nilgiri Leaf Monkey

Principal synonyms *Semnopithecus cucullatus* I. Geoffroy, 1831:38, atlas, pl. 1; *Semnopithecus jubatus* Wagner, 1840:305.

Type localities Of *johnii*, 'in India orientali (Tellicheri)'; of *cucullatus*, 'Les montagnes des Gates et Bombay'; of *jubatus*, southern India.

Taxonomic note In 1928, Pocock treated *P. johnii* as a race of *Presbytis vetulus*, but in 1939 he gave it specific status within the genus *Kasi* Reichenbach, 1862, following Hill (1934).

Range Western Ghats from Coorg (*c.* 12°30′N) to the Tinnevelly (= Tirunelveli) Hills (*c.* 8°45′N), usually above 3000ft.

Dimensions Adult specimens in the museum collections have the following field measurements (see Appendix 2 for comparative table):

h & b 5 ♂♂ 570 (508-645)
tail 5 ♂♂ 860 (756-965)
Weights, cited by Leigh (1926), are as follows: 4 ♂♂ 12 712 g (11 804-13 620); 1 ♀ 10 896 g. Hill (1934) gives an average weight of 25 lbs (11 350 g).

Species To draw attention to the close relationship between *P.johnii* and the following species, *P. vetulus*, their keys are united in a single table (p. 73) from which it can be seen that *P. johnii* resembles *P. vetulus* particularly closely in its brown crown, black white-ticked coat and shorter paler rump hairs, but differs in lacking the striking white cheek whiskers of its Sri Lankan counterpart. It is also considerably larger and heavier than any *P. vetulus* subspecies.

Locality	Date of Collection	Sex	Age	Nature of Specimen	Accession no.	References, Remarks, etc.
Presbytis johnii (Fischer 1829)						
INDIA						
Anaimalai [Hills], Mysore, [Kerala]	—	F	Ad	Skull	RCS(OM) A70.371	
„ „ „	—	I	Juv	Skull	RCS(OM) A70.372	
Kurg [=Coorg], S.India	—	I	[Ad]	Skin	ZD.1891.10.7.7	
Nr. Kuriarkutti, Parambikulam Valley, Cochin State, [Kerala], 10°25′N 76°43′E, 1600ft	Nov 1937	M	[Ad]	Skin	ZD.1953.730	
S.India	—	M	Ad	Skin, Skull	ZD.1879.11.21.591	[Cited] Horsfield,T 1851: 8, Cited Hill,W.C.Osman 1936: 125
„ „ „	—	F	Juv	Skin, Skull	ZD.1879.11.21.592	[Cited] Horsfield,T. 1851: 8
„ „ „	—	[M]	[Ad]	Skin	ZD.1979.982	
—	—	[M]	[Juv]	Skin	ZD.1979.981	[Cited] Lindsay,H.M. 1926: 592
[INDIA]						
Anamaad, [S. of Palghat], S.Malabar, [Kerala], 3200ft	25 Apr 1921	M	Ad	Skin, Skull	ZD.1921.11.5.5	Cited Lindsay,H.M. 1926: 592, Cited Pocock,R.I. 1928: 504, Cited Pocock,R.I. 1939: 148, 157
Cotengady Estate, [24 miles S. of Palghat], Cochin, [Kerala], 3500ft	27 Apr 1921	M	Ad	Skin, Skull	ZD.1921.11.5.3	Cited Lindsay,H.M. 1926: 592, Cited Pocock,R.I. 1939: 148, 157

Locality	Date of Collection	Sex	Age	Nature of Specimen	Accession no.	References, Remarks, etc.
Presbytis johnii *continued*						
" " "	8 Apr 1921	M	Inf	Skin, Skull	ZD.1933.7.29.10	Cited Lindsay,H.M. 1926: 592, Cited Hill,W.C.Osman 1936: 125-126
Kumblacodie, [S. of Palghat], Cochin, [Kerala], 3500ft	29 Mar 1921	M	Ad	Skin, Skull	ZD.1921.11.5.4	Cited Lindsay,H.M. 1926: 592, Cited Pocock,R.I. 1928: 504, Cited Pocock,R.I. 1939: 157
Nilghiri [Hills], [c.11°25′N 76°45′E]	—	F	Ad	Skin	ZD.1845.8.12.2	Associated with next specimen (1845.9.20.4)
" " "	—	F	Ad	Skull	ZD.1845.9.20.4	Listed Gerrard,E. 1862: 8, Listed Gray,J.E. (unpubl.): 10
Palagapandy, [S. of Palghat], S.Malabar, [Kerala], 3300ft	23 May 1921	M	Inf	Skin, Skull	ZD.1933.7.29.11	Cited Lindsay,H.M. 1926: 592, Cited Hill,W.C.Osman 1936: 125-126
Seetagundy, [S. of Palghat], Malabar, [Kerala], 3500ft	8 May 1921	M	Ad	Skin, Skull	ZD.1921.11.5.6	Cited Lindsay,H.M. 1926: 592
Nr. Srimangala, (Brahmagiri Hills), S.Coorg, [Karnataka], [c.12°01′N 75°58′E], 2780-5135ft	11 Feb 1913	M	Ad	Skin, Skull	ZD.1913.8.22.1	Cited Ryley,K.V. 1913b: 492, Cited Pocock,R.I. 1928: 504, Figured Pocock,R.I. 1939: 148, 157, Fig.45
—	—	I	Juv	Skull, Skeleton, Hyoid	ZD.1857.10.19.1	Listed Gray,J.E. (unpubl.): 10
—	—	M	[Juv]	Skin	GCMAG	
—	—	M	Ad	Skull	RCS(OM) A70.37	
Presbytis johnii? (Fischer 1829)						
[INDIA]						
—	—	F	Ad	Skull	RCS(OM) A68.93	
—	—	M	Ad	Skull	CMZ E7220B	(*Semnopithecus* sp.)

PRESBYTIS VETULUS (Erxleben, 1777)

Purple-faced Leaf Monkey
(4 subspecies)

Taxonomic notes The change in the scientific name of the species from *senex* to *vetulus* has been brought about by the discovery (Brandon-Jones, personal communication, 1979) that in 1926, Phillips selected *vetulus* Erxleben, 1777, as the prior name. He considered the name *senex*, published simultaneously by Erxleben in 1777, referred to an albino which might have come from any of the geographic regions whence subspecies had at that time been described. As first reviser, Phillips' decision is binding (Article 24 of the International Code of Zoological Nomenclature).

Most unfortunately, Pocock was unaware of Phillips' action when, in 1928, he selected *senex* as the name of the taxon on the basis of page priority, a decision to which he adhered in his *Fauna of British India* (1939). It was also accepted by Ellerman and Morrison-Scott in their *Checklist of Palaearctic and Indian Mammals* (1951), and thus the name *Presbytis senex* for the purple-faced leaf-monkeys has been perpetuated in the literature of the last 30 years.

At present the situation is not absolutely clearcut for two reasons:

1. the identity of *Presbytis vetulus* is still unsettled. Based on the purple-faced leaf-monkey and the lion-tailed macaque, it will need the designation of a lectotype to make its identification unequivocal; and
2. the name *Presbytis senex* is based on albino monkeys from Sri Lanka reported by Robert Knox in 1681. They were probably observed in the zone of contact between two subspecies, *nestor* and *philbricki*.

Probably the best course is to establish the identity of *P. vetulus* as a purple-faced leaf monkey by designating a suitable lectotype; to eliminate *senex* as subspecifically indeterminable, and to revert to the nomenclature adopted by Hill (1934) and Phillips (1935) as a basis for the taxonomy of the group.

The difficulties involved in attempting to preserve the name *senex* for the species are twofold: it is impossible to identify *senex* subspecifically, neither can its identity be fixed arbitrarily, as suggested by Pocock (1939), since an abnormal (albino) specimen cannot be designated a neotype.

Presbytis albinus Kelaart, 1851, placed by Pocock (1928) in the synonymy of *P. senex*, is another albinistic leaf-monkey whose provenance is in doubt. Kelaart (1851) stated that the species was found near Kandy, but later (1852) he noted that the holotype had been forwarded to him from Matale. The place where it was collected is not stated, and as both the above-mentioned localities are in the zone of contact between *nestor* and *philbricki*, it is impossible to allocate it to either with any confidence.

Dimensions Adult specimens in the museum collections have the following field measurements (see Appendix 2 for comparative table):

P.v.vetulus			
h & b	1 ♂	540	: 2 ♀♀ 509 (492, 525)
tail	1 ♂	730	: 2 ♀♀ 718 (687, 750)
weight	1 ♂	5675	: 2 ♀♀ 5108 (4994, 5221)
P.v.monticola			
h & b	1 ♂	595	: 1 ♀ 540
tail	1 ♂	618	: 1 ♀ 666
weight	2 ♂♂	9421 (9080, 9761)	: 1 ♀ 7491
P.v.nestor			
h & b	1 ♂	495	: 4 ♀♀ 518 (490-560)
tail	1 ♂	712	: 4 ♀♀ 702 (604-770)
P.v.philbricki			
h & b	2 ♂♂	608 (600, 616)	: 1 ♀ 484
tail	2 ♂♂	853 (820, 885)	: 1 ♀ 820

Subspecies As mentioned above, the close relationship of *P. vetulus* to the preceding species is shown in the following key, adapted from Hill (1934).

Key to the species *P. johnii* and *P. vetulus*, and the subspecies of the latter

	P. johnii	*P.v.vetulus*	*P.v.nestor*	*P.v.monticola*	*P.v.philbricki*
Range	W. Ghats, India	SW. Sri Lanka	W. Sri Lanka	Central highlands, Sri Lanka	NE. Sri Lanka
Size (from Hill, 1934)	Very large (11 350g or more)	Medium (up to 6810g)	Small (3632-4540g)	Large (9080g)	Very large (9988g)
Upper back	Black, sparsely ticked with whitish; smooth coat	Black, ticked with whitish: smooth coat	Greyish-brown with whitish ticking in some; smooth coat	Greyish-brown, no ticking; long shaggy coat	Greyish-brown, no ticking; coat shorter and less shaggy than *monticola*
Lower back and rump	Silvery short-haired rump-patch, not sharply defined from upper back; silvery ticking extends to tail	Whitish short-haired rump-patch, sharply defined from upper back, extending to thighs and tail	Beige short-haired rump-patch, sharply defined from upper back, extending to thighs and tail	Rump-patch not sharply defined; hair shorter but scarcely paler than upper back	Rump-patch not sharply defined; hair shorter but scarcely paler than upper back
Limbs	As upper back; white ticking most pronounced on thighs, fading distally; hands and feet black	As upper back; whitish ticking most pronounced on thighs, fading distally; hands and feet black	As upper back, except thighs which are like rump, darkening distally; hands and feet dark brown	Usually darker than upper back; hands and feet dark brown or black.	Usually darker than upper back; hands and feet dark brown
Tail	Long, black distally; ticking extends to tailtip in some	Long; middle portion greyish-brown, becoming paler towards tip	Long; middle portion light greyish-brown becoming paler towards tip	Short but thick; like back but paler at tip in some	Long; greyish-brown with white tip
Underparts	Black	Black	Greyish-brown	Greyish-brown	Greyish-brown
Crown and nape	Light brown, contrasting with back; hair long, longer and yellower on nape	Brown, contrasting with back; hair short	Light brown, contrasting with back; hair short, longer and lighter on nape	Scarcely paler than back; hair long	Scarcely paler than back; hair short, longer on nape
Whiskers	Long, brown, merging with crown hairs to form complete brown hood	Moderate, dirty white; separated from crown by narrow tract of black hairs	Moderate; cream; separated from crown by narrow tract of black hairs	Very long and full; dirty white; separated from crown by triangle of black hairs which tapers towards ear	Long and full; cream; separated from crown by broad tract of black hairs which tapers towards ear

Presbytis vetulus vetulus (Erxleben, 1777)

South-western Purple-faced Leaf Monkey, Black Wanderoo

Principal synonyms *Cercopithecus kephalopterus* Zimmermann, 1780:185; *Simia (Cercopithecus) silenus purpuratus* Kerr, 1792:65; *Simia porphyrops* Link, 1795:62; *Simia latibarba* Temminck, 1807:3; *Cercopithecus latibarbatus* E. Geoffroy, 1812:94; *Cercopithecus*(?) *leucoprymnus* Otto, 1825:503.

Type localities Of *vetulus*, 'Zeylona', restricted by Hill (1934) to the district near Galle, Southern Province; of *kephalopterus*, 'Ceilon'; of *purpuratus*, 'Ceylon'; of *porphyrops*, none cited; of *latibarba* 'Ceylan'; of *latibarbatus*, unknown; of *leucoprymnus*, unknown.

Presbytis vetulus monticola Kelaart, 1850
Highland Purple-faced Leaf Monkey, Great Wanderoo, Bear Monkey

Principal synonym *Presbytis ursinus* Blyth, 1851:155.

Type localities Of *monticola,* Nuwara Eliya, Central Province; of *ursinus*, Nuwara Eliya.

Presbytis vetulus nestor (Bennett, 1833)
Western Purple-faced Leaf Monkey, Dusky Wanderoo

Principal synonyms *Semnopithecus kelaartii* Schlegel, 1876:52; *Pithecus vetulus phillipsi* Hinton, 1923:510.

Type localities Of *nestor,* unknown, assumed to be from India; of *kelaartii*, 'La partie méridionale de l'île de Ceylan'; of *phillipsi*, Gonapola, Panadura District, Ceylon.

Presbytis vetulus philbricki (Phillips, 1927)
North-eastern Purple-faced Leaf Monkey

Principal synonym *Presbytis senex harti* Deraniyagala, 1954 (p. 9) is provisionally assigned here, following Eisenberg and McKay (1970).

Type localities Of *philbricki*, Kantalai, near Trincomalee, Eastern Province; of *harti*, Thunuikai (= Tunnukkai), Northern Province.

Locality	Date of Collection	Sex	Age	Nature of Specimen	Accession no.	References, Remarks, etc.
Presbytis vetulus vetulus (Erxleben 1777)						
[SRI LANKA]						
Anasigalla, Matugama, Kalutara, [W.P.], Ceylon, [6°29'N 80°03'E], 150ft	15 Dec 1919	F	Ad	Skin, Skull	ZD.1920.5.1.1	Cited Hinton,M.A.C. 1923: 509, Cited Pocock,R.I. 1928: 501, Cited Hill,W.C.Osman 1934: 60
Anasigalla, Matugama, Kalutara, [W.P.], Ceylon, [6°29'N 80°03'E], 100ft	7 Apr 1920	F	Ad	Skin, Skull, Cast of lower L molars	ZD.1920.9.26.1	Cited Hinton,M.A.C. 1923: 509, 514, Cited Hill,W.C.Osman 1934: 60
Anasigalla, Matugama, Kalutara Dist., [W.P.], Ceylon, [6°29'N 80°03'E], 100ft	7 Dec 1921	M	Juv	Skin, Skull	ZD.1923.1.18.1	(*Pithecus kephalopterus*), Cited Hinton,M.A.C. 1923: 509, Cited Hill,W.C.Osman 1934: 60
,, ,, ,,	23 Dec 1921	M	Juv	Skin, Skull	ZD.1923.1.18.2	,, ,, ,, ,,
Anasigalla, Matugama, Kalutara Dist., [W.P.], Ceylon, [6°29'N 80°03'E], 50ft	29 Dec 1921	M	Ad	Skin, Skull	ZD.1923.1.18.3	(*Pithecus kephalopterus*), Cited Hinton,M.A.C. 1923: 509, 513, Cited Pocock,R.I. 1928: 501, Cited Hill,W.C.Osman 1934: 60

Locality	Date of Collection	Sex	Age	Nature of Specimen	Accession no.	References, Remarks, etc.
Presbytis vetulus vetulus *continued*						
„ „ „	3 Oct 1920	F	[Juv]	Skin	ZD.1923.1.18.4	(*Pithecus kephalopterus*), Cited Hinton,M.A.C. 1923: 509, Cited Hill,W.C.Osman 1934: 60
Anasigalla, Matugama, Kalutara, [W.P.], Ceylon, [6°29′N 80°03′E]	28 Sep 1922	F	Juv	Skull	ZD.1966.5545	
Anasigalla, Matugama, Kalutara, [W.P.], Ceylon, [c.6°29′N 80°03′E]	13 Feb 1920	M	Juv	Skull	ZD.1979.993	(*Pithecus ursinus*), Cited Hinton,M.A.C. 1923: 509, Cited Hill,W.C.Osman 1934: 60
Galle, [S.P.], Ceylon, [6°01′N 80°13′E]	16 Nov 1940	F	[Juv]	Skin	ZD.1950.5	
Hiyare, Nr. Galle, [S.P.], [6°04′N 80°18′E]	—	I	Juv	Skull, Skeleton	ZE.1950.7.17.4	Background Hill,W.C.Osman 1936: 147
„ „ „	—	F	Ad	Skull, Skeleton	ZE.1950.7.17.5	„ „ „ „
„ „ „	—	M	Ad	Skull, Skeleton	ZE.1950.7.17.6	„ „ „ „
„ „ „	—	[M]	[Juv]	Skull, Skeleton	ZE.1950.7.17.7	„ „ „ „
„ „ „	—	[M]	[Juv]	Skull, Skeleton	ZE.1950.7.17.8	„ „ „ „
Kottawa, S.P., Ceylon, [6°06′N 80°18′E]	12 Apr 1913	F	Juv	Skin, Skull	ZD.1915.3.1.7	Cited Hinton,M.A.C. 1923: 508-509
Matugama, [Kalutara], Ceylon, [6°32′N 80°06′E]	—	[M]	[Ad]	Skin	ZD.1923.2.6.1	
„ „ „	—	[M]	[Ad]	Skin	ZD.1923.2.6.2	
[Matugama], [Kalutara], [W.P.], [c.6°32′N 80°06′E]	—	M	Ad	Skull	ZD.1975.1806	
„ „ „	—	M	Ad	Skull	ZD.1975.1807	
Pasdon Corola, Ceylon, [c.6°31′N 80°07′E]	—	[M]	Juv	Skin, Skull	ZD.1879.9.5.1	(*Semnopithecus leucoprymnus*), Cited Hinton,M.A.C. 1923: 508-509
Sinharaja Forest, Nr. Ketala-Patela, Ceylon, [6°23′N 80°25′E], 1000-2000ft	18 Aug 1936	F	[Ad]	Skin	ZD.1936.11.26.1	
S.P., Ceylon	19 Sep 1938	F	[Ad]	Skin	ZD.1950.6	
Presbytis vetulus monticola Kelaart 1850						
[SRI LANKA]						
Hakgalla [Hill], Uva, Ceylon, [6°55′N 80°49′E], 5600ft	6 Apr 1914	F	[Ad]	Skin	ZD.1915.3.1.6	(*Presbytis ursinus*), Cited Wroughton,R.C. 1915b: 84, Cited Pocock,R.I. 1928: 503, Cited Hill,W.C.Osman 1934: 69, Cited Pocock,R.I. 1939: 156, 158
Hakgalla, [Uva], C.P., Ceylon, [6°55′N 80°49′E]	—	I	[Ad]	Skin	ZD.1979.1440	(*Kasi senex monticola*)
Newara Eliya, [C.P.], Ceylon, [6°58′N 80°46′E], 6000ft	30 Jan 1944	F	[Ad]	Skin	ZD.1950.8	
[Newara Eliya], [C.P.], Ceylon, [6°58′N 80°46′E]	—	[F]	[Juv]	Skin, Skull	ZD.1852.5.9.16	(*Semnopithecus leucoprymnus*), Cited Pocock,R.I. 1928: 503, Cited Hill,W.C.Osman 1936: 128, Cited Colyer,J.F. 1936: 35
Ohiya, [Horton Plains], [C.P.], Ceylon, [6°49′N 80°50′E]	Oct 1942	M	[Ad]	Skin	ZD.1949.514	
Pattipola, C.P., Ceylon, [6°51′N 80°50′E], 6210ft	12 Mar 1914	M	Ad	Skin, Skull	ZD.1915.3.1.5	Cited Wroughton,R.C. 1915b: 84, Cited Hinton,M.A.C. 1923: 514, Cited Pocock,R.I. 1928: 503, Cited Hill,W.C.Osman 1934: 69, Cited Pocock,R.I. 1939: 156, 158
Ceylon	—	[M]	Ad	Skin, Skull	ZD.1855.12.24.295	(*Semnopithecus leucoprymnus*)
Presbytis vetulus monticola? Kelaart 1850						
[SRI LANKA]						
Adam's Peak, Ceylon, [6°49′E 80°30′E]	1908	M	Ad	Skin, Skull	ZD.1911.9.9.1	(*P.ursinus?*), Cited Hinton,M.A.C. 1923: 514, Cited Pocock,R.I. 1928: 502-503, Cited Hill,W.C.Osman 1934: 69, Cited Pocock,R.I. 1939: 156
Presbytis vetulus nestor (Bennett 1833)						
SRI LANKA						
Bandaragama, Panadura Dist., [W.P.], Ceylon, [6°43′N 79°59′E]	12 Dec 1921	M	Juv	Skin, Skull	ZD.1923.1.19.2	(*Pithecus vetulus phillipsi*), Background Hinton,M.A.C. 1923, Cited Hill,W.C.Osman 1934: 65, 67
Godigamuwa, Panadura Dist., [W.P.], Ceylon, [6°45′N 79°59′E]	15 Dec 1921	F	Ad	Skin, Skull	ZD.1923.1.19.4	Paratype *Pithecus vetulus phillipsi* Hinton 1923, Cited Hinton,M.A.C. 1923: 510, 513, Cited Hill,W.C.Osman 1934: 66
„ „ „	15 Dec 1921	F	Ad	Skin, Skull	ZD.1923.1.19.5	„ „ „ „
Gonabendaduwa, [=Mahagonaduwa], Panadura Dist., [W.P.], Ceylon, [6°40′N 79°58′E]	16 Dec 1921	F	Ad	Skin, Skull	ZD.1923.1.19.3	Paratype *Pithecus vetulus phillipsi* Hinton 1923, Cited Hinton,M.A.C. 1923: 510, 513-514, Cited Pocock,R.I. 1928: 502, Cited Hill,W.C.Osman 1934: 66
Gonapola, Panadura Dist., Ceylon, [6°46′N 80°01′E]	14 Dec 1921	M	Ad	Skin, Skull	ZD.1923.1.19.1	Holotype *Pithecus vetulus phillipsi* Hinton 1923, Cited Hinton,M.A.C. 1923: 510, 513-514, Cited Pocock,R.I. 1928: 502, Cited Hill,W.C.Osman 1934: 63, 65
Homogama, W.P., Ceylon, [6°51′N 80°E]	19 Dec 1935	M	[Juv]	Skin	ZD.1950.2	

Locality	Date of Collection	Sex	Age	Nature of Specimen	Accession no.	References, Remarks, etc.
Presbytis vetulus nestor *continued*						
Horana, Panadura Dist., [W.P.], Ceylon, [6°42′N 80°04′E]	13 Dec 1921	F	Ad	Skin, Skull	ZD.1923.1.19.6	Paratype *Pithecus vetulus phillipsi* Hinton 1923, Cited Hinton,M.A.C. 1923: 513, Cited Pocock,R.I. 1928: 502, Cited Hill,W.C.Osman 1934: 66
Polgahawela, Ceylon, [7°20′N 80°19′E], [below 150m]	—		Ad	2 skulls	RCS(OM) S70.3	Cited Colyer,J.F. (unpubl.): 243
Raygam Korale, [W.P.], Ceylon, [6°43′N 80°03′E]	—	M	Ad	Skin, Skull	ZD.1879.9.5.2	(*Semnopithecus leucoprymnus*), Cited Hinton,M.A.C. 1923: 510
" " "	—	[F]	Juv	Skin, Skull	ZD.1879.9.5.3	" " " "
Raygam Korale, W.P., Ceylon, [6°43′N 80°03′E]	8 Jun 1943	M	[Juv]	Skin	ZD.1950.1	
" " "	21 Apr 1938	F	[Ad]	Skin	ZD.1950.3	
" " "	Died 22 Apr 1935	F	[Juv]	Skin	ZD.1950.4	
Ceylon	—	I	Juv	Skin, Skull	ZD.1855.12.24.12	Holotype *Semnopithecus nestor* Bennett 1833, Cited Bennett,E.T. 1833: 67, Cited Hinton,M.A.C. 1923: 508, 510, Cited Hill,W.C.Osman 1934: 63
" " "	—	F	Ad	Skin, Skull	ZD.1881.4.29.1	(*Presbytis* v. *nestor*)(*Semnopithecus leucoprymnus* var. *kelaarti* Schlegel)
—	—	[F]	Juv	Skin, Skull	ZD.1846.6.15.29	
—	—	[F]	Juv	Skin, Skull	ZD.1939.1147	
—	—	[F]	[Juv]	Skin	ZD.1939.1150	(*Pithecus*)
—	Died 29 Aug 1942	F	Ad	Wet, kidneys	ZD.1972.1314	
—	—	F	Juv	Wet, kidneys	ZD.1972.1317	
—	Died 7 Feb 1938	F	[Ad]	Cast of brain	ZD.1979.1403	(*Kasi senex nestor*)
"India"	—	I	Juv	Skin, Skull	ZD.1979.1406	
Presbytis vetulus nestor? (Bennett 1833)						
[SRI LANKA]						
Kala Oya, [Boundary of N.W. and N.C.P.], Ceylon	Feb 1935	F	[Ad]	Skin	ZD.1950.7	(*Kasi vetulus philbricki*), [Cited] Hill,W.C.Osman 1936: 130-132
Kala Oya, [Boundary of N.W. and N.C.P.], Ceylon	Apr 1935	F	[Ad]	Skin	ZD.1935.3.30.1	(*P.senex philbricki*), Cited Pocock,R.I. 1939: 155
" " "	Apr 1935	F	[Ad]	Skin	ZD.1935.3.30.2	" " " "
Kala Oya, [Boundary of N.W. and N.C.P.]	Jan 1935	M	[?Ad]	Wet, skin	ZD.1972.930	
" " "	—	F	Inf	Wet, cadaver	ZD.1972.931	[Cited] Hill,W.C.Osman 1936: 130-132
Kala Oya, [Boundary of N.W. and N.C.P.], Ceylon, [c.8°12′N 80°06′E], [below 150m]	—			9 skulls	RCS(OM) S70.32	(*Presbytis senex senex*), Cited Colyer,J.F. (unpubl.): 30r
Presbytis vetulus philbricki (Phillips 1927)						
[SRI LANKA]						
Kantalai, Nr. Trincomalee, [E.P.], Ceylon, [8°22′N 81°E], [200ft]	14 Sep 1925	M	Ad	Skin, Skull	ZD.1928.7.12.1	Holotype *Pithecus philbricki* Phillips, 1927, (*P.senex senex*), Cited Phillips,W.W.A. 1927: 57, Cited Pocock,R.I. 1939: 154-155
Kantalai, [20 miles W. of Trincomalee], [E.P.], Ceylon, [8°22′N 81°E], [200ft]	15 Sep 1925	M	Juv	Skin, Skull	ZD.1928.7.12.2	Paratype *Pithecus philbricki* Phillips, 1927, (*P.senex senex*), Cited Phillips,W.W.A. 1927: 57, Cited Pocock,R.I. 1939: 155
" " "	16 Sep 1925	F	Ad	Skin, Skull	ZD.1928.7.12.3	" " " "
Kantalai, Nr. Trincomalee, E.P., Ceylon, [8°22′N 81°E], [below 150m]	—			7 skulls	RCS(OM) S70.32	(*Presbytis senex senex*), Cited Colyer,J.F. (unpubl.): 30r, Cited Pocock,R.I. 1939: 155
Kantalai Tank, E.P., Ceylon, [8°22′N 81°E], 200ft	2 Aug 1925	M	Ad	Skin, Skull, L and R radii, ulnae, tibiae, fibulae	ZD.1966.5544	(*Presbytis senex senex*)
Nitre Cave Dist., [C.P.], Ceylon, [c.7°25′N 80°52′E], 900m	—			3 skulls	RCS(OM) S70.3	(*Presbytis senex nestor*), Correspondence Phillips,W.W.A. (unpubl.), Cited Hill,W.C.Osman 1934: 78
Trincomalee, E.P., Ceylon, [8°34′N 81°13′E], [below 150m]	—			3 skulls	RCS(OM) S70.32	(*Presbytis senex senex*), Cited Colyer,J.F. (unpubl.): 22r
Presbytis vetulus philbricki? (Phillips 1927)						
[SRI LANKA]						
—	—	F	[Ad]	Skin	ZD.1979.994	(*Presbytis senex senex*)
Presbytis vetulus (Erxleben 1777)						
[SRI LANKA]						
—	—	F	Ad	Skin, Skull	ZD.1855.12.24.10	(*Kasi? senex senex*)
—	—	F	Ad	Skull	ZD.1855.12.26.61	(*Presbytis cephalopterus*), Cited Gerrard,E. 1862: 8, Listed Anon. (unpubl.)a, Listed Gray,J.E. (unpubl.): 1106
—	—	[M]	[Ad]	Skin	ZD.1979.995	

Locality	Date of Collection	Sex	Age	Nature of Specimen	Accession no.	References, Remarks, etc.
Presbytis vetulus *continued*						
Ceylon	Died 20 Dec 1904	M	Juv	Skull	ZD.1979.1404	(*Semnopithecus cephalopterus*)
" " "	—	I	Inf	Skull	ZD.1979.1405	" " " "
" " "	—	F	[Ad]	Skin	GCMAG 02-11a	(*P.cephalopterus*)
" " "	—	F	Ad	Skull	RCS(OM) A70.32	(*Presbytis senex senex*)
—	—		Ad	Skull	RCS(OM) G68.872	(*P.senex senex*), Cited Colyer,J.F. (unpubl.)
Presbytis vetulus? (Erxleben 1777)						
[SRI LANKA]						
—	—			Wet, hands and feet	CMZ E7262A	(*Semnopithecus leucoprymnus*)

Presbytis entellus group

PRESBYTIS ENTELLUS (Dufresne, 1797)

Entellus Langur, Hanuman Langur
(15 subspecies)

Taxonomic notes Pocock's 1939 arrangement of subspecies is followed, except in one instance involving the *P.e.schistacea* group from the Himalayas and northern India. Confusion over the colour of the Himalayan Langur was initiated by Hodgson himself when, in his original description (1841), he referred to *Presbytis schistacea* as 'dark slaty'. As Pocock (1928) has commented, not one of Hodgson's specimens which came to the Museum could conceivably be described as dark slaty, most of them being 'emphatically brown'. However Blanford's (1888) description: 'earthy or greyish brown, frequently with a slight purplish tinge' is as accurate a verbal description of the colour as can reasonably be expected, and with minor variations it characterises all the races of *P. entellus.*

Later, in 1846, Hodgson described three varieties of *P. schistacea*:
'Var. 1. Slate-grey, head white, fur moderate.
'Var. 2. Paler, sides and head white; fur long, rather curling.
'Var. 3. Nearly white, head rather whiter; fur long and rather curling.'

Hodgson's Var. 3 is represented in the collections by an almost completely white juvenile (no. 1845. 1. 8.221) which Pocock considered an albino. However, as juveniles of the *schistacea* group are markedly lighter in colour than adults, it probably pertains to 'Var. 2'. In other words there appear to be at least two types of greyish brown langur in Nepal, one distinctly darker than the other.

That this was indeed Hodgson's view is made clear in his unpublished manuscripts and drawings, particularly those at the Zoological Society of London. Here he illustrates, under the name *Semnopithecus schistaceus*, a mature male head, hand and foot (p. 4) and a dark greyish brown female and juvenile (p. 6), describing them as 'the large variety'. The mature male is called the 'Bara Langoor' (Big Langur) and is stated to be from 'the Kachár, the northern region of mountains near the snows'. On p. 5 he illustrates a pale grey langur from the Tarai, calling it '*Semnopithecus* ? *petrophilus*, the lesser nipalese langur', indicating that he was considering giving taxonomic status to this smaller pale grey form. In fact he never did so, and *P. petrophilus* (Hodgson, 1846) remains a nomen nudum.

The arrival of Hodgson's huge collections of mammals and birds from Nepal in the early 1840s caused considerable confusion at the British Museum. Gray's many manuscript lists and the numbers of entries in the register indicate the difficulty he experienced in coping with the sheer numbers of new and unique specimens, many of which were given to other museums in Britain and Europe. Some of Hodgson's primate specimens cannot now be accounted for, but the two specimens first registered by Gray are still extant, and are particularly pertinent to the present problem. The first, no. 1843. 1.12.1, an adult male langur which although faded and worn, is clearly of the darker (and browner) kind (Var. 1), was selected as lectotype by Thomas (1918). Thomas wrote on the label that it came from the Tarai Forest in spite of the fact that Gray had noted in the register that it came from the Kachár. Subsquently, Wroughton (1918) cited as 'cotype' (= paralectotype) no. 1843. 1.12.2, an adult female which is equally clearly of the paler greyer kind (Var. 2). From the measurements of museum specimens (see below), there does appear to be a slight difference in size, but neither Thomas nor Wroughton seems to have been aware that they were dealing with two different races of Nepalese langur.

Most unfortunately, Thomas's lectoytpe designation was completely ignored by Hinton and Fry (1923) in their report on the mammals of Nepal collected by the Bombay Natural History Society's Mammal Survey of India in 1920-21. They identified the only langur collected, a pale grey female (no. 1922. 5.16.1) from the lowlands of Nepal, as true *schistacea*. In fact this specimen closely resembles Wroughton's supposed paralectotype of *P.e.schistacea* and also the langurs of Kumaun which were later described by Pocock (1928) as *P.e.hector*. In this paper, Pocock also ignored Thomas's lectotype designation and called the pale grey langurs of Nepal *P.e.schistacea*, at the same time restricting the type locality to the Tarai Forest (i.e. the lower altitudes of Nepal). For the darker browner langurs from the higher altitudes, he proposed a new name, *P.e.achilles*, to which taxon he attributed Hodgson's brown specimens.

In order to rectify this contradictory state of affairs, which through the wide acceptance of Pocock's work has been perpetuated in the literature of the last 50 years, it will be necessary to distinguish the Nepalese races of langur as follows:

Presbytis entellus schistacea (Hodgson, 1841), with synonym *P.e.achilles* (Pocock, 1928), for the darker browner langurs from the higher altitudes, and *P.e.hector* (Pocock, 1928) for the smaller paler greyer langurs of the lesser hills and Tarai.

P.e.ajax (Pocock, 1928) from N. Punjab and Kashmir and *P.e.lania* (Elliot, 1909) from Chumbi in Tibet are retained as possibly distinct forms, following Pocock (1928) and Roonwal (1979).

Cercopithecus albocinereus Desmarest, 1822, is probably a synonym of *P. entellus* in view of its grey coloration, darker in lumbar region, with blackish hands and feet, black face with long stiff projecting brow hairs, and large, angled, bare, black ears.

Range In the north, from Kashmir to the Chumbi Valley, Tibet. Its existence in Bhutan is doubtful and is based on Griffiths' (1840) sightings of langurs at Tongsa and Bulphai which were probably *P. geei* (Brandon-Jones, personal communication, 1979). Further south *P. entellus* has been observed to the east of the Hooghly River (Gittin & Akonda, 1982); in the west it reaches Kathiawar, and in the south is found throughout peninsular India and Sri Lanka, except in the southwest quadrant of the latter. Altitudinal range from sea level to 12 000ft.

Dimensions Field measurements of adult specimens in the museum collections indicate that the northern races are, on average, larger and heavier than the central and southern forms. *P.e.thersites* from Sri Lanka is the smallest of all the races.

P.e.schistacea group (except *P.e.hector*, see taxonomic notes above):

h & b	4 ♂♂	766	(760-780)	:	6 ♀♀	619	(520-680)
tail	4 ♂♂	887	(810-950)	:	5 ♀♀	761	(693-820)
weight	2 ♀♀	20 000	(19 545-20 455)	:	3 ♀♀	15 303	(12 727-17 273)

P.e.hector (see taxonomic notes above):

h & b	1 ♂	690		:	3 ♀♀	592	(580-605)
tail	1 ♂	978		:	3 ♀♀	853	(823-900)
weight					2 ♀♀	13 636	(13 182, 14 091)

P.e.entellus, *hypoleucos* and *priam* groups (except *P.e.thersites*):

h & b	15 ♂♂	644	(575-775)	:	19 ♀♀	596	(483-650)
tail	15 ♂♂	955	(830-1080	:	19 ♀♀	909	(785-1016)
weight	9 ♂♂	13 041	(9091-18 182)	:	11 ♀♀	9886	(7727-12 273)

P.e.thersites (from Sri Lanka):

h & b	8 ♂♂	568	(510-622)	:	5 ♀♀	490	(406-558)
tail	8 ♂♂	836	(714-918)	:	5 ♀♀	789	(752-868)
weight	8 ♂♂	10 641	(7955-13 409)	:	4 ♀♀	6705	(5455-8182)

Subspecies The subspecies of *P. entellus* can be divided into four groups, roughly equivalent to the four species recognised by Blanford (1888): the *P.e.schistacea* group (including *ajax, hector* and *lania*) from India north of the R. Ganges and the southern slopes of the Himalayas; the *P.e.entellus* group (including *anchises*) from south of the R. Ganges and northeastern parts of the peninsula; the *P.e.hypoleucos* group including *achates, aeneas, hypoleucos* and *iulus* from western peninsular India, mainly the western slopes of the Western Ghats; and the *P.e.priam* group (including *elissa* and *thersites*) from southeastern peninsular India and Sri Lanka. The position of *P.e.priamellus* is not entirely clear but it seems to belong with the *P.e.priam* group.

The groups can be separated in the following key:

1.	Whiskers long and thick, nearly or completely hiding ears.	**P.e.schistacea** group
1a.	Whiskers shorter, not or only partly covering ears	2
2.	Crest on crown	**P.e.priam** group
2a.	No crest on crown	3
3.	Arms and legs dark, blending with black or darkbrown hands and feet	**P.e.hypoleucos** group
3a.	Arms and legs pale greybrown to whitish, contrasting with black hands and feet (black fingers and toes only in *anchises*)	**P.e.entellus** group

The *P.e.schistacea* group are often stated to be easily differentiated from the others by the marked contrast between the pale head and darker body. However this contrast is seen in all groups to a greater or lesser extent and thus is not a particularly reliable character.

Key to the *P.e.schistacea* group of subspecies

1.	Colour of back pale greybrown with pinkish tinge; short coat	**P.e.hector**
1a.	Colour of back earthy greybrown; long shaggy coat	2
2.	Forearms and hands dark brown or black, contrasting with body; tailtip white	**P.e.ajax**
2a.	Forearms and hands same as back or only slightly darker; tail as back but tip occasionally pale	**P.e.schistacea**

The type and only specimen of *P.e.lania* lacks its tail; it cannot be distinguished from *P.e.schistacea* from Sikkim except by its peculiar woolly coat.

The *P.e.entellus* group includes *P.e.anchises* which, with reduced amounts of black on hands and feet, is transitional between *P.e.entellus* and the completely pale-footed *P.e.priam*. Individual specimens from the Madhya Pradesh (*c.* 22°N 78°E) show characters intermediate between *entellus* and *anchises* (Eastern Ghats) and *entellus* and *achates* (Western Ghats). They have been catalogued as *P.e.entellus*?

The *P.e.hypoleucos* group is the darkest and is found mainly in the wettest region of southern India, the western slopes of the Western Ghats. The races can be distinguished in the following key:

1.	Back dark chocolate brown; head golden, darker on vertex and nape; dark smudge between eye and ear in some specimens; entire outer aspect of limbs dark brown or black	**P.e.aeneas**
1a.	Back dark greybrown; head yellowish, lacking dark patch on vertex; dark smudge between eye and ear; outer aspect of legs greybrown, arms darker	2
2.	Legs greybrown, paler distally, the paler shade encroaching on to feet, toes black; arms greybrown, darker distally, the darker shade encroaching on to hands, fingers black	**P.e.achates**
2a.	Legs greybrown, darker distally, entire foot black; arms darker, blackish distally, entire hand black	**P.e.iulus**

Two subspecies, *P.e.hypoleucos* and *P.e.dussumieri*, are known only from their descriptions and type specimens which are held in Calcutta and Paris respectively. *P.e.hypoleucos* is described as having a dark patch on the vertex like *P.e.aeneas*, but lacking the dark smudge between eye and ear. *P.e.dussumieri*, though much faded, resembles *P.iulus* most closely except that it too lacks the dark smudge (Olson, personal communication, 1979). Neither has a precise type locality: *P.e.hypoleucos* is said to have come from Travancore and *P.e.dussumieri* from the 'Côte de Malabar'.

The *P.e.priam* group is generally distinguished by a crest or tuft on the vertex, though this feature is occasionally absent.

Key to the subspecies of the *P.e.priam* group:

1.	Crown (and nape) entirely yellowish-white, contrasting with pale greybrown back; outer aspect of limbs like back, becoming darker distally, hands and feet dark brown to black	**P.e.elissa**
1a.	Crown bicoloured, brows anterior to whorl yellowish-white, crown pale greybrown like back: outer aspect of limbs like back, becoming paler distally	2
2.	Nape conspicuously paler than crown patch and back; hands and feet whitish	**P.e.priam**
2a.	Nape scarcely paler than crown patch and back; hands like back or paler, feet whitish	**P.e.thersites**

P.e.priamellus (Pocock, 1928) from Shernelly in the Nelliampathy Hills is based on a single specimen without a crest. Hill (1937) placed it in the synonym of *P.e.priam* which it most closely resembles except that the arms become darker distally rather than paler. He suggested that the absence of the tuft might be due to moulting or to the crown hairs having slipped.

Subspecific variation in tail carriage has been recorded by Roonwal (1979).

Presbytis entellus entellus (Dufresne, 1797)

Type locality Bengal.

Presbytis entellus achates (Pocock, 1928)

Type locality Hawsbhavi, S. Dharwar, at 2000ft, India.

Presbytis entellus aeneas (Pocock, 1928)

Type locality Makut, S. Coorg, at 250ft, India.

Presbytis entellus ajax (Pocock, 1928)

Type locality Deolah, Chamba, at 6000ft, Punjab, India.

Presbytis entellus anchises (Blyth, 1844)

Type locality Deccan, India.

Presbytis entellus dussumieri (I. Geoffroy, 1842)

Dussumier's Langur

Type locality 'Côte de Malabar'.

Presbytis entellus elissa (Pocock, 1928)

Type locality Nagarhole, SE. Coorg, at 2600ft, India.

Presbytis entellus hector (Pocock, 1928)

Principal synonym *Semnopithecus petrophilus* Hodgson, 1846:1 (nomen nudum).

Type localities Of *hector*, Sitabani, Ramnagar, at 2000ft, Kumaun, N. India: of *petrophilus*, none cited.

Presbytis entellus hypoleucos (Blyth, 1841)

Type locality Travancore, SW. India.

Presbytis entellus iulus (Pocock, 1928)

Type locality Jog, Gersoppa Falls, at 1300ft, Kanara-Mysore border, India.

Presbytis entellus lania Elliot, 1909

Type locality Chumbi, Chumbi Valley, at 10 000ft, SE. Tibet.

Presbytis entellus priam (Blyth, 1844)

Principal synonyms *Semnopithecus pallipes* Blyth, 1844:312 (nomen nudum); *Presbytis priamus* Blyth, 1847:732; *Semnopithecus albipes*, I. Geoffroy, 1851:14.

Type localites Of *priam*, Coromandel coast, India; of *pallipes*, Southern India; of *priamus*, as for *priam*, but range extended to include Ceylon (Sri Lanka); of *albipes*, Manilla.

Presbytis entellus priamellus (Pocock, 1928)

Type locality Shernelly (Nelliampathy Plateau), Cochin, India.

Presbytis entellus schistacea (Hodgson, 1941)

Nepalese Langur, Himalayan Langur

Principal synonyms *Semnopithecus nipalensis* Hodgson, 1841:1212; *Pithecus entellus achilles* Pocock, 1928:478.

Type localities Of *schistacea*, restricted by Pocock (1928) to the Nepal Terai, but lectotype probably came from the Kachár (see taxonomic notes); of *nipalensis*, Nepal; of *achilles*, Satthar Hill, Ghorka, at 12 000ft, 50 miles NW. of Katmandu, Nepal.

Presbytis entellus thersites Blyth, 1847

Type locality Trincomalee, Sri Lanka. The provenance of the holotype has been discussed by Hill (1937*a*).

Locality	Date of Collection	Sex	Age	Nature of Specimen	Accession no.	References, Remarks, etc.
Presbytis entellus entellus (Dufresne 1797)						
INDIA						
Pachmarhi, Hoshangabad, [Madhya Pradesh], C.P., [22°29′N 78°26′E], 3300ft	11 Mar 1912	M	Ad	Skin, Skull	ZD.1914.11.18.14	
—	Died 24 Aug 1900	F	[Ad]	Skin	ZD.1939.1138	(*Semnopithecus entellus*)
[INDIA]						
Bengal	—	M	Ad	Skull	ZD.1856.5.6.10	
Calcutta, [West Bengal], [22°30′N 88°20′E]	—	I	Inf	Skin, Skull	ZD.1845.8.12.1	
Deesa, [Gujarat], [24°14′N 72°13′E]	—	M	Ad	Skull	ZD.1979.525	Cited Ryley,K.V. 1914: 688
Hazaribagh, Bengal, [Bihar], [24°N 85°23′E]	—	M	Ad	Skull	ZD.1856.5.6.11	(*P.priamus*)
Hoshangabad, [Madhya Pradesh], C.P., [22°44′N 77°45′E]	—	F	Ad	Skin, Skull	ZD.1912.2.7.1	
Lohra, Bengal, Hazaribagh, Bihar, 1000ft	27 Apr 1914	M	Ad	Skin, Skull	ZD.1915.4.3.3	Cited Wroughton,R.C. 1915c: 100
Midnapore, [West Bengal], [22°25′N 87°24′E], 200ft	6 Sep 1914	F	Ad	Skin, Skull	ZD.1915.4.3.4	(*Pithecus entellus*), Cited Wroughton,R.C. 1915c: 100
Pudari, Jeypore Agency, Vizag Dist., [Orissa]	15 Sep 1927	M	[Ad]	Skin	ZD.1928.3.7.1	(*Pithecus entellus*)
Ramagiri, Jeypore Agency, Vizag Dist., [c.18°46′N 82°15′E]	25 Aug 1927	F	[Inf]	Skin	ZD.1928.3.7.2	" " " "
Talala, Kathiawar, Junagadh, [Gujarat], [21°N 70°39′E], 200ft	23 Oct 1912	M	Ad	Skin, Skull	ZD.1914.11.18.18	(*Presbytis entellus anchises*)
" " "	22 Oct 1912	F	Ad	Skin, Skull	ZD.1914.11.18.19	" " " "
Presbytis entellus entellus? (Dufresne 1797)						
INDIA						
Lahi, Seoni-Malwa, Hoshangabad, [Madhya Pradesh], C.P., [c.22°29′N 77°31′E]	—	[M]	[Ad]	Skin	ZD.1912.2.8.4	(*P.entellus*)
" " "	20 Dec 1909	F	Ad	Skin, Skull	ZD.1912.2.8.6	(*Pithecus entellus* aff. *anchises*)
[INDIA]						
Chikalda, Amraoti Dist., Berar, [Maharashtra], [21°29′N 77°21′E], 3660ft	21 May 1911	M	[Ad]	Skin	ZD.1913.7.5.1	Cited Wroughton,R.C. 1912a: 821
Hewra, Nimar, [Madhya Pradesh], C.P., 1000ft	—	F	Ad	Skin, Skull	ZD.1914.11.18.13	(*Pithecus entellus anchises*), Cited Wroughton,R.C. 1912b: 845
Lahi, Seoni-Malwa, Hoshangabad, [Madhya Pradesh], C.P., [c.22°29′N 77°31′E]	22 Dec 1909	F	Juv	Skin, Skull	ZD.1912.2.8.5	(*Semnopithecus entellus anchises?*)
Sembadoh, Berar, [Maharashtra], C.I., [c.21°30′N 77°20′E], 1800ft	30 May 1911	F	[Ad]	Skin	ZD.1913.7.5.2	(*Pithecus entellus* aff. *anchises*), Cited Wroughton,R.C. 1912a: 821
Shendurni, East Kandesh, [Maharashtra], [c.20°39′N 75°36′E]	—	M	Juv	Skull	ZD.1914.11.18.20	Cited Wroughton,R.C. 1912: 397
" " "	—	F	Ad	Skull	ZD.1914.11.18.21	" " " "
" " "	—	F	Ad	Skull	ZD.1914.11.18.22	" " " "
Presbytis entellus achates (Pocock 1928)						
INDIA						
Karwar, Bombay Presidency, North Kanara, [Karnataka], [14°50′N 74°09′E]	1909	M	Ad	Skin, Skull	ZD.1912.2.8.1	Figured Elliot, D.G. 1913: Pl.8
" " "	1909	F	Ad	Skin, Skull	ZD.1912.2.8.2	
" " "	1909	F	Ad	Skin, Skull	ZD.1912.2.8.3	
[INDIA]						
Alnavar, S.Mahratta, Dharwar Dist., [Karnataka], [c.15°25′N 74°45′E]	2 Nov 1911	F	Ad	Skin, Skull	ZD.1913.7.5.3	Cited Wroughton,R.C. 1912c: 1176, Cited Pocock,R.I. 1928: 489-490
Barchi, 10 miles E. of Supa, 4 miles N. of R.Kalinadi, North Kanara, [Karnataka], [c.15°18′N 74°36′E]	30 Jan 1912	M	Inf	Skin, Skull	ZD.1933.7.29.7	Cited Wroughton,R.C. 1913: 32, Cited Pocock,R.I. 1939: 104

Locality	Date of Collection	Sex	Age	Nature of Specimen	Accession no.	References, Remarks, etc.
Presbytis entellus achates *continued*						
Devikop, 26 miles S. of Dharwar, edge of Kanara Forest, South Mahratta, [Karnataka], [c.15°08′N 74°56′E], 2000ft	26 Nov 1911	M	Ad	Skin, Skull	ZD.1913.7.5.4	Cited Wroughton,R.C. 1912c: 1176, Cited Pocock,R.I. 1928: 489
„ „ „	22 Nov 1911	M	Ad	Skin, Skull	ZD.1914.11.18.4	„ „ „ „
„ „ „	26 Nov 1911	F	Ad	Skin, Skull	ZD.1914.11.18.5	„ „ „ „
Hawsbhavi, S.Dharwar, South Mahratta, [Karnataka], 2000ft	23 Feb 1912	M	Ad	Skin, Skull	ZD.1914.11.18.1	Holotype *Pithecus entellus achates* Pocock 1928, Cited Wroughton,R.C. 1912c: 1176, Figured Pocock,R.I. 1928: 488, Fig.7B, Cited Pocock,R.I. 1939: 104
Helwåk, Satara Dist., [Maharashtra], [c.17°23′N 73°43′E]	8 Dec 1914	M	[Ad]	Skin	ZD.1951.610	(*Semnopithecus entellus anchises*), Cited Wroughton,R.C 1916a: 311
Potoli, S.E. of Supa, 6 miles S. of R.Kalinadi, North Kanara, [Karnataka], [c.15°10′N 74°33′E], 1800ft	18 Jan 1912	M	Ad	Skin, Skull	ZD.1914.11.18.2	Cited Wroughton,R.C. 1913: 32, Cited Pocock,R.I. 1928: 490
„ „ „	18 Jan 1912	F	Ad	Skin, Skull	ZD.1914.11.18.3	Cited Wroughton,R.C. 1913: 32, Cited Pocock,R.I. 1939: 119
„ „ „	18 Jan 1912	M	Inf	Skin, Skull	ZD.1933.7.29.8	Cited Wroughton,R.C. 1913: 32, Cited Pocock,R.I. 1939: 104
Potoli, S.E. of Supa, 6 miles S. of R.Kalinadi, North Kanara, [Karnataka], [c.15°10′N 74°33′E]	18 Jan 1912	F	Inf	Skin, Skull	ZD.1933.7.29.9	„ „ „ „
Samasgi, Kanara boundary, S.W. Dharwar, South Mahratta, [Karnataka], 2000ft	12 Mar 1912	M	Juv	Skin, Skull	ZD.1914.11.18.10	Cited Wroughton,R.C. 1912c: 1176, Cited Pocock,R.I. 1928: 489, Cited Pocock,R.I. 1939: 104
Samasgi, Kanara boundary, S. Dharwar, South Mahratta, [Karnataka]	19 Mar 1912	M	Ad	Skin, Skull	ZD.1914.11.18.11	„ „ „ „
„ „ „	12 Mar 1912	F	Ad	Skin, Skull	ZD.1914.11.18.12	Cited Wroughton,R.C. 1912c: 1176, Cited Pocock,R.I. 1928: 489, Cited Pocock,R.I. 1939: 104, 119
Samasgi, S.W. Dharwar, South Mahratta, [Karnataka], 2000ft	22 Mar 1912	I	Inf	Skin, Skull	ZD.1933.7.29.6	Cited Wroughton,R.C. 1912c: 1176, Cited Pocock,R.I. 1939: 104
Vijayanagar [=Hampi], [R. bank of R.Tungabhadra], Bellary, [Karnataka], [15°19′N 76°28′E], 1500ft	16 Aug 1912	M	Ad	Skin, Skull	ZD.1914.11.8.8	(*Presbytis entellus anchises*), Cited Wroughton,R.C. and Ryley,K.V. 1913a: 60, Cited Pocock,R.I. 1928: 489, Cited Pocock,R.I. 1939: 104
„ „ „	4 Aug 1912	F	Ad	Skin, Skull	ZD.1914.11.8.9	„ „ „ „
Presbytis entellus achates? (Pocock 1928)						
INDIA						
S.India, Kanara	—	F	Ad	Skin, Skull, Cast of lower Is and L molars	ZD.1929.9.21.1	(*Semnopithecus entellus*)
NO DATA						
—	—	[F]	Inf	Skin, Skull	ZD.1939.1148	(*Pithecus* sp.)
Presbytis entellus aeneas (Pocock 1928)						
[INDIA]						
Makut, [8 miles S. of Wotekolli], below Ghats, South Coorg, [Karnataka], 250ft	12 Jan 1913	M	Ad	Skin, Skull	ZD.1914.11.18.23	Holotype *Pithecus entellus aeneas* Pocock 1928, Cited Ryley,K.V. 1913b: 493, Cited Pocock,R.I. 1928: 492, Cited Pocock,R.I. 1939: 106
Wotekolli, South Coorg, [Karnataka], 2000ft	1 Jan 1913	F	Ad	Skin, Skull	ZD.1914.11.18.24	Paratype *Pithecus entellus aeneas* Pocock 1928, (*Pithecus hypoleucus*), Cited Ryley,K.V. 1913b: 493, Cited Pocock,R.I. 1928: 492
„ „ „	2 Jan 1913	M	Juv	Skin, Skull	ZD.1914.11.18.25	(*Pithecus hypoleucos*), Cited Ryley,K.V. 1913b: 493
„ „ „	27 Dec 1912	M	[Juv]	Skin	ZD.1979.529	Cited Ryley,K.V. 1913b: 493, Background Pocock,R.I. 1928: 492
„ „ „	1 Jan 1913	M	[Juv]	Skin	ZD.1979.530	„ „ „ „
Presbytis entellus ajax (Pocock 1928)						
INDIA						
Bara, Tissa, Chamba, [Himachal Pradesh], [c.32°52′N 76°09′E], 7500ft	10 Dec 1922	M	Ad	Skin, Skull	ZD.1923.9.1.2	Paratype *Pithecus entellus ajax* Pocock 1928, Cited Lindsay,H.M. 1926a: 599, Cited Pocock,R.I. 1928: 481, Cited Pocock,R.I. 1939: 97
Chandra Nullah, Chamba, [Himachal Pradesh]	—	M	Juv	Skull	ZD.1939.851	
„ „ „	—	I	Juv	Skull, Atlas	ZD.1939.852	
Chatri, Chamba, [Himachal Pradesh], 6000ft	2 Dec 1922	M	Ad	Skin, Skull	ZD.1923.9.1.1	Paratype *Pithecus entellus ajax* Pocock 1928, Cited Lindsay,H.M. 1926a: 599
„ „ „	1 Dec 1922	F	Juv	Skin, Skull	ZD.1928.7.11.4	„ „ „ „
Chichian, Kangra Valley, [Kangra], [Punjab], 9000ft	20 Feb 1922	F	Ad	Skin, Skull	ZD.1933.12.1.1	Cited Lindsay,H.M. 1926a: 599, Cited Pocock,R.I. 1928: 481, Cited Pocock,R.I. 1939: 97, Figured Colyer,J.F. 1936: 35, Fig.44
Deolah, Chamba Valley, Chamba, [Himachal Pradesh], 6000ft	7 Jan 1923	M	Ad	Skin, Skull	ZD.1928.7.11.5	Holotype *Pithecus entellus ajax* Pocock 1928, Cited Lindsay,H.M. 1926a: 599, Figured Pocock,R.I. 1928: 480, Pl.11, Figured Pocock,R.I. 1939: 97, Pl.6

Locality	Date of Collection	Sex	Age	Nature of Specimen	Accession no.	References, Remarks, etc.
Presbytis entellus ajax *continued*						
Kangra, Punjab, 2400ft	1 Apr 1922	F	Ad	Skin, Skull	ZD.1923.9.1.119	Cited Lindsay,H.M. 1926a: 599, Cited Pocock,R.I. 1928. 481
,, ,, ,,	2 Apr 1922	M	Inf	Skin, Skull	ZD.1928.7.11.2	Cited Lindsay,H.M. 1926a: 599
Kangra, [Punjab], 2500ft	29 Mar 1921	F	Ad	Skull	ZD.1979.976	
,, ,, ,,	20 Mar 1921	F	Inf	Skull	ZD.1979.977	
Kangra Fort, [Kangra], Punjab, [c.32°06′N 76°16′E], 2450ft	31 Mar 1922	M	Ad	Skin, Skull	ZD.1928.7.11.1	Cited Lindsay,H.M. 1926a: 599, Cited Pocock,R.I. 1928: 481, Cited Pocock,R.I. 1939: 97
Rahla, Kulu Valley, Kangra, [Punjab], [c.32°22′N 77°12′E], 9800ft	4 Jun 1922	F	Ad	Skin, Skull	ZD.1928.7.11.3	Cited Lindsay,H.M. 1926a: 599, Cited Pocock,R.I. 1928: 481, Cited Pocock,R.I. 1939: 98
Presbytis entellus ajax? (Pocock 1928)						
INDIA						
[Simla], [Himachal Pradesh], [31°07′N 77°09′E]	—		Ad	4 skulls	RCS(OM) S70.16	(*Presbytis entellus ajax*)
[INDIA]						
Simla or Mussoorie	—	M	Ad	Skull	ZD.1879.11.21.215	(*Presbytis entellus*), Cited Pocock,R.I. 1928: 483, Cited Pocock,R.I. 1939: 98
,, ,, ,,	—	M	Ad	Skull	ZD.1879.11.21.216	,, ,, ,, ,,
KASHMIR						
By Butoch La, Kishtwar, [c.33°19′N 75°46′E], 6000ft	1 Nov 1898	M	[Ad]	Skin, Skull in skin	PCM Kashmir	(*Semnopithecus schistaceus*)
,, ,, ,,	1 Nov 1898	M	[Ad]	Skin, Skull in skin	PCM Kashmir	,, ,, ,, ,,
Gugai Nala, Kishen Ganga R.,Gurais, [c.34°37′N 74°53′E], 10,000ft	23 May 1906	I	[Ad]	Skin	ZD.1907.2.15.1	(*Presbytis schistaceus*)
Gugga Nullah, 8-12,000ft	May 1909	M	Ad	Skull	ZD.1910.1.18.65	,, ,, ,, ,,
Presbytis entellus anchises (Blyth 1844)						
[INDIA]						
Diguvametta, [Nallamalai Range, near Nandikanama Pass], Kurnool Dist., [Andhra Pradesh], [c.15°24′N 78°49′E], [1036ft]	23 Apr 1930	F	Ad	Skin, Skull	ZD.1933.7.29.2	Listed Baptista,N.A. (unpubl.)
,, ,, ,,	23 Apr 1930	F	Ad	Skin, Skull	ZD.1933.7.29.3	,, ,, ,, ,,
,, ,, ,,	28 Apr 1930	F	Ad	Skin, Skull	ZD.1933.7.29.4	,, ,, ,, ,,
Diguvametta, [Nallamalai Range], E.Ghats, Kurnool Dist., [Andhra Pradesh], [c.15°24′N 78°49′E], [1036ft]	17 Apr 1930	F	Juv	Skin, Skull	ZD.1937.5.26.5	,, ,, ,, ,,
Nallamalai Range, South Kurnool, [Andhra Pradesh], 2000ft	3 Nov 1929	F	Ad	Skin, Skull	ZD.1933.7.29.5	,, ,, ,, ,,
[Nallamalai Range], [South Kurnool], [Andhra Pradesh]	—	[M]	Ad	Skull	ZD.1979.539	
Presbytis entellus dussumieri (I.Geoffroy 1842)						
NO SPECIMENS						
—	—	—	—	—		
Presbytis entellus elissa (Pocock 1928)						
INDIA						
Honnametti Estate, Attikan, S.Mysore	—	[F]	Ad	Skull	RCS(OM) A70.19	
[INDIA]						
Nagarhole, S.E.Coorg, [Karnataka], 2600ft	7 Feb 1913	M	Ad	Skin, Skull	ZD.1914.11.18.15	Paratype *Pithecus entellus elissa* Pocock 1928, (*Presbytis entellus anchises*), Cited Ryley,K.V. 1913b: 494, Cited Pocock,R.I. 1928: 493
,, ,, ,,	16 Feb 1913	F	Ad	Skin, Skull	ZD.1914.11.18.16	Holotype *Pithecus entellus elissa* Pocock 1928, Cited Ryley,K.V. 1913b: 494, Cited Pocock,R.I. 1928: 493, Cited Pocock,R.I. 1939: 113
,, ,, ,,	16 Feb 1913	F	Ad	Skin, Skull	ZD.1914.11.18.17	Paratype *Pithecus entellus elissa* Pocock 1928, Cited Ryley,K.V. 1913b: 494, Cited Pocock,R.I. 1928: 493
,, ,, ,,	6 Feb 1913	M	Juv	Skin, Skull	ZD.1937.5.26.4	,, ,, ,, ,,
Presbytis entellus hector (Pocock 1928)						
INDIA						
Champawat, Almora, Kumaun, [Uttar Pradesh], [29°20′N 80°06′E], 6000ft	21 Feb 1914	M	Ad	Skin, Skull	ZD.1914.7.10.12	(*Presbytis schistaceus*), Cited Wroughton,R.C. 1914: 286, Cited Pocock,R.I. 1928: 482
Dela, [c.8 miles W. of] Ramnagar, Kumaun, [Uttar Pradesh], [29°26′N 79°E], 1500ft	7 Jan 1914	F	Ad	Skin, Skull	ZD.1914.7.10.11	(*Presbytis schistaceus*), Cited Wroughton,R.C. 1914: 286, Cited Pocock,R.I. 1928: 482
Ramnagar, [30 miles W. of Naini Tal], Kumaun, [Uttar Pradesh], [29°25′N 79°08′E], [1100ft]	—	M	Ad	Skull	ZD.1914.7.10.14	(*Presbytis schistaceus*), Cited Wroughton,R.C. 1914: 286

Locality	Date of Collection	Sex	Age	Nature of Specimen	Accession no.	References, Remarks, etc.
Presbytis entellus hector *continued*						
Ratighat, [On road between Almora and] Naini Tal, Kumaun, [Uttar Pradesh], 3700ft	6 Nov 1913	F	Ad	Skin, Skull	ZD.1914.7.10.13	(*Presbytis schistaceus*), Cited Wroughton,R.C. 1914: 286, Cited Pocock,R.I. 1928: 482, Figured James,W.Warwick 1960: 226-227
Sitabani, [c.20 miles W. or S.W. of Naini Tal], Kumaun, [Uttar Pradesh], 2000ft	21 Nov 1913	M	Ad	Skin, Skull	ZD.1914.7.10.10	Holotype *Pithecus entellus hector* Pocock 1928, Cited Pocock,R.I. 1928: 481, Cited Wroughton,R.C. 1914: 286, Figured Pocock,R.I. 1939: Pl.6
NEPAL						
Hazaria, [in terai to N. of Bairagria, on Bagmati R.], Patherghatta c.600ft	15 Feb 1921	F	Ad	Skin, Skull	ZD.1922.5.16.1	(*Pithecus schistaceus*), Cited Hinton,M.A.C. and Fry,T.B. 1923: 404, Figured Pocock,R.I. 1928: 477-478, Fig.4B, Figured Pocock,R.I. 1939: 93, Fig.23
—	—	I	Juv	Skin, Skull	ZD.1845.1.8.3	(*S.schistaceus*), Background Hodgson,B.H. 1841, Background Hodgson,B.H. 1846
—	—	I	[Juv]	Skin	ZD.1845.1.8.221	(*Pithecus entellus achilles*), Background Hodgson,B.H. 1846
[NEPAL]						
Himalaya	—	F	Ad	Skin, Skull in skin	ZD.1843.1.12.2	Paralectotype *Semnopithecus schistaceus* Hodgson 1841, Background Hodgson,B.H. 1841, Background Hodgson,B.H. 1846, Cited Wroughton,R.C. 1918: 559, Background Thomas,O. 1918
NO DATA						
—	—	[F]	[Ad]	Skin	ZD.1972.1346	
Presbytis entellus hector? (Pocock 1928)						
[INDIA]						
Bhinga, Bahraich, [Uttar Pradesh], [c.27°42′N 81°56′E], 1000ft	20 Mar 1893	M	Ad	Skull	ZD.1926.10.8.6	Cited Pocock,R.I. 1928: 483, Cited Pocock,R.I. 1939: 95
Garhwal, [Uttar Pradesh]	—	M	Ad	Skull	ZD.1926.10.8.3	(*Pithecus entellus*), Cited Pocock,R.I. 1928: 483, Cited Pocock,R.I. 1939: 95
,, ,, ,,	—	M	Ad	Skull	ZD.1926.10.8.4	(*Pithecus entellus*), Cited Pocock,R.I. 1928: 483,Cited Pocock,R.I. 1939: 95, Figured James,W.Warwick 1960: 224-225
,, ,, ,,	—	M	Ad	Skull	ZD.1926.10.8.5	(*Pithecus entellus*), Cited Pocock,R.I. 1928: 483, Cited Pocock,R.I. 1939: 95
Presbytis entellus hypoleucos (Blyth 1841)						
NO SPECIMENS						
—	—	—	—	—	—	
Presbytis entellus iulus (Pocock 1928)						
[INDIA]						
Jog [Falls], Gersoppa Falls, Kanara-Mysore boundary, [Karnataka], [14°13′N 74°45′E], 1300ft	4 May 1912	M	Ad	Skin, Skull	ZD.1914.11.18.6	Paratype *Pithecus entellus iulus* Pocock 1928, Cited Wroughton,R.C. 1913: 32, Figured Pocock,R.I. 1928: 490, Fig.8A, Figured Pocock,R.I. 1939: 105
Jog [Falls], Gersoppa Falls, Kanara-Mysore boundary, North Kanara, [Karnataka], [14°13′N 74°45′E], 1300ft	4 May 1912	F	Ad	Skin, Skull	ZD.1914.11.18.7	Holotype *Pithecus entellus iulus* Pocock 1928, Cited Wroughton,R.C. 1913: 32, Cited Pocock,R.I. 1928: 490, Cited Pocock,R.I. 1939: 105
Jog [Falls], [Gersoppa Falls, Kanara-Mysore boundary], North Kanara, [Karnataka], [14°13′N 74°45′E], [1300ft]	1912	F	Ad	Skin, Skull	ZD.1937.5.26.6	Cited Wroughton,R.C. 1913: 32, Cited Pocock,R.I. 1928: 490
Presbytis entellus lania Elliot 1909						
[CHINA]						
Chumbi, [Chumbi Valley], Tibet, [27°28′N 89°E], 10,000ft	9 Apr 1909	F	Ad	Skin, Skull	ZD.1909.7.16.1	Holotype *Presbytis lania* Elliot 1909, Cited Elliot,D.G. 1909: 273, Cited Pocock,R.I. 1928: 480, Cited Colyer,J.F. 1936: 35
Presbytis entellus priam (Blyth 1844)						
[INDIA]						
Dasarladoddi, [10 miles W. of Chintarajupalli], Palkonda Hills [E.Ghats], [Cuddapah], [Andhra Pradesh]	[29 Jun 1930]	F	Ad	Skull	ZD.1933.7.29.13	(*Pithecus entellus*), Listed Baptista,N.A. (unpubl.)
Dasarladoddi, [10 miles W. of Chintarajupalli], Palkonda Hills, E.Ghats, [Cuddapah], [Andhra Pradesh], 1600ft	29 Jun 1930	F	[Ad]	Skin	ZD.1930.11.1.10	(*Pithecus entellus pallipes*), Listed Baptista,N.A. (unpubl.)

Locality	Date of Collection	Sex	Age	Nature of Specimen	Accession no.	References, Remarks, etc.
Presbytis entellus priam *continued*						
Hogainakal, Dharmapuri Range, North Salem, [Tamil Nadu], [c.12°07′N 77°46′E], 850ft	26 Aug 1930	M	Ad	Skin, Skull	ZD.1932.4.14.3	„ „ „ „
Muthur, [Dharmapuri Range], [North] Salem, [Tamil Nadu], [950ft]	2 Oct 1930	M	Juv	Skin, Skull	ZD.1932.4.14.1	„ „ „ „
Muthur, Dharmapuri Range, North Salem, [Tamil Nadu], 950ft	30 Sep 1930	F	Ad	Skin, Skull	ZD.1932.4.14.2	„ „ „ „
Nilghiris, [Nilgiri Hills], [Tamil Nadu]	—	M	Ad	Skull	ZD.1888.2.5.15	
„ „ „	—	F	Ad	Skull	ZD.1888.2.5.16	
Tirthamalai, E.Ghats, Salem Dist., [Tamil Nadu], [c.12°06′N 78°36′E], 1000ft	10 Jul 1929	M	Ad	Skull	ZD.1930.5.24.7	(*Pithecus pallipes*), Listed Baptista,N.A. (unpubl.)
„ „ „	10 Jul 1929	M	Ad	Skin	ZD.1933.7.29.1	(*Pithecus entellus pallipes*), Listed Baptista,N.A. (unpubl.)
Presbytis entellus priamellus (Pocock 1928)						
INDIA						
Shernelly, [Nelliampathy Plateau], Cochin, [Kerala]	—	M	Ad	Skin, Skull	ZD.1921.11.5.7	Holotype *Pithecus entellus priamellus* Pocock 1928, Cited Lindsay,H.M. 1926: 592, Figured Pocock,R.I. 1928: 494, Figs.8B and 9B, Figured Pocock,R.I. 1939: 112, Fig.33
Presbytis entellus schistacea (Hodgson 1841)						
INDIA						
Lachen, Sikkim, [27°46′N 88°36′E], 8800ft	13 Dec 1914	F	Ad	Skin, Skull	ZD.1928.7.11.7	(*Pithecus schistaceus achilles*), Cited Wroughton,R.C. 1916b: 476, Cited Pocock,R.I. 1928: 480, Cited Pocock,R.I. 1939: 96
Sikkim	—	M	Ad	Skin, Skull	ZD.1891.10.7.6	(*Pithecus entellus achilles*), Cited Pocock,R.I. 1928: 479
Tsunthang, Sikkim, [27°38′N 88°35′E], 5350ft	18 Dec 1914	F	Ad	Skin, Skull	ZD.1915.9.1.6	(*Pithecus entellus achilles*), Cited Wroughton,R.C. 1916b: 476, Cited Pocock,R.I. 1928: 480, Cited Pocock,R.I. 1939: 96
„ „ „	19 Dec 1914	M	Juv	Skin, Skull	ZD.1928.7.11.6	(*Pithecus entellus achilles*), Cited Wroughton,R.C. 1916b: 476, Cited Pocock,R.I. 1928: 480, Cited Pocock,R.I. 1939: 96
NEPAL						
Satthar Hill, Gorkha, Central Nepal, [c.28°01′N 84°37′E], 12,000ft	[18 Jan 1921]	M	Ad	Skin, Skull	ZD.1926.12.3.1	Holotype *Pithecus entellus achilles* Pocock 1928, Figured Pocock,R.I. 1928: 478, Pl.11, Figured Pocock,R.I. 1939: 96, Pl.6
—	—	F	Ad	Skin, Skull	ZD.1842.5.26.3	Background Hodgson,B.H. 1841, Background Hodgson,B.H. 1846
—	—	M	Ad	Skin	ZD.1843.1.12.1	Lectotype *Semnopithecus schistaceus* Hodgson 1841, Cited Hodgson,B.H. 1841: 1212, Cited Thomas,O. 1918: 370, Cited Wroughton,R.C. 1918: 559
—	—	M	Ad	Skull	ZD.1845.10.8.4	„ „ „ „
—	—	M	Ad	Skin, Skull	ZD.1853.8.16.1	(*S.schistaceus*), Background Hodgson,B.H. 1841, Background Hodgson,B.H. 1846
—	—	F	Juv	Skin, Skull	ZD.1879.11.21.593	Background Hodgson,B.H. 1841, Background Hodgson,B.H. 1846
Presbytis entellus schistacea? (Hodgson 1841)						
INDIA						
—	—	M	Ad	Skull	ZD.1895.3.21.1	Listed Anon (unpubl.)a
NEPAL						
—	—	M	Ad	Skull	ZD.1845.1.8.2	Background Hodgson,B.H. 1841
[NEPAL]						
—	—	F	Ad	Skull	ZD.1845.1.8.1	„ „ „ „
NO DATA						
—	—	M	Ad	Skin, Skull in skin	ZD.1972.1345	
—	—	I	[Juv]	R limb bones	ZD.1979.978	
Presbytis entellus thersites Blyth 1847						
[INDIA]						
50 miles S. of Trivandrum, Travancore, [Tamil Nadu]	—	M	Ad	Skin, Skull, Skeleton	ZD.1910.10.13.1	(*Presbytis priamus*)
[SRI LANKA]						
Arugam Bay, E.P., Ceylon, [c.6°51′N 81°50′E]	22 Aug 1913	F	Ad	Skin, Skull	ZD.1930.11.1.9	Cited Wroughton,R.C. 1915b: 84, Cited Pocock,R.I. 1939: 115, 118
Cheddikulam, N.P., Ceylon, [8°40′N 80°18′E]	16 Nov 1913	M	Ad	Skin, Skull	ZD.1915.3.1.12	Cited Wroughton,R.C. 1915b: 84, Cited Pocock,R.I. 1939: 115

Locality	Date of Collection	Sex	Age	Nature of Specimen	Accession no.	References, Remarks, etc.
Presbytis entellus thersites *continued*						
,, ,, ,,	3 Dec 1913	M	Ad	Skin, Skull	ZD.1915.3.1.13	,, ,, ,, ,,
About 10 miles S. of Elahera, C.P., Ceylon, [c.7°44′N 80°47′E], 1000ft?	Dec 1930	M	[Ad]	Skin	ZD.1931.9.11.1	
Hakgalla[?], Uva[?], Ceylon, 5600ft[?]	—	M	Ad	Skull	ZD.1979.980	(*Pithecus ursinus*), Background Wroughton,R.C. 1915b: 79-84
Hambantota Dist., [S.P.], Ceylon	—	M	Ad	Skin, Skull	ZD.1879.9.5.4	(*Semnopithecus priamus*)
,, ,, ,,	—	M	Ad	Skin, Skull	ZD.1879.9.5.5	,, ,, ,, ,,
Kala Oya, about midway between Puttalam and Anuradhapura, Boundary between N.W. and N.C.P., [c.8°12′N 80°06′E], [below 150m]	—		Ad	5 skulls	RCS(OM) S70.2	
Kala Oya, [about midway between Puttalam and Anuradhapura], [Boundary between N.W. and N.C.P.], Ceylon, [c.8°12′N 80°06′E]	Apr 1935	F	Ad	Skin, Skull	ZD.1935.3.30.3	
,, ,, ,,	Apr 1935	F	Ad	Skull	ZD.1979.694	(*Pithecus senex philbricki*)
Kala Oya, [Boundary between N.W. and N.C.P.], Ceylon, [c.8°12′N 80°06′E]	—	F	Ad	Skin, Skull	ZD.1930.11.1.8	Cited Wroughton,R.C. 1915b: 84
Kala Oya, Ceylon, [c.8°12′N 80°06′E], [below 150m]	—		Ad	2 skulls	RCS(OM) S70.32	(*Presbytis senex senex*), Cited Colyer,J.F. (unpubl.): 30r
[Kala Oya], [c.8°12′N 80°06′E]	—	F		Wet, legs and tail	ZD.1972.932	
,, ,, ,,	—	F		Wet, legs and tail	ZD.1972.933	
,, ,, ,,	—	F		Wet, legs	ZD.1972.934	
Kumbukkan, [Uva], [c.6°48′N 81°18′E], [below 150m]	—			11 Skulls	RCS(OM) S70.2	
Mankeni, E.P., Ceylon, [8°01′N 81°29′E]	8 Oct 1913	M	Ad	Skin, Skull	ZD.1915.3.1.8	Cited Wroughton,R.C. 1915b: 84, Background Pocock,R.I. 1939: 115-116
,, ,, ,,	9 Sep 1913	F	Ad	Skin, Skull	ZD.1915.3.1.9	(*Presbytis priamus*), Cited Wroughton,R.C. 1915b: 84, Background Pocock,R.I. 1939: 115-116
,, ,, ,,	11 Sep 1913	F	Ad	Skin, Skull	ZD.1915.3.1.10	,, ,, ,, ,,
,, ,, ,,	28 Sep 1913	F	Ad	Skin, Skull	ZD.1915.3.1.11	Cited Wroughton,R.C. 1915b: 84, Background Pocock,R.I. 1939: 115-116
,, ,, ,,	1 Sep 1913	M	Ad	Skin, Skull	ZD.1930.11.1.1	,, ,, ,, ,,
,, ,, ,,	6 Sep [1913]	M	Ad	Skin, Skull	ZD.1930.11.1.2	,, ,, ,, ,,
,, ,, ,,	6 Oct 1913	M	Ad	Skin, Skull, Cast of lower L M3	ZD.1930.11.1.3	,, ,, ,, ,,
,, ,, ,,	8 Oct 1913	M	Ad	Skin, Skull	ZD.1930.11.1.4	,, ,, ,, ,,
,, ,, ,,	1 Sep 1913	F	Ad	Skin, Skull	ZD.1930.11.1.5	,, ,, ,, ,,
Monaragala, [Uva], [6°52′N 81°22′E], [below 150m]	—			35 Skulls	RCS(OM) S70.2	
Tammannewa, N.C.P., Ceylon, [8°23′N 80°35′E]	3 May 1914	M	Ad	Skin, Skull	ZD.1930.11.1.6	Cited Wroughton,R.C. 1915b: 84
Telulla, [10 miles S. of Wellawaya], Uva, Ceylon, [6°36′N 81°07′E]	[3 Feb 1914]	F	Ad	Skull	ZD.1920.2.8.1	
Telulla, [10 miles S. of Wellawaya], Uva, Ceylon, [6°36′N 81°07′E], 300ft	10 Apr 1928	F	Ad	Skin, Skull	ZD.1928.7.18.1	
Telulla, [10 miles S. of Wellawaya], Uva, Ceylon, [6°36′N 81°07′E]	10 Feb 1967	F	[Ad]	Skin	ZD.1971.775	
Welligatta, [c.9 miles N.E. of Hambantota], [S.P.], Ceylon, [6°12′N 81°12′E]	24 Jun 1913	M	Ad	Skin, Skull	ZD.1930.11.1.7	Cited Ryley,K.V. 1914a: 704
Ceylon	—			Skull	CMZ E7257A	(*Semnopithecus priamus*)
Trincomalee, N.P., Ceylon, [8°34′N 81°13′E], [below 150m],	—		Ad	5 Skulls	RCS(OM) S70.32	(*Presbytis senex senex*), Cited Colyer,J.F. (unpubl.): 22r
—	—	M	Ad	Skull	RCS(OM) A70.18	(*Presbytis entellus priam*)
Presbytis entellus thersites? Blyth 1847						
[INDIA]						
Aramboli, Foot of Mahendragiri Range, Travancore, [Tamil Nadu], [c.8°15′N 77°33′E]	7 Oct 1915	M	Ad	Skin, Skull	ZD.1923.1.20.1	(*Pithecus priam*), Cited Wroughton,R.C. 1921: 550, Cited Pocock,R.I. 1928: 496
NO DATA						
—	Died 1 Apr 1942	M	[Ad]	Skin	ZD.1979.996	(*Semnopithecus priam*)
Presbytis entellus (Dufresne 1797)						
INDIA						
Bombay	—	M	Juv	Skin, Skull	ZD.1837.6.10.29	Listed Anon. (unpubl.)b: 16
Gwalior, [Madhya Pradesh]	—	M	[Ad]	Skin, Skull in skin	GCMAG 76-1b	
—	—	M	Ad	Skull	ZD.13q	(*P.entellus entellus*), Listed Anon. (unpubl.)b: 16
—	—	F	Juv	Skin, Skull	ZD.1837.6.10.28	
—	—	F	Juv	Skull	ZD.1856.5.6.15	
—	—	[F]	[Ad]	Skin	ZD.1939.3466	(*Semnopithecus entellus*)
[INDIA]						
Bombay Presidency	—	M	Ad	Skull	ZD.1939.252	
Meerut, [Uttar Pradesh], [29°N 77°46′E]	1875			Skull	CMZ E7254C	(*Semmnopithecus entellus*)
—	—	M	Ad	Skull	ZD.1858.5.4.190	Listed Gerrard,E. 1862: 8

Locality	Date of Collection	Sex	Age	Nature of Specimen	Accession no.	References, Remarks, etc.
Presbytis entellus *continued*						
NEPAL						
—	—	I	[Ad]	Skeleton	ZD.1845.1.12.453	Background Hodgson,B.H. 1841
NO DATA						
—	—	I	Juv	Skull	ZD.1848.8.14.21	(*Colobus temmincki*)
—	—	M	Ad	Skull	ZD.1848.9.23.1	
—	—	[F]	Juv	Skull, Skeleton	ZD.1859.2.8.1	(*Cercopithecus entellus*)
—	—	M	Ad	Skull, Skeleton	ZD.1865.10.9.11	
—	—	M	Ad	Skull	ZD.1931.1.2.1	(*Colobus p.kikuyuensis*)
—	—	M	Ad	Skull	ZD.1931.1.2.2	,, ,, ,, ,,
—	—	M	[Ad]	Wet,cadaver	ZD.1931.3.11.1	
—	—	F	Juv	Skin, Skull	ZD.1939.1151	(*Presbytis entellus x Presbytis pileatus* hybrid)
—	—	I	Ad	R talus	BM(NH)A M10869	Figured Cautley and Falconer 1840: 501, Figured Cautley and Falconer 1868: 293, Figs.6-9, Cited Lydekker,R. 1885: 3, Cited Lydekker,R. 1886: 5
—	—	I	Juv	Skull	ZD.1977.592	
—	—	F	Ad	Skull	ZD.1979.979	
—	—			Skull, Skeleton	CMZ E7254A	(*Semnopithecus entellus*)
—	—	M	Ad	Skull, Skeleton	DLPA 71.0.1	
—	—			Skull, Skeleton	EDA	(*Semnopithecus butollus*)
—	—			Skull, Skeleton	EDZ XH6-25.1	
—	—			6 skulls	RCS(OM) S70.1	(*Presbytis entellus entellus*)
Presbytis sp.						
BORNEO						
[Cave deposit], [Sarawak]	—	I	[Ad]	Upper R P2, Lower R P1, Upper R M, Lower L M	BM(NH)A M20557c	(Subfossil)(*?Macaca nemestrina*), Background Everett,A.H. 1879: 149-155
,, ,, ,,	—	I	[Ad]	R maxillary fragment with M1-3	BM(NH)A M20557d	,, ,, ,, ,,
,, ,, ,,	—	I	[Juv]	Upper L M1-2, P2 erupting, Upper R M	BM(NH)A M20557f	(Subfossil)(*?Macaca nemestrina*)
,, ,, ,,	—		[Ad]	Molars (5)	BM(NH)A M20557h	(Subfossil)(*?Macaca nemestrina*), Background Everett,A.H. 1879: 149-155
Sarawak	—			Fragment of L mandibular corpus with P4-M2	BM(NH)A M1971	(Subfossil)(*Semnopithecus* sp.)
INDIA						
Billa Surgam Caves, [3 miles E.S.E. Betamcherla, near Banganapalle], Kurnool, [Andhra Pradesh]	—	M	Ad	Lower L C, P3-4, M1, M2, Upper M3	BM(NH)A M2963a	(Pleistocene)(*Semnopithecus priamus*), Cited Lydekker,R. 1886a: 28
,, ,, ,,	—	M	Ad	Lower L C	BM(NH)A M2963b	,, ,, ,, ,,
,, ,, ,,	—	F	Ad	Lower L C	BM(NH)A M2963c	,, ,, ,, ,,
,, ,, ,,	—	I	Juv	Lower dP	BM(NH)A M2963d	,, ,, ,, ,,
NO DATA						
—	—	I	Juv	Skull	ZD.1980.863	
—	—	F	Ad	Skull	ZG.B522b	
—	—	F	Ad	Skull	ZG.B522c	
—	—	F	Ad	Skull, Cervical vertebrae (6)	ZG.B512b	
—	—			2 crania	RCS(OM) S70.8(A)	
?Presbytis						
INDIA						
Siwalik Hills	—	I	Ad	Fragment of R mandible with P4-M3, Radiograph	BM(NH)A 15710	Lectotype *Semnopithecus palaeindicus* Lydekker 1884, (Lower Pliocene), Figured Falconer,H. and Cautley,P.T. 1837: 357, Pl.23,fig.5, Figured Falconer,H. and Cautley,P.T. 1868: 302, pl.24,figs.5-6, Cited Lydekker,R. 1884: 123, Cited Lydekker,R. 1885: 2, Correspondence Delson,E. (unpubl.)a
Siwalik Hills, [Probably Tatrot zone]	—	I	Ad	R mandibular fragment with M3	BM(NH)A 15711	Paralectotype *Semnopithecus palaeindicus* Lydekker 1884, (Lower Pliocene), Figured Falconer,H. and Cautley,P.T. 1837: 357, pl.23,fig.4, Figured Falconer,H. and Cautley,P.T. 1868: 302, pl.24,figs.7-8, Cited Lydekker,R. 1884: 123, Cited Lydekker,R. 1885: 5, Cited Lydekker,R. 1886: 5, Listed Delson,E. (unpubl.): 4
Siwalik Hills	—	I	[Ad]	Casts of R talus (3)	BM(NH)A M1539	(Lower Pliocene)(*Semnopithecus palaeindicus*), Figured Cautley,P.T. and Falconer,H. 1840: 501, Figured Cautley,P.T. and Falconer,H. 1868: 293, figs.6-9, Cited Lydekker,R. 1885: 3, Cited Lydekker,R. 1886: 5

PYGATHRIX E. Geoffroy, 1812

(2 subgenera: *Pygathrix* E. Geoffroy, 1812 and *Rhinopithecus* Milne Edwards 1872)

Taxonomic note *Rhinopithecus* is treated as subgenus of *Pygathrix*, following Groves (1970) and Thorington & Groves, 1970.

Morphology Large monkeys with typical colobine characters; sacculated stomach and no cheek pouches. Distinguished from all other southeast Asian leaf monkeys by the conformation of the nose: two small flaps of skin on the upper border of the nostrils lie flat in the subgenus *Pygathrix* but stand erect in the subgenus *Rhinopithecus*. Arms are nearly as long as legs, with intermembral indices of 90 (subgenus *Pygathrix*) and 95 (subgenus *Rhinopithecus*), similar to *Macaca* (92) or *Papio* (95). Unlike *Presbytis* and *Nasalis*, ischial callosities are separate in both males and females. Facial portions of skull short and broad; orbits with strong brow ridges; interorbital space very broad, but nasals reduced, sometimes obsolete. Posterior nares wide and deep, unlike *Presbytis*. Always a hypoconulid on M3, but the presence of a 6th cusp is variable.

Range China and Indo-China. The four species replace each other geographically.

Subgenus: PYGATHRIX E. Geoffroy, 1812

Douc
(1 species: type species *Simia nemaeus* Linnaeus, 1771)

Morphology Distinguished from subgenus *Rhinopithecus* by the colour pattern of the fur which is very sharply defined, giving rise to the German name 'Kleideraffe' (Clothes Monkey) since the nominate subspecies looks like a football team wearing white shirts, grey pullovers, black shorts and red socks. The tail is pure white, as is the short-haired rump-patch which is triangular in shape, the apex directed caudally. The inner thigh is also white except that in adult females it becomes suffused with red during oestrus and pregnancy; in adult males the red hue is also present but only when the female is in oestrus (Lippold, 1977). Though sexual dimorphism is not marked in size (see dimensions, below), males can be distinguished from females by the white spots at the anterior angles of the white rump-patch (Lippold, 1977). The ischial callosities are separated in males by a narrow strip of white furred skin. Limb proportions are close to those of *Nasalis* and *Macaca*, suggesting a terrestrial habitus, but observations in the wild so far indicate that douc is never seen on the ground, except momentarily when fleeing. Large males were occasionally observed to brachiate (Van Peenen *et al*., 1971).

The subgenus *Pygathrix* differs from subgenus *Rhinopithecus* in having a narrower face, and the palate is shorter than the distance from basion to staphylion. A 6th cusp is present on M_3 in 20% of skulls (n=9) in *P.n.nemaeus* and 60% (n=10) in *P.n.nigripes* (Groves, 1970, pp. 566-567). Chromosome diploid no. 44 (Chiarelli, 1973).

Range Laos, S. Vietnam and ?Hainan. Lippold (1977) reports the possible existence of *P.n.nigripes* near the Vietnam-Cambodian border near Tinh Bien (10°30′N 105°E), and Pfeffer (1969) observed it in NE. Cambodia.

Ecology The only field of study of *Pygathrix nemaeus* is that of the nominate race by Lippold (1977) on the Mt Sontra peninsula near Da Nang, S. Vietnam (16°07′N 108°18′E). Here the rain forest that once clothed the peaks has been much disturbed in recent years. Some remains on the northern slopes, also some secondary forest from which the taller trees have been removed for charcoal burning. Doucs were most often seen in these two types of forest though they occasionally enter banana plantations. They eat young rather than mature leaves, also buds, fruits and flowers. No insects were eaten and no water was drunk, though dew was licked from leaves. No predators were observed except man; doucs are hunted by natives for their meat. Observations in Cambodia of *P.n.nigripes* by Pfeffer (1969) reveal its preference for gallery forest or isolated wooded hills (or 'phnoms') characteristic of the area.

Social behaviour Reports of group size vary from Osgood's (1932) estimate of 30-50 animals in the Col des Nuages to the small groups of 3-11 observed on Mt Sontra by van Peenen *et al*. (1971), Gochfeld (1974) and Lippold (1977). Both one-male and multi-male groups have been recorded, with females outnumbering males by at least 2 to 1. There is some evidence of a dominance hierarchy in both males and females. Self-grooming is seen less frequently than social grooming, in which females play a more active role than males. Social grooming sessions can last up to an hour, indicating the importance of the activity in the douc's way of life. Both hands and mouth are used to remove particles from the fur.

Reproduction The menstrual cycle lasts 28-30 days. At oestrus the white inguinal region becomes red, and remains so throughout pregnancy. Gestation is thought by Lippold and Brockman (1974) to last 180-190 days. Hick (1975) estimates it as 165 days. In captivity however copulation continues during pregnancy, so precise information is not available, The newborn infant has a black face with two pale stripes beneath the eyes, As in *Presbytis*, females other than the mother handle and groom the infant during the first days of its life. Lippold (1977:528) lists the characters by which age can be estimated in doucs. Males become adult at about 5 years, females at about 4 years.

PYGATHRIX (PYGATHRIX) NEMAEUS (Linnaeus, 1771)

Douc
(2 subspecies)

Range Groves (1970) gives the following localities: *P.n.nemaeus* from Nape in Laos, and Col des Nuages and Mt Sontra in S. Vietnam; *P.n.nigripes* from Bien Hoa, Da Lat, Djiring (=Di Linh), Nui Lang Bian, Nha Trang, Saigon and Trang Bom in S. Vietnam. A specimen of *P.n.nemaeus* was reported from Hainan by Meyer in 1892, but there are no subsequent records.

Dimensions The subgenus *Pygathrix* is remarkable amongst colobines for its lack of sexual dimorphism. Adult specimens in the museum collections have the following field measurements (see p. 99 for comparative tables)

h & b	4 ♂♂ 586 (550-630)	:	1 ♀ 597
tail	4 ♂♂ 680 (600-735)	:	1 ♀ 597

Subspecies Two are recognised. Both have distinctively speckled grey fur with black and white annulations, more closely resembling that of De Brazza's Monkey (*Cercopithecus neglectus*) than any other leaf monkey. Both races have a white neck, cheeks, short-haired triangular rump-patch and tail. In *P.n.nemaeus*, the face is pale terracotta; the forearms and wrists are white, contrasting with the grey upper arm and black fingers, while the legs and ankles are red, contrasting with the black thighs and toes. There is also a semi-circular red pectoral band, outlined with black. In *P.n.nigripes*, the face is black; the red areas are reduced to a narrow pectoral band; the forearm is grey like the upper arm and the entire leg is black. The genitalia are as striking as the pelage: the penis is bright red in both races, contrasting with the white scrotum of *P.n.nemaeus* and the blue scrotum of *P.n.nigripes*.

Pygathrix (Pygathrix) nemaeus nemaeus (Linnaeus, 1771)

Type locality Cochin China.

Pygathrix (Pygathrix) nemaeus nigripes (Milne Edwards, 1871)

Principal synonym *Presbytis nemaeus moi* Kloss, 1926:214.

Type localities Of *nigripes*, 'environs de Saigon. Forêts qui bordent le Mékong vers son embouchure'; of *moi*, Lang Bian Peak, S. Vietnam at 5500-6500ft.

Locality	Date of Collection	Sex	Age	Nature of Specimen	Accession no.	References, Remarks, etc.
Pygathrix (Pygathrix) nemaeus nemaeus (Linnaeus 1771)						
LAOS						
Nape, [18°18'N 105°07'E], 2500ft	27 Jan 1928	M	Ad	Skin, Skull	ZD.1928.7.1.2	Listed Lowe,W.P. (unpubl.), Cited Pocock,R.I. 1935: 959-960, Cited Thomas,O. 1929: 832
Nape, [18°18'N 105°07'E]	31 May 1927	I	Juv	Skin, Skull	ZD.1928.7.1.3	
[S.VIETNAM]						
Annam	[Died] Aug 1936	F	[Ad]	Wet, cadaver	ZD.1937.3.4.3	
Col des Nuages, Annam, [16°11'N 108°08'E]	3 Feb 1926	M	Ad	Skin, Skull	ZD.1926.10.4.4	Listed Lowe,W.P. (unpubl.), Cited Thomas,O. 1927: 43, Cited Pocock,R.I. 1935: 959-960, Figured James,W.Warwick 1960: 232-233, figs.57a-c
" " "	20 Nov 1925	F	Ad	Skin, Skull	ZD.1926.10.4.5	Listed Lowe,W.P. (unpubl.), Cited Thomas,O. 1927: 43, Cited Pocock,R.I. 1935: 959-960
NO DATA						
—	—	F	[Ad]	Skin	ZD.1855.12.24.9	
Pygathrix (Pygathrix) nemaeus nigripes (Milne Edwards 1871)						
[S.VIETNAM]						
Bali, near Nha Trang, Annam, [250ft]	10 Nov 1905	M	Ad	Skin, Skull, Radiograph of penis	ZD.1906.11.6.1	
Djiring [=Di Linh], S.Annam, [11°38'N 108°07'E], 3500ft	2 Mar 1927	M	Ad	Skin, Skull	ZD.1927.12.1.8	Listed Lowe,W.P. (unpubl.), Cited Thomas,O. 1928: 141
" " "	1 Mar 1927	M	[Inf]	Skin	ZD.1927.12.1.9	" " " "
" " "	2 Mar 1927	F	Ad	Skin, Skull	ZD.1927.12.1.10	" " " "
Djiring [=Di Linh], S.Annam, [11°38'N 108°07'E], 3000ft	4 [Mar] 1927	F	[Inf]	Skin	ZD.1927.12.1.11	
Djiring [=Di Linh], S.Annam, [11°38'N 108°07'E], 3500ft	1 Mar 1927	M	Juv	Skin, Skull, [Sketch]	ZD.1927.12.1.14	Listed Lowe,W.P. (unpubl.), Cited Thomas,O. 1928: 141
Djiring [=Di Linh], S.Annam, [11°38'N 108°07'E]	18 Mar 1927	M	Inf	Skin, Skull	ZD.1927.12.1.15	Cited Thomas,O. 1928: 141
Lang Bian, Annam, [c.12°05'N 108°30'E], 1500ft	10 Jun 1908	M	Ad	Skin, Skull	ZD.1908.11.1.2	
" " "	—	M	Ad	Skin, Skull, Radiograph of penis	ZD.1908.11.1.3	Figured Elliot,D.G. 1913: Pl.9
" " "	10 Jun 1908	F	Ad	Skin, Skull	ZD.1908.11.1.4	
Lang Bian Peak, S.Annam, [12°03'N 108°27'E], [55-6500ft]	23 Aug 1919	M	Ad	Skin, Skull	ZD.1947.1501	Holotype *Presbytis nemaeus moi* Kloss 1926, Cited Kloss,C.Boden 1926: 214, Cited Pocock,R.I. 1935: 960
Trang Bom, 25 miles E. of Saigon, Cochin China, [10°57'N 107°E]	2 Jun 1918	M	Ad	Skin, Skull	ZD.1928.2.24.1	Cited Pocock,R.I. 1935: 959-960
[?S.VIETNAM]						
[Cambodia or Cochin China]	—	M	[Juv]	Skin, Skull in skin	ZD.1878.6.17.3	

Subgenus: RHINOPITHECUS E. Geoffroy, 1812

Snub-nosed Leaf Monkeys
(3 species: type species *Semnopithecus roxellana* Milne Edwards, 1870)

Principal synonyms *Presbytiscus* Pocock, 1924:330 (type species: *Rhinopithecus avunculus* Dollman, 1912).

Taxonomic notes Three species are recognised, following Thorington & Groves, 1970. *Presbytiscus* was distinguished by Pocock (1924) from *Rhinopithecus* on account of its longer digits, particularly the hallux.

Morphology The colour pattern of the fur is much less sharply demarcated than in subgenus *Pygathrix*; the front of the head and the inner and underparts are generally lighter than the upper parts. The snub nose is set well back from the prominent rounded muzzle, with the nostrils pointing directly forward. Two flaps of skin on the upper rim of the nostrils stand erect, forming two little peaks which in adulthood approach the forehead. The lateral extensions of these skin flaps tend to curl inwards slightly, at least in dried skins. Ischial callosities are separated in males by a wide tract of furred skin. In skull characters, the subgenus *Rhinopithecus* differs from *Pygathrix* in having a broader face, and the palate is longer than the distance from basion to staphylion. A 6th cusp on M_3 is present in 21% of *P. roxellana* skulls (n=14); in 100% of *P. avunculus* skulls (n=7), and in 100% of *P. brelichi* skulls (n=3) (Groves, 1970). Limb proportions are like those of *Nasalis*, resembling those of terrestrial cercopithecines like *Papio* rather than other colobines. As in *Nasalis* and *Simias*, the thumb is longer in relation to the rest of the hand than in other leaf monkeys, and considerably longer than in subgenus *Pygathrix* (Jouffroy & Lessertisseur, 1977). Chromosome diploid no. 2n=44 (Chen *et al.*, 1978).

Range Western and central China and North Vietnam.

Ecology *P. roxellana*: coniferous forests, bamboo jungle and rhododendron thickets in high mountains of Szechwan and Yunnan, where snow covers the ground for about half the year. Observed at 8000ft in the Min Shan, N. Szechwan, near the boundary with Kansu (Chen Ho-yi, 1974), and at 3350-4000ft in Yun-ling Mts (*P.r.bieti*) (Li *et al.*, 1980). Allen (1938) notes their nomadic way of life, spending the summer at 10 000ft, and descending to cultivated levels in winter. This is confirmed by P'eng *et al.* (1968) for *P. brelichi* on the SE. slopes of Fan-ching-shan: here they descend to the valleys in the spring for wild cherries and in the autumn for wild pears and cucumbers. *P. avunculus*: little information. It is found exclusively within the biogeographical province named Thailand Monsoon Forest (Udvardy, 1975).

Social behaviour Little information. Allen (1938) mentions large groups of 100 or more. In dense jungle in Moupin, the type locality, the Roosevelts (1929) killed 9 specimens from a single group from which it can be inferred that the group was at least twice that size, and probably larger.

Reproduction No information.

Key to the Species of *Pygathrix (Rhinopithecus)*

Species	Head	Face	Back	Limbs	Cheeks & neck	Underparts	Tail
roxellana (Szechwan, Kansu and NW. Yunnan)	Dark crest on vertex, short in *roxellana*, long in *bieti*	Muzzle white; circumocular region bluish or greenish	Greyish brown in *roxellana*, dark grey in *bieti*	Light patch on back of thighs; hands and feet yellowish in *roxellana*, black in *bieti*	Orange (*roxellana*) or white (*bieti*) extending around neck to form collar	Yellowish white in *roxellana*; white in *bieti*	Shorter than h + b length (86-95%*; colour as back; curly hairs at base in some
brelichi (NE. Kweichow)	No crest on vertex	Skin pigmented; some white fur on cheeks (skin imperfect)	Dark brown; white inter-scapular patch	Lighter on back of thighs; hands and feet dark	As dark or darker than back	As back	Longer than h + b length (133%)*; black with curly hairs at base; small white tip
avunculus (N. Vietnam)	No crest on vertex	Forehead bluish white; cheeks and mouth blue-black; chin flesh pink†	Dark brown, darkening towards rump	Black outside; yellowish white inside; hands and feet black	Orange patch on throat; chin yellowish white	Yellowish white	Much longer than h + b length (143-148%)*; black and strawcoloured curly hairs give a 'corkscrew' effect; white tip.

* Cited from Groves (1970)
† Cited from field notes

PYGATHRIX (RHINOPITHECUS) ROXELLANA (Milne Edwards, 1870)

Snub-nosed Leaf Monkey, Golden Snub-nosed Leaf Monkey, Snow Monkey
(2 subspecies)

Taxonomic note Although Milne Edwards altered the specific name *roxellana* to *roxellanae* in 1872, there is no indication in the original description that *roxellana*, a noun standing in apposition to the generic name, was an inadvertent error. It should not therefore be amended (Art. 32 (a) (ii) of the International Code of Zoological Nomenclature). Li *et al.* (1980) give specific rank to *bieti*, and supply notes on its range and ecology.

Range Szechwan, S. Kansu and N. Yunnan, China.

Dimensions None recorded.

Subspecies *P.r.roxellana* has the upper and outer parts a greyish brown, the back ornamented with long golden strands. The underparts are yellowish white which becomes golden with increasing age, particularly on the head and neck. The short hairs on the backs of the thighs are yellowish white, and the hands and feet yellow. *P.r.bieti* has dark smoky grey to blackish upper and outer parts with white underparts. The white hairs on the backs of the thighs are long and bushy, and the hands and feet black.

Pygathrix (Rhinopithecus) roxellana roxellana (Milne Edwards, 1870)
Golden Snub-nosed Leaf Monkey, Snow Monkey

Principal synonym *Rhinopithecus roxellanae* Milne Edwards, 1872:233, pls. 36-37.

Type locality 'Principauté de Moupin' (=Pao-hsing, 30°24′N 102°50′E), Central Szechwan, China.

Pygathrix (Rhinopithecus) roxellana bieti Milne Edwards, 1897
Biet's Snub-nosed Leaf Monkey

Type locality Kiape (*c.* 28°25′N 98°55′E), a day's journey southwards from Atentse, L. bank of upper R. Mekong, NW. Yunnan, China (Milne Edwards & Pousargues, 1898:124).

Locality	Date of Collection	Sex	Age	Nature of Specimen	Accession no.	References, Remarks, etc.
Pygathrix (Rhinopithecus) roxellana roxellana (Milne Edwards 1870)						
CHINA						
Szechwan, N.W.China	—	F	Juv	Skin, Skull, Skeleton	ZD.1908.10.9.5	
Tsao Po′, Ken Yuen Chiao, Szechwan, W.China	27 May 1938	[F]	Inf	Skin, Skull	ZD.1955.608	
W.China	—	M	Ad	Skin, Skull, Skeleton	ZD.1908.10.9.1	Figured Elliot,D.G. 1913: Pl.10, Figured James,W.Warwick 1960: 234-235, figs.58a-c
" " "	—	F	Juv	Skin, Skull, Skeleton	ZD.1908.10.9.2	
" " "	—	F	Juv	Skin, Skull, Skeleton	ZD.1908.10.9.3	
" " "	—	I	[Juv]	Skin, Skeleton	ZD.1908.10.9.4	
Yang-liu-pa, Szechwan	—	F	Ad	Skin, Skull	ZD.1899.3.1.2	
[CHINA]						
[Szechwan]	—	M	Ad	Skin, Skull	ZD.1899.3.1.1	
NO DATA						
—	[Died] 21 Feb 1939	F	[Juv]	Skin	ZD.1939.271	
Pygathrix (Rhinopithecus) roxellana bieti (Milne Edwards 1897)						
NO SPECIMENS						
—	—	—	—	—	—	

PYGATHRIX (RHINOPITHECUS) AVUNCULUS (Dollman, 1912)

Tonkin Snub-nosed Monkey

Type locality Yen-Bay (21°43′N 104°54′E), R. Song-koi, N. Vietnam.

Taxonomic note *P. avunculus* is the type species of the subgenus *Presbytiscus* which Pocock (1924:330) separated on account of its elongated digits which resemble those of *Nasalis*. He noted that the tip of the hallux of *P. avunculus* reaches the distal end of the 1st phalanx of the 2nd digit, while in *Rhinopithecus* and *Presbytis* it does not reach even the distal margin of the plantar pad.

Range Known only from the type locality and Bac Kan (22°09′N 105°50′E), N. Vietnam.

Dimensions Adult specimens in the museum collections have the following field measurements:

h & b	2 ♂♂ 590 (560, 620)	: 2 ♀♀ 515 (510, 520)
tail	2 ♂♂ 871 (822, 920)	: 2 ♀♀ 695 (660, 730)

Locality	Date of Collection	Sex	Age	Nature of Specimen	Accession no.	References, Remarks, etc.
Pygathrix (Rhinopithecus) avunculus (Dollman 1912)						
[N.VIETNAM]						
Bac Kan, Tonkin, [22°09′N 105°50′E], 500ft	11 Dec 1926	M	Ad	Skin, Skull	ZD.1927.12.1.2	(*Presbytiscus avunculus*), Listed Lowe,W.P. (unpubl.), Cited Thomas,O. 1928: 140
" " "	11 Dec 1926	M	[Inf]	Skin	ZD.1927.12.1.3	" " " "
" " "	15 Jan [1927]	M	Ad	Skin, Skull, Radiograph of penis	ZD.1927.12.1.4	" " " "
" " "	20 Jan 1927	M	[Ad]	Skin, Sketch	ZD.1927.12.1.5	" " " "
" " "	27 Dec 1926	F	Ad	Skin, Skull	ZD.1927.12.1.6	(*Presbytiscus avunculus*), Listed Lowe,W.P. (unpubl.), Cited Thomas,O. 1928: 140, Figured Colyer,J.F. 1936: 222, fig.265
" " "	8 Jan 1927	F	Juv	Skin, Skull	ZD.1927.12.1.7	(*Presbytiscus avunculus*), Listed Lowe,W.P. (unpubl.), Cited Thomas,O. 1928: 140
Yen Bay, R.Song koi, Tonkin, [21°43′N 104°54′E]	14 Sep 1911	F	Ad	Skin, Skull	ZD.1912.4.21.1	Holotype *Rhinopithecus avunculus* Dollman 1912, Cited Dollman,G. 1912: 18, Cited Dollman,G. 1912a: 503, Cited Thomas,O. 1928: 140
" " "	14 Sep 1911	F	Inf	Skin, Skull	ZD.1912.4.21.2	

PYGATHRIX (RHINOPITHECUS) BRELICHI (Thomas, 1903)

Brelich's Snub-nosed Leaf Monkey

Type Locality Van Gin Shan Range (= Fan-ching Shan, 27°57′N 108°50′E, Peng *et al.*, 1968), N. Kweichow, Central China.

Taxonomic note *Rhinopithecus tingianus* Matthew & Granger, 1923, from the Middle Pleistocene of Yenchingkou, a village near the right bank of the R. Yangtse opposite Wan-hsien (30°48′N 108°17′E), SE. Szechwan, is provisionally classified as a subspecies of *P. brelichi* by Groves (1970:569).

Range Region of NE. Kweichow, south of the Middle Yangtse River.

Dimensions The following measurements were recorded by Thomas (1903:224) from the type specimen, a flat native pelt (a female): h & b 730, tail 970 (with hairs 1040).

Locality	Date of Collection	Sex	Age	Nature of Specimen	Accession no.	References, Remarks, etc.
Pygathrix (Rhinopithecus) brelichi (Thomas 1903)						
CHINA						
Van Gin Shan Mts. [=Fan-ching Shan], N.Kweichow, [27°57′N 108°50′E]	—	F	[Ad]	Skin	ZD.1903.3.14.1	Holotype *Rhinopithecus brelichi* Thomas 1903, Figured Thomas,O. 1903: 224, pl.21
Van Gin Shan range [=Fan-ching Shan], Kweichow, [27°57′N 108°50′E], 7-8000ft	Jan 1905	F	Juv	Skin, Skull	ZD.1905.3.24.1	

NASALIS E. Geoffroy, 1812

Proboscis Monkey
(1 species: type species *Cercopithecus larvatus* Wurmb, 1784)

Taxonomic notes The classification of the Proboscis Monkey of Borneo and the Pigtailed Leaf Monkey of the Mentawai Islands is still uncertain. Groves (1970) united them in the genus *Nasalis* on account of the many skull characters which they share, but they differ in almost every point of external appearance. The long bulbous nose (in the adult male) and long well-furred tail of *Nasalis* are in striking contrast with the snubnose and bare pig-tail of *Simias*; the colour and disposition of the fur is completely different in the two except, as pointed out by Groves, the banding of individual hairs is concentrated on the upper back and arms in both. However the light-coloured short-haired rump-patch and adpressed backswept whiskers of *Nasalis* are not in any way duplicated in *Simias*. Much more striking is the superficial resemblance of the latter to *Macaca nemestrina pagensis* from Siberut; both are almost black with striking white cheek patches, and both have short pig-tails. On ecological grounds, Medway (1970:543) considered that there was no necessity to associate the two. Their social behaviour has also been shown to differ (Tilson, 1977). *Nasalis* is retained as a separate genus, mainly on account of its extraordinary pendant fleshy nose, an organ unique among the primates with the exception perhaps of man. The resemblance between the two is sufficiently marked for the natives of Borneo to have nicknamed the proboscis monkey 'Orang Blanda', or Dutchman.

No subspecies are recognised by Groves (1970:563); he noted differing frequencies of paler or more intense coloration in three different localities.

Morphology Large colobine monkeys with marked sexual dimorphism in size and weight. Males have a huge bulbous nose overhanging the mouth; in females the nose is prominent but does not overhang the mouth while in infants it is even slightly upturned. The facial skin is deep blue in infancy and pinkish brown in adult life. The general fur colour is brick red, darkest on crown where there is a distinct central whorl. The dark area continues down the nape on to the shoulders and contrasts with the cream-coloured cheeks and ruff. Arms, legs, hands, feet and tail are pale grey, as is the short-haired triangular rump patch. Ischial callosities are contiguous in males but separate in females.

Two characters differentiate the skull from other southeast Asian colobines and ally it with *Simias*: the narrow interorbital breadth and the long nasal bones. Verheyen (1962, p. 132) gives the interorbital width/muzzle length index* as 14.4-19.6 (n = 15) in *Nasalis* and 13.3-15.1 (n = 2) in *Simias*, compared with 18.8-32.9 (n = 104) in *Presbytis*. However, the nasals are broader in *Nasalis* with a distinct median projection (Miller, 1903).

Limb proportions are those of terrestrial quadrupedal monkeys, the intermembral index being 93, strikingly high for a colobine but close to the other 'odd-nosed' leaf monkeys of southeast Asia, *Pygathrix* and *Simias* (see Appendix 3, p. 101). The thumb is longer in relation to the index finger than in any other colobine except *Simias* (see p. 90). The opposability index† is 46 (n = 4) (Napier & Napier, 1967). The subgenus *Rhinopithecus* also has a relatively long thumb (Jouffroy & Lessertisseur, 1977).

Chromosome diploid no. 48 (Chiarelli, 1973). All other colobines so far examined have a diploid no. of 44.

* The interorbital width/muzzle length index is obtained as follows:

$$\frac{\text{anterior interorbital width} \times 100}{\text{glabella-prosthion}} \quad \text{(Verheyen, 1962)}$$

† The opposability index is estimated as follows:

$$\frac{\text{Thumb length} \times 100}{\text{Index ray length}} \quad \text{(Napier \& Napier, 1967)}$$

Range Borneo.

Ecology Observations of *Nasalis* have been made mainly in Sabah, N. Borneo by Davis (1962), Kern (1964), and Kawabe & Mano (1972). *Nasalis* live in tidal mangrove on the coast, in the nipa palm and mangrove association of the estuaries, and in lowland rain forest further inland. They eat leaves and shoots of mangrove and pedada trees, also some fruits and flowers. They are mainly arboreal, but in rain forest where the ground is dry they come to the ground to feed, even venturing into the open on the banks of rivers which they can cross by swimming (Kawabe & Mano, 1972). In the mangrove the day is spent in the lower shadier parts, while at night the taller pedada trees provide protection from predators such as the clouded leopard (Davis, 1962; Kern, 1964, pl. 4).

Social behaviour *Nasalis* lives in multi-male groups of about 19 members (11-32: n = 8). Kawabe & Mano (1972), describe the groups as well integrated. They were able to list 14 different vocalisations which can be divided into three groups: (1) growls or snorts; (2) honks, and (3) shrieks which are invariably given by females and young. Adult males utter growls and honks, the former when calming the rather excitable females and young, and the latter when responding to an external disturbance. Adult males protect the group by facing the source of danger, honking loudly, thus enabling the females and young to retreat into the depths of the forest. A dominance hierarchy was observed, a dominant male displacing a subordinate male at a feeding ground without obvious antagonism.

Reproduction Observations in captivity indicate that copulation is initiated by the female, not only at oestrous but also at other periods in the cycle as well as during pregnancy. Gestation is thought to be about 166 days (Schultz, 1942). In captivity, infants are held by other females besides the mother, and even by adult males (Hollihn, 1973). The face is deep blue at birth, becoming slate-grey at about 3 months. The permanent dentition is fully erupted at 7 years (Schultz, 1942).

NASALIS LARVATUS (Wurmb, 1784)

Principal synonyms *Cercopithecus nasica* Lacépède, 1799:4; *Nasalis recurvus* Vigors & Horsfield, 1829:110; *Nasalis larvatus orientalis* Chasen, 1940:84.

Type localities Of *larvatus*, Pontianak, W. Borneo: of *nasica*, none cited: of *recurvus*, probably Pontianak or Sambas, about 100 miles to the north, according to Chasen (1940:83); of *orientalis*, Salim Batu, Bulungan, NE. Kalimantan.

Dimensions Adult specimens in the museum collections have the following field measurements (see Appendix 2, p. 99 for comparative table):

h & b	2 ♂♂ 745 (730, 760)	: 3 ♀♀ 620 (610-640)
tail	2 ♂♂ 665 (660, 670)	: 3 ♀♀ 573 (550-620)

Schultz (1942) gives the following weights: 10 ♂♂ 20 344 (14 062-23 587) : 15 ♀♀ 9873 (8165-11 794), giving a female/male weight ratio of less than 50%.

Locality	Date of Collection	Sex	Age	Nature of Specimen	Accession no.	References, Remarks, etc.
Nasalis larvatus (Wurmb 1784)						
BORNEO						
Buntok, R.Barito, South Central Borneo, [1°45′S 114°47′E]	—	M	Ad	Skin, Skull	ZD.1910.4.5.1	
Buntok, R.Barito, South Central Borneo, [1°45′S 114°47′E], 20ft	4 Oct 1909	M	Ad	Skin, Skull, Photograph, Sketches	ZD.1910.4.5.2	(*Nasalis*), Figured James,W.Warwick 1960: 238-239, figs.60a-c
„ „ „	5 Oct 1909	M	Ad	Skin, Skull	ZD.1910.4.5.4	(*Nasalis*)
„ „ „	5 Oct 1909	M	Ad	Skin, Skull, Sketch	ZD.1910.4.5.5	(*Nasalis*), Figured Elliot,D.G. 1913: Pl.13
„ „ „	2 Oct 1909	F	Juv	Skin, Skull	ZD.1910.4.5.6	(*Nasalis*)
„ „ „	2 Oct 1909	F	Ad	Skin, Skull, Sketch	ZD.1910.4.5.7	„ „ „ „
„ „ „	3 Oct 1909	F	Ad	Skin, Skull, Sketch	ZD.1910.4.5.8	„ „ „ „
„ „ „	10 Oct 1909	F	Ad	Skin, Skull	ZD.1910.4.5.9	„ „ „ „
„ „ „	6 Oct 1909	F	Inf	Skin, Skull	ZD.1910.4.5.10	„ „ „ „
„ „ „	—	F	Inf	Skull	ZD.1910.4.5.11	„ „ „ „
Lumbidan [=?Lambidan I.], N.W.Borneo, [?5°23′N 115°21′E]	—	M	Ad	Skull, Skeleton	ZD.1894.6.8.5	
Probably N.W. coast, [c.1°22′N 111°10′E]	—	[F]	Juv	Skull	ZD.1844.12.17.4	Listed Gray,J.E. (unpubl.): 6
[R.Lawas], [N.W.Borneo], [c.4°58′N 115°25′E]	—	F	Ad	Skin, Skull	ZD.1923.7.1.1	
Santubong, [Sarawak], [c.1°43′N 110°18′E]	—	M	Juv	Skull	CMZ E7281C	
Sarawak	—	[M]	Ad	Skull, Skeleton	CMZ E7281B	
—	—	M	Ad	Skull	ZD.1846.8.8.1	Listed Gray,J.E. (unpubl.): 6
—	—	[M]	Juv	Skin, Skull	ZD.1850.10.24.1	
—	Apr 1897	M	[Inf]	Skin	ZD.1899.5.2.1	
—	—	M	Ad	Skin	ZD.1939.3479	
—	—			Skin	GZD	(*Presbytes recurvus*)
—	—	M	Ad	Skull, Skeleton	DLPA 74.0.1	
—	—	M	Ad	Skull, Skeleton	DLPA 74.0.2	

Locality	Date of Collection	Sex	Age	Nature of Specimen	Accession no.	References, Remarks, etc.
Nasalis larvatus *continued*						
—	—			Skin	CMZ E7281D	
—	—	[M]	Ad	Skull	CMZ E7281A	
[BORNEO]						
Sadong, Sarawak, [c.1°33′N 110°45′E]	—	I	Juv	Skull, Skeleton	CDA M44	
" " "	—	M	Ad	Skull, Skeleton	CDA M45	
Sadong Dist., Sarawak, [c.1°33′N 110°45′E]	—	M	[Ad]	Skin, Radiograph of penis	ZD.1939.1152	
" " "	—	M	[Ad]	Skin, Radiograph of penis	ZD.1939.1153	
" " "	—	F	[Ad]	Skin	ZD.1939.1154	
—	—	M	Ad	Skull	ZD.1846.10.27.11	
—	—	F	Ad	Skull, Skeleton	ZD.1855.12.26.242	Listed Anon (unpubl.), Listed Gray,J.E. (unpubl.): 6, Cited Martin,W.C.L. 1838a: 72, Figured Martin,W.C.L. 1841: 456, fig.282
—	—	M	Ad	Skull	ZD.1867.4.12.34	
—	—	M	[Ad]	Wet, head and larynx	ZD.1972.939	
—	—	M	[Ad]	Skin	ZD.1980.881	
—	—	F	[Ad]	Skin	ZD.1980.882	
—	—	F	[Juv]	Skin	ZD.1980.883	
—	—	I	[Inf]	Skin	ZD.1980.884	
—	—	M	Ad	Skull	RCS(OM) A70.9	
—	—	F	Ad	Skull	RCS(OM) A70.91	
—	—			Skull	EDZ XH6-23.1	
—	—	I	Juv	Wet, cadaver	CDA M46	

SIMIAS Miller, 1903

Simakobu, Mentawai Islands Snub-nosed Leaf Monkey, Pagai Islands Snub-nosed Leaf Monkey, Pigtailed Leaf Monkey.

(1 species: type species *Simias concolor* Miller, 1903)

Taxonomic note *Simias* is retained as a separate genus (see *Nasalis*, taxonomic notes, p. 88).

Morphology Medium-sized monkeys more closely resembling the pigtailed macaque than any other colobine in the relatively short limbs and bare pig-tail. Dark brown to black fur, longer and slightly grizzled on nape, shoulders and upper back due to the greyish buff annulations of individual hairs. The only adornments are the white cheek patches which are also present in *Macaca nemestrina pagensis*. Cream-buff variants have been recorded: in Siberut these amounted to 37% of the 91 animals observed (Tilson, 1977:205), and Miller (1903:69) reported 2 out of a total of 10 specimens from S. Pagai Island. (For photographs of both forms, see Tilson, 1977:204-205.)

Simias' nose most closely resembles that of *Rhinopithecus*, but instead of two upturned triangular flaps of skin above the nostrils, the superior nasal margin is lengthened to form a slightly elevated transverse ridge, giving a snub-nosed appearance. There is little sexual dimorphism in body size. The skull is closest to *Nasalis* with a similar interorbital breadth/muzzle length index (13.3-15.1, n = 2, see *Nasalis* morphology, p. 89), but it is smaller, with the rostrum less developed. The nasal bones are narrower and probably never form a distinct median projection as in *Nasalis* (Miller, 1903:67). As in *Nasalis*, ischial callosities are contiguous in males, separate in females. Limb proportions, based on a single specimen in the United States National Museum, are closest to *Papio* and *Rhinopithecus* though the limbs are relatively shorter (see Appendix III, p. 101). As in *Nasalis*, the thumb is long in relation to the index finger (opposability index 46: n = 1, Napier, unpublished data). The chromosome diploid no. of *Simias* has so far not been investigated.

Range Siberut, Sipora, N. and S. Pagai Islands, Mentawai Islands (Feiler & Eck, 1979).

Ecology From Tilson's (1977) account of the Simakobu in Siberut, the only place where it has been studied, it lives in rain forest, preferably in the inland hilly regions, rather than in mangrove swamp or coastal lowlands, as implied by Chasen & Kloss (1927). Simakobu are leaf-eaters, but also eat fruits & berries. There are no large raptors or felids in Siberut, though the reticulated python is found there (Tenaza, 1975). Their main predator is man by whom they are hunted for their meat. *Simias* are inconspicuous in coat colour and body build, the absence of a long pendant tail aiding in this respect. Their concealment in the treetops is enhanced by their silence, their small groups (usually not more than 5 individuals), and their quiet quadrupedal progression in single file through the canopy. They feed for long periods in a single tree, selecting the densest patches of foliage.

Social behaviour (from Tilson, 1977 and Watanabe, 1981): Simakobu generally exhibit the same monogamous 'family' group structure as both Kloss's Gibbon (*Hylobates klossii*) and the Mentawai Island Leaf Monkey (*Presbytis potenziani*); an adult male and female and their offspring. Most groups number five individuals or less, but a minority contain more than one adult female and number as many as twenty. Groups forage within areas of approximately 25-30 hectares. When two groups meet near the boundary of their home ranges, the encounters lack the noisy display element described by Tilson & Teneza (1976) for *P. potenziani*. Each adult male gives a single loud vocalisation before retreating rapidly.

Simakobu vocalisations are limited: no predawn spacing calls are heard, nor 'contact' calls to ensure group cohesion. Adult males very occasionally give a loud nasal bark which is thought to be associated with courtship, either by attracting a female or repelling other males.

Reproduction Little information. Tilson (1977) notes that births appeared to occur only during the months of June and July.

SIMIAS CONCOLOR Miller, 1903

(2 subspecies)

Range As for the genus.

Dimensions Adult specimens in the museum collection have the following field measurements (see Appendix II for comparative table):

	Simias concolor siberu	*Simias concolor concolor*
h & b	1 ♂ 515	: 1 ♀ 500
tail	1 ♂ 155	: 1 ♀ 140
weight		: 1 ♀ 7144

Subspecies *Simias concolor siberu* was described by Chasen & Kloss (1927) as black in general coloration with more grizzling upon the shoulders and upper back than in the dusky brown *S. concolor concolor* from the Pagai Islands. The white cheek patches are also much more conspicuous in the darker *S.c.siberu* than they are in the paler *S.c.concolor*.

Simias concolor concolor Miller, 1903

Type locality S. Pagai Island, Mentawai Islands.

Simias concolor siberu Chasen & Kloss, 1927

Type locality Siberut Island, Mentawai Islands.

Locality	Date of Collection	Sex	Age	Nature of Specimen	Accession no.	References, Remarks, etc.
Simias concolor concolor Miller 1903						
SUMATRA						
Vickin's Bay, S.Pagai I., Mentawai Is., [3°12′S 100°26′E]	3 Dec 1902	F	Ad	Skin, Skull	ZD.1904.5.4.2	Paratype *Simias concolor* Miller 1903, Cited Miller,G.S. 1903: 69, Figured James,W.Warwick 1960: 236-237, figs.59a-c
Simias concolor siberu Chasen and Kloss 1927						
SUMATRA						
Siberut I., Mentawai Is.	25 Sep 1924	M	Ad	Skin, Skull	ZD.1928.11.2.2	Holotype *Simias concolor siberu* Chasen and Kloss 1927, Cited Chasen,F.N. and Kloss,C.Boden 1927: 813, pl.2

European Colobinae
MESOPITHECUS Wagner, 1839 (extinct)

(2 species: type species *Mesopithecus pentelici* Wagner, 1839)

Taxonomic note Gaudry's (1862) spelling of the name '*pentelici*', although an unjustified emendation of the original '*pentelicus*', has been followed by all subsequent authors, and is continued here. Two species are recognised, following Szalay & Delson (1979), to whom reference should be made for further details.

Principal synonym *Anthropodus* Lapouge, 1894.

Morphology Small to medium-sized colobines, the cranium resembling similar-sized *Presbytis* specimens rather than the larger *Colobus* or *Nasalis*. The post-cranial skeleton of *M. pentelici* shows a mosaic of arboreal and terrestrial features, with limb proportions closest to those of *Macaca sylvanus*. Fragmentary post-cranial material of *M. monspessulanus* suggests greater arboreality.

Range Southern and Central Europe between 40-50°N and 0-30°E, and Iran; also south-eastern England if the Red Crag molar is correctly referred here (see below). *Age* Late Miocene to late Pliocene.

MESOPITHECUS PENTELICI Wagner, 1839 (extinct)

Principal synonym *Mesopithecus major* Roth & Wagner, 1854:385.

Type localities Of *pentelici*, Pikermi, near Athens; of *major*, Pikermi.

Referred material From southern and central European sites, and Iran, the largest number of specimens having been recovered from Pikermi. *Age* Late Miocene (*c.* 10-5.5 MYA).

Locality	Date of Collection	Sex	Age	Nature of Specimen	Accession no.	References, Remarks, etc.
Mesopithecus pentelici Wagner 1839 (extinct)						
GREECE						
Pikermi	—	[F]	Ad	Cast of face with L M2-3, R C-M3	BM(NH)A M32287	
,, ,, ,,	—	I	Juv	Cast of R dP3-M2	BM(NH)A M32288	
,, ,, ,,	—	M	Ad	Cast of mandible	BM(NH)A M32289	
[GREECE]						
Pikermi, Attica	1901	[M]	Ad	Skull, Cast of mandible	BM(NH)A M8944	
,, ,, ,,	1901	M	Ad	Skull	BM(NH)A M8945	Figured Szalay,F.S. and Delson,E. 1979: 407, fig.206D
,, ,, ,,	1901	M	Ad	Skull	BM(NH)A M8946	
,, ,, ,,	1901	M	Ad	Skull, [Radiograph]	BM(NH)A M8947	
,, ,, ,,	1901	M	Ad	Skull	BM(NH)A M8948	
,, ,, ,,	1901	F	Ad	Skull	BM(NH)A M8949	
,, ,, ,,	1901	I		Partial dentition	BM(NH)A M8950	
,, ,, ,,	1901	M	Ad	L maxilla	BM(NH)A M8951	
,, ,, ,,	1901	[F]	Ad	Partial maxilla, Isolated teeth	BM(NH)A M8952	
,, ,, ,,	1901	[M]	Ad	Partial mandible, Cast of same	BM(NH)A M8953	
,, ,, ,,	1901	I	[Ad]	Partial mandible	BM(NH)A M8954	
,, ,, ,,	1901	I	Ad	Partial mandible, Cast of same	BM(NH)A M8955	
,, ,, ,,	1901	I	Ad	Fragment of L mandible with M1-3	BM(NH)A M8956	
,, ,, ,,	1901	I	Juv	Fragment of R mandible with I2, dP3-4, M1, Cast of same	BM(NH)A M8957	
,, ,, ,,	—			2 mandibular fragments, 1 molar	BM(NH)A M2279	
,, ,, ,,	—		M	Cast of skull	CMZ E7201A	
,, ,, ,,	—		F	Cast of skull	CMZ E7201B	
Pikermi, [near Athens]	—	I	Ad	Cast of cranium	BM(NH)A 49671	Figured Wagner,A. 1857: 113, pl.3, fig.2
,, ,, ,,	—	[F]	Ad	Cast of cranium	BM(NH)A 49672	Figured Wagner,A. 1857: 112, pl.3, fig.1
,, ,, ,,	—	I	Juv	Cast of mandible	BM(NH)A 49672a	
,, ,, ,,	—			2 mandibular fragments	BM(NH)A 49719	
[UKRAINA, U.S.S.R.]						
Grebeniki	—	I	Ad	Cast of upper L P4-M3	BM(NH)A M32290	(*Mesopithecus ukrainicus*)

MESOPITHECUS MONSPESSULANUS (Gervais, 1849) (extinct)

Principal synonym *Anthropodus rouvillei* Lapouge, 1894:202.

Type localities Of *monspessulanus*, Montpellier, SW. France; of *rouvillei*, between Celleneuve and La Mosson, W. of Montpellier, SW. France.

Referred material From southern and central European sites. *Age* Late Miocene to late Pliocene (*c.* 7-2.25 MYA). The upper molar crown from Red Crag in Suffolk, England (BM(NH)A M9171), previously catalogued as *Macaca* sp. by Napier (1981:35) is referred here by Szalay & Delson (1979). Its inclusion in Part 3 of the *Catalogue of Primates* draws attention to this considerable extension of the range of the subfamily Colobinae. Previously the only monkeys thought to have reached the British Isles have been identified as *Macaca* (subfamily: Cercopithecinae) and dated to Pleistocene times. Fossils in the Pleistocene Red Crag deposits are all derived from earlier deposits and could easily be Pliocene in age.

Locality	Date of Collection	Sex	Age	Nature of Specimen	Accession no.	References, Remarks, etc.
Mesopithecus monspessulanus (Gervais 1849) (extinct)						
ENGLAND, BRITISH ISLES						
Red Crag, Suffolk	—	I	Ad	Crown of upper L M3	BM(NH)A M9171	(*Semnopithecus monspessulanus*)(*Macaca florentina*)(*Hyracotherium?*), Cited Szalay, F.S. and Delson,E 1979: 405
FRANCE						
Montpellier, Herault	—		[Ad]	Casts of 5 isolated teeth	BM(NH)A M1893	Figured Gervais,P. 1859: 10, pl.1, figs.7-11
[ITALY]						
Casino, [near Siena], [Tuscany]	—		Ad	Isolated R and L M3	BM(NH)A M7155	(*Semnopithecus monspessulanus*), Figured Ristori,G. 1890: 195, pl.8, figs.11-12, 15-16
?Mesopithecus (extinct)						
ITALY						
?Monte Bamboli, [Grossetto]	—			Tooth fragment	BM(NH)A M11555	(*?Semnopithecus monspessulanus*)

DOLICHOPITHECUS Depéret, 1889 (extinct)

(1 species: type species *Dolichopithecus ruscinensis* Depéret, 1889)

Taxonomic note A species attributed to *Dolichopithecus* was described by Depéret in 1929 as *Dolichopithecus arvernensis*. It has since been referred to the cercopithecine genus *Paradolichopithecus* Necrasov, Samson & Rădulesco, 1961, by Delson (1975).

Morphology Moderately large colobine with considerable sexual dimorphism, and typical colobine dentition. The post-cranial elements indicate extreme adaptation to terrestrial life, the estimated limb proportions being similar to those of *Papio* and *Theropithecus*. *Dolichopithecus* is probably more closely related to *Mesopithecus* than to any other known colobine (Szalay & Delson, 1979).

Range Southern and central-eastern Europe. *Age* Latest Miocene to late Pliocene (*c*. 6.5-2.5 MYA).

DOLICHOPITHECUS RUSCINENSIS Depéret, 1889 (extinct)

Type locality Serrat d'en Vacquer, (*c*. 42°40′N 2°53′E), near Perpignan, Pyrénées Orientales, France.

Range As for the genus.

Locality	Date of Collection	Sex	Age	Nature of Specimen	Accession no.	References, Remarks, etc.
Dolichopithecus ruscinensis Depéret 1889 (extinct)						
FRANCE						
Perpignan, Pyrénées Orientales, [c.42°40′N 2°53′E]	—	I	[Ad]	Cast of R femur	BM(NH)A M47106	
[FRANCE]						
[Perpignan], [c.42°40′N 2°53′E]	—	I	Juv	Cast of lower R dP4-M1	BM(NH)A M32284	Listed Delson,E. (unpubl.)c: 13
" " "	—	I	Juv	Cast of lower L dP3-M1	BM(NH)A M32285	" " " "
" " "	—	I		Cast of L molar	BM(NH)A M32286	Listed Delson,E. (unpubl.)c: 14
[Serrat d'en Vacquer], [nr. Perpignan], [Pyrénées Orientales], [c.42°40′N 2°53′E]	—	I	[Juv]	Cast of lower R dI1-dP4	BM(NH)A M32283	Figured Depéret,C. 1890: Pl.II, fig.3

Cercopithecidae *incertae sedis*
PROHYLOBATES Fourtau, 1918 (extinct)

(2 species: type species *Prohylobates tandyi* Fourtau, 1918)

Morphology Small to medium-sized (*P. tandyi*) or large (*P. simonsi*) Old World Monkeys with incomplete bilophodonty. Known only from mandibular fragments, the former is characterised by having M3 only very slightly larger than M2, whereas in *P. simonsi* it is significantly larger.

Range Egypt and Libya. *Age* Early to middle Miocene, *c.* 19-15 MYA.

PROHYLOBATES TANDYI Fourtau, 1918 (extinct)

Principal synonym ?*Dryopithecus mogharensis* Fourtau, 1918:95.

Type localities Of *tandyi* and *mogharensis*, Wadi Moghara (30°28′N 29°E), northern Egypt.

Referred material Known only from the type site. *Age* Early to middle Miocene, *c.* 19-15 MYA.

Locality	Date of Collection	Sex	Age	Nature of Specimen	Accession no.	References, Remarks, etc.
Prohylobates tandyi Fourtau 1918 (extinct)						
EGYPT						
Hateyet el Moghara, Libyan Desert, [30°28′N 29°E]	—		Ad	R mandibular fragment with P4-M3	BM(NH)A M20494	Holotype *Prohylobates tandyi* Fourtau 1918, Figured Fourtau,R. 1920: 93, fig.63
" " "	—		Ad	Cast of R mandibular fragment with P4-M3	BM(NH)A M26219	
" " "	—		Ad	Cast of R mandibular fragment with M1-3 (broken)	BM(NH)A M20492	(?*Dryopithecus mogharensis*), Figured Fourtau,R. 1920: 95, fig.64a
" " "	—		Ad	Cast of R mandibular fragment with M1-3 (broken)	BM(NH)A M26220	" " " "
" " "	—			Cast of L mandibular fragment with P4-M2	BM(NH)A M20493	(?*Dryopithecus mogharensis*), Figured Fourtau,R. 1920: 95, fig.64b

PROHYLOBATES SIMONSI Delson, 1979 (extinct)

Type locality Jebel Zelten (*c.* 28°37′N 19°43′E), Libya.

Referred material Known only from the type site. *Age* Early to middle Miocene, *c.* 19-15 MYA.

Dimensions Weight *c.* 19kg, estimated from a regression equation based on the length of the second lower molar (Kay & Simons, 1980:23).

Locality	Date of Collection	Sex	Age	Nature of Specimen	Accession no.	References, Remarks, etc.
Prohylobates simonsi Delson 1979 (extinct)						
LIBYA						
6km E. of Ora and 15km W. of edge of Jebel Zelten, ?Lower Marada Formation, [c.28°37′N 19°43′E]	1967		Ad	Cast of L mandibular fragment with M2-3	BM(NH)A M47255	Holotype *Prohylobates simonsi* Delson 1979, (Dichobunid artiodactyl), Figured Delson,E. 1979: 725-733, pl.1, figs.1-3, 7-10

VICTORIAPITHECUS Von Koenigswald, 1969 (extinct)

(2 species: type species *Victoriapithecus macinnesi* Von Koenigswald, 1969)

Taxonomic notes The two included species are probably not members of the same genus but are treated together for convenience. Hence the specific name is written *'Victoriapithecus' leakeyi*, to indicate that, in the event of a revision, a new generic name might have to be proposed for it. The set of casts in the collections follow Von Koenigswald's identifications, with the exception of BM(NH)A M32301 which has been assigned to *'V'. leakeyi* by Delson (1975:189, fig 8c) and Szalay & Delson (1979:421, fig 214c) to whom reference should be made for further details. The latter have referred the early Miocene molar tooth from Napak V, Karamoja, Uganda, to *Victoriapithecus* sp., thus associating it with *V. macinnesi* rather than with *'V'. leakeyi*. The cast, BM(NH)A M35191, has been catalogued as Cercopithecoidea, fam., gen. et sp. indet.

Morphology Small cercopithecids. *V. macinnesi* possesses fully bilophodont molar teeth, whereas *'V'. leakeyi* has a crista obliqua on the type molar. Delson (1975:191) has suggested that the divergence between the cercopithecine and colobine subfamilies is foreshadowed here. *V. macinnesi* shows the tooth characters predicted for the ancestral colobine stock and *'V'. leakeyi* those predicted for the ancestral cercopithecines.

Range East Africa. *Age* Middle Miocene (18.5-15 MYA) (Szalay & Delson, 1979:420).

VICTORIAPITHECUS MACINNESI Von Koenigswald, 1969 (extinct)

Type locality Maboko Island (0°10′S 34°36′30″E), N. of Kavirondo Gulf, Kenya.

Referred material Loperot, Ombo and Maboko, Kenya. *Age* Middle Miocene (*c.* 15 MYA).

Locality	Date of Collection	Sex	Age	Nature of Specimen	Accession no.	References, Remarks, etc.
Victoriapithecus macinnesi Von Koenigswald 1969 (extinct)						
KENYA						
Maboko, [N. of Kavirondo Gulf], [0°10′S 34°36′30″E]	—		[Ad]	Cast of R mandibular corpus with P4-M3	BM(NH)A M32298	Holotype *Victoriapithecus macinnesi* Von Koenigswald 1969, Figured Von Koenigswald,G.H.R. 1969: 41, pl.1, figs.1a-d, Figured Delson,E. 1975: 189, fig.8b,f,g
,, ,, ,,	—		Ad	Cast of L mandibular fragment with M2-3	BM(NH)A M32300	Figured Von Koenigswald,G.H.R. 1969: 41, fig.3
,, ,, ,,	—		[Ad]	Cast of L distal humerus	BM(NH)A M32307	Figured Delson,E. 1975: 190, fig.9d, Figured Szalay,F.S. and Delson,E. 1979: 424, fig.216C,M
,, ,, ,,	1951		[Ad]	Cast of lower L M3	BM(NH)A M32311	
,, ,, ,,	—			Cast of upper R M3[?]	BM(NH)A M32312	Figured Von Koenigswald,G.H.R. 1969: Pl.1, fig.5
,, ,, ,,	—		Ad	Cast of lower L M3	BM(NH)A M32313	
,, ,, ,,	—		Ad	Cast of lower R M3	BM(NH)A M32314	
,, ,, ,,	—			Cast of upper R M	BM(NH)A M32320	Figured Von Koenigswald,G.H.R. 1969: Pl.1, fig.12
,, ,, ,,	—			Cast of upper R M	BM(NH)A M32325	
,, ,, ,,	—			Cast of upper R M1	BM(NH)A M32326	
,, ,, ,,	—		Ad	Cast of upper [L] M3	BM(NH)A M32327	
,, ,, ,,	—		Ad	Cast of lower R M3	BM(NH)A M32328	
,, ,, ,,	—		[Ad]	Cast of lower L C	BM(NH)A M32329	
,, ,, ,,	—		[Ad]	Cast of lower R C	BM(NH)A M32330	
[KENYA]						
[Maboko]	—		[Ad]	Upper R C	BM(NH)A M32303	

'VICTORIAPITHECUS' LEAKEYI Von Koenigswald, 1969 (extinct)

Type locality Maboko Island (0°10′S 34°36′30″E), N. of Kavirondo Gulf, Kenya.

Referred material Maboko and ?Ombo, Kenya. *Age* Middle Miocene (*c.* 15 MYA).

Locality	Date of Collection	Sex	Age	Nature of Specimen	Accession no.	References, Remarks, etc.
"Victoriapithecus" leakeyi Von Koenigswald 1969 (extinct)						
KENYA						
Maboko, [N. of Kavirondo Gulf], [0°10′S 34°36′30″E]	—		Ad	Cast of L mandibular fragment with M2-3	BM(NH)A M32301	(*Victoriapithecus macinnesi*), Figured Von Koenigswald,G.H.R. 1969: 41, pl.1, fig.2, Figured Delson,E. 1975: 189, fig.8c,h,i, Figured Szalay,F.S. and Delson,E. 1979: 421, fig.214c
,, ,, ,,	—		[Ad]	Cast of proximal fragment R ulna	BM(NH)A M32306	Figured Von Koenigswald,G.H.R. 1969: 43, pl.1, fig.19
,, ,, ,,	—			Cast of L mandibular fragment with M1-2	BM(NH)A M32308	Figured Von Koenigswald,G.H.R. 1969: 43, pl.1, fig.11, Figured Delson,E. 1975: 189, fig.8d
,, ,, ,,	—		Ad	Cast of lower R M3	BM(NH)A M32310	Figured Von Koenigswald,G.H.R. 1969: 43, pl.1, fig.7
,, ,, ,,	—			Cast of upper L M2[?]	BM(NH)A M32315	Holotype *Victoriapithecus leakeyi* Von Koenigswald 1969, Figured Von Koenigswald,G.H.R. 1969: 43, pl.1, fig.8, Figured Szalay,F.S. and Delson,E. 1979: 421, fig.214h
,, ,, ,,	—			Cast of upper R M	BM(NH)A M32316	
,, ,, ,,	—			Cast of upper R M	BM(NH)A M32317	
,, ,, ,,	—			Cast of upper L M1	BM(NH)A M32319	
,, ,, ,,	—			Cast of lower R M	BM(NH)A M32322	Cited Von Koenigswald,G.H.R. 1969: 43
,, ,, ,,	—		[Ad]	Cast of upper L C	BM(NH)A M32331	
,, ,, ,,	—			Cast of R mandibular fragment with P4-M2	BM(NH)A M32299	Figured Von Koenigswald,G.H.R. 1969: 43, pl.1, fig.10
Victoriapithecus sp.indet. (extinct)						
KENYA						
Maboko, [N. of Kavirondo Gulf], [0°10′S 34°36′30″E]	—			Cast of upper P[?]	BM(NH)A M32318	
,, ,, ,,	—			Cast of upper R M	BM(NH)A M32321	
,, ,, ,,	—			Cast of lower L M	BM(NH)A M32323	
,, ,, ,,	—			Cast of lower L M	BM(NH)A M32324	
,, ,, ,,	—			Upper L dC	BM(NH)A M32304	
,, ,, ,,	—			Lower L P3	BM(NH)A M32305	
[KENYA]						
Bur-Siala, Ombo, [0°04′S 34°34′E]	—			Lower R M1 or M2	BM(NH)A M20209	(*?Mesopithecus*), Listed Delson,E. (unpubl.)c

Locality	Date of Collection	Sex	Age	Nature of Specimen	Accession no.	References, Remarks, etc.
Victoriapithecus sp. indet. *continued*						
[Maboko], [N. of Kavirondo Gulf], [0°10′S 34°36′30″E]	—	—	—	Cast of lower R M1	BM(NH)A M32332	
,, ,, ,,	—	—	—	Cast of upper R M	BM(NH)A M32333	
,, ,, ,,	—	—	—	Cast of upper L C	BM(NH)A M32334	
—	—	—	—	Lower L P3	BM(NH)A M32302	
Cercopithecidae gen. et sp. indet.						
KENYA						
Gamble's Cave II, Elmenteita, Gt. Rift Valley, [c.0°30′S 36°05′E]	—	—	—	Lower L P4	BM(NH)A M20610	(Upper Pleistocene *(*Gamblian*)*)
Cercopithecoidea fam., gen. et sp. indet.						
[EGYPT]						
—	—	—	—	Casts of four teeth	BM(NH)A M32292	Figured Stromer,E. 1920: 352-354, figs.3,6-8
KENYA						
Gamble's Cave II, Elmenteita, Gt. Rift Valley, [0°30′S 36°05′E]	—	—	—	14Ms, 1 lower R I2	BM(NH)A M34880	(Upper Pleistocene *(*Gamblian*)*) (Primate)
,, ,, ,,	—	—	—	4 proximal radius fragments	BM(NH)A M34881	,, ,, ,, ,,
,, ,, ,,	—	—	—	Maxillary fragment with 2Ms, 2 mandibular fragments with Ms, mesial portion of upper L C	BM(NH)A M34882	,, ,, ,, ,,
,, ,, ,,	—	—	—	2 metatarsals	BM(NH)A M34883	,, ,, ,, ,,
,, ,, ,,				3 metacarpals (IV)	BM(NH)A M34884	,, ,, ,, ,,
,, ,, ,,	—	—	—	17 canines	BM(NH)A M34885	,, ,, ,, ,,
,, ,, ,,	—	—	—	5 premolars	BM(NH)A M20612	,, ,, ,, ,,
,, ,, ,,	—	—	[Ad]	R metatarsal I	BM(NH)A M20613	(Upper Pleistocene (Gamblian) (?Cercopithecoid)
,, ,, ,,	—	—	[Ad]	R metatarsal I	BM(NH)A M20614	,, ,, ,, ,,
,, ,, ,,	—	—	[Ad]	L metatarsal I	BM(NH)A M20615	,, ,, ,, ,,
,, ,, ,,	—	—	[Ad]	L metatarsal I	BM(NH)A M20616	,, ,, ,, ,,
,, ,, ,,	—	—	[Ad]	Metacarpal III	BM(NH)A M20618	(Upper Pleistocene (Gamblian) (Primate ?Cercopithecus)
,, ,, ,,	—	—	[Ad]	Proximal head of L humerus	BM(NH)A M20622	(Upper Pleistocene (Gamblian) (Primate)
,, ,, ,,	—	—	[Ad]	Distal head of R humerus	BM(NH)A M20623	,, ,, ,, ,,
,, ,, ,,	—	—	[Ad]	3 1st metatarsals	BM(NH)A M20627	(Upper Pleistocene (Gamblian) (?Cercopithecus)
,, ,, ,,	—	—	[Ad]	3 metacarpals	BM(NH)A M20628	(Upper Pleistocene (Gamblian)
Lukeino, [N.W. of L. Baringo], [c.0°50-55′N 35°55′E]	—	—	—	Cast of upper molar	BM(NH)A M34469	(Miocene) (Monkey)
Ngorora, [c.0°54′N 35°51′E]	—	—	—	Cast of lower I1	BM(NH)A M34468	,, ,, ,, ,,
,, ,, ,,	—	—	—	Cast of lower R P4	BM(NH)A M34470	,, ,, ,, ,,
PAKISTAN						
Kotal Kund, [32°46′N 73°18′E]	—	[M]	Ad	Cast of fused maxilla and mandible	BM(NH)A M35128	(Upper Miocene)
[S.AFRICA]						
Hrdlicka's cave, Buxton Limeworks, Taung, S.African Bechuanaland, [Cape Province], [c.27°32′S 24°48′E]	1953	—	—	29 cheek teeth, 7 associated postcranial fragments (unidentified)	BM(NH)A M47301	(Lower Pleistocene) (Primate)
UGANDA						
Napak [Site V], [Karamoja], [2°05′N 34°14′E]	—	—	—	Cast of upper molar	BM(NH)A M35191	(Early Miocene), Cited Pilbeam,D. and Walker,A.C. 1968: 657-660
?Cercopithecoidea						
[MALAWI]						
Mwenirondo, [W. of N. end of L.Malawi], (Chiwondo Beds), Karongo Dist., Nyasaland	1963	—	—	Head of R femur	BM(NH)A M25299	(Pleistocene)

Appendix I

Abbreviations employed for other museums and departments

Other departments in the museum and other museums in the British Isles are indicated in the Accession Number column by the following abbreviations:

BDS	= Dental School, University of Bristol, Lower Maudlin Street, Bristol BS1 2LY.
BM(NH)A	= Sub-Department of Anthropology, Department of Palaeontology, British Museum (Natural History).
CDA	= Department of Anatomy, University of Cambridge, Downing Street, Cambridge.
CMZ	= University Museum of Zoology, Downing Street, Cambridge.
Cuming Mus.	= The Cuming Museum (Borough of Southwark), Walworth Road, London, SE17.
DLPA	= Duckworth Laboratory of Physical Anthropology, Faculty of Archaeology and Anthropology, Downing Street, Cambridge.
EDA	= Department of Anatomy, University Medical School, Teviot Place, Edinburgh 8.
EDZ	= Department of Zoology, The Ashworth Laboratory, West Mains Road, Edinburgh 9.
GCMAG	= Glasgow City Museum and Art Gallery, Glasgow G3 8AG.
GU(HM)	= Hunterian Museum, Department of Zoology, The University, Glasgow W2.
GZD	= Zoology Department, University College, Galway.
HM	= The Hancock Museum, Newcastle upon Tyne NE2 4PT.
LM	= Leicester Museum and Art Gallery, New Walk, Leicester LE1 6TD.
MHMS(DA)	= Department of Anatomy, Middlesex Hospital Medical School, Cleveland Street, London W1P 6BD.
NMI	= National Museum of Ireland, Kildare Street, Dublin 2.
NMW	= National Museum of Wales, Cathays Park, Cardiff CF1 3NP.
PCM	= The Powell-Cotton Museum, Quex Park, Birchington, Kent.
RCS(HM)	= Hunterian Museum, Royal College of Surgeons, Lincoln's Inn Fields, London WC2A 3PN.
RCS(OM)*	= Odontological Museum, Royal College of Surgeons, Lincoln's Inn Fields, London WC2A 3PN.
RSM	= Royal Scottish Museum, Chambers Street, Edinburgh.
SM	= Sheffield Museum, Weston Park, Sheffield S10 2TP.

*The large holdings of skulls housed in the Odontological Museum were mainly collected under the aegis of Sir Frank Colyer, who made extensive studies of dental development and dental abnormalities in mammals. His primate collections have recently been catalogued and details of the dentition, etc. entered in a computer-based retrieval system by Dr Caroline Grigson. Many skins pertaining to these skulls are in the British Museum (Natural History), the Powell-Cotton Museum, and the National Museum of Victoria, Melbourne. For further information, application should be made to the Odontological Museum, Royal College of Surgeons.

Appendix II

Field measurements of adult specimens of Colobinae in the British Museum (Natural History). Linear measurements are expressed in millimetres, weight in grams.

		Male			Female		
		Mean	Range	No.	Mean	Range	No.
Colobus polykomos	head & body	633	584-680	10	600	570-648	8
	tail	874	718-1000	10	886	806-959	8
	total	1507	1302-1660		1486	1410-1569	
	weight	8420	6600-10 000	5	7133	6100-8500	6
C. angolensis	head & body	590	553-660	7	529	483-593	7
	tail	829	762-920	7	706	629-763	7
	total	1419	1315-1560		1235	1149-1346	
	weight	9679	9072, 10 285	2	7403	6804-8895	5
C. guereza	head & body	615	543-699	20	576	521-673	22
	tail	667	521-826	20	687	528-797	22
	total	1282	1085-1499		1263	1073-1402	
	weight	9327	7970-11 790	8	7826	5448-10 230	12
C. satanas	head & body	670		1	635		1
	tail	800		1	800		1
	total	1470			1435		
C. badius	head & body	569	470-690	30	528	470-584	20
	tail	665	540-800	30	665	515-750	20
	total	1234	1020-1438		1193	1038-1315	
	weight	9348	6300-11 340	7	6770	5500-9072	7
C. kirkii	head & body	450		1	490	465-506	3
	tail	595		1	671	644-715	3
	total	1045			1161	1120-1215	
	weight	5800		1	5463	5100-6100	3
Procolobus verus	head & body	480	455-510	6	466	380-546	14
	tail	559	500-610	6	573	430-710	14
	total	1039	960-1086		1039	880-1220	
	weight	4000		1	3500	2900, 4100	2
Presbytis melalophos	head & body	492	423-565	24	494	420-555	22
	tail	711	618-824	24	711	636-820	22
	total	1203	1077-1389		1205	1075-1375	
	weight	7031		1	6464		1
P. comata	head & body	505	430-595	3	535	475-570	3
[= *aygula*]	tail	655	560-724	3	659	590-720	3
	total	1160	990-1275		1194	1143-1290	
P. frontata	head & body	600		1	545	525-570	3
	tail	740		1	726	720-730	3
	total	1340			1271	1253-1300	
P. hosei	head & body	533	520, 545	2	550		1
	tail	665	660, 670	2	650		1
	total	1198	1190, 1205		1200		
P. potenziani	head & body	500		1			
	tail	580		1			
	total	1080					
P. rubicunda	head & body	520		1	495	470-508	3
	tail	730		1	702	675-725	3
	total	1250			1197	1175-1233	
P. thomasi	head & body	550		1	550		1
	tail	690		1	760		1
	total	1240			1310		
P. cristata	head & body	556	465-650	19	506	420-630	15
	tail	737	671-838	19	701	560-795	15
	total	1293	1174-1430		1207	1003-1400	
	weight	8626		1			

		Male			Female		
		Mean	Range	No.	Mean	Range	
P. francoisi	head & body	548	485-635	4	570	550-590	3
	tail	849	820-872	4	852	830-887	3
	total	1397	1330-1493		1422	1410-1437	
P. geei	head & body				490		1
	tail				713		1
	total				1203		
P. obscura	head & body	534	420-675	27	519	425-595	38
	tail	734	570-790	27	723	635-813	38
	total	1268	1060-1439		1242	1090-1360	
	weight	7319	6129-9080	12	6602	4994-8626	22
P. phayrei	head & body	542	420-600	19	516	442-570	8
	tail	770	745-858	19	764	719-795	8
	total	1312	1165-1400		1280	1161-1345	
	weight	7931	5675-9080	8	6946	6356-7491	5
P. pileata	head & body	618	533-710	12	543	490-660	6
	tail	946	859-1025	12	867	830-955	6
	total	1564	1392-1730		1410	1330-1615	
	weight	5774	4310-6630	4	5139	4752, 5525	2
P. vetulus [= *senex*]	head & body	569	495-616	5	514	484-560	8
	tail	753	618-885	5	717	640-820	8
	total	1322	1207-1485		1231	1130-1330	
	weight	8172	5675-9761	3	5902	4994-7491	3
P. johnii	head & body	570	508-645	5			
	tail	860	756-965	5			
	total	1430	1264-1512				
P. entellus	head & body	641	501-780	28	584	406-680	33
	tail	912	714-1080	28	862	693-1016	32
	total	1553	1215-1750		1444	1208-1651	
	weight	12 748	7945-20 430	19	9421	5448-17 252	20
Pygathrix nemaeus	head & body	586	550-630	4	597		1
	tail	680	600-735	4	597		1
	total	1266	1230-1320		1194		
P. avunculus	head & body	590	560, 620	2	515	510, 520	2
	tail	871	822, 920	2	695	660, 730	2
	total	1461	1382, 1540		1210	1180, 1240	
P. brelichi	head & body				730		1
	tail				970		1
	total				1700		
Nasalis larvatus	head & body	745	730, 760	2	620	610-640	3
	tail	665	660, 670	2	573	550-620	3
	total	1410	1400, 1420		1193	1160-1230	
Simias concolor	head & body	515		1	500		1
	tail	155		1	140		1
	total	670			640		
	weight				7144		1

Appendix III

Comparative Table of Characters

Field measurements are taken from Appendix II unless otherwise stated. Dimensions are cited in the following order: the mean, the number of observations (in parentheses) and the range (in square brackets). Superscript figures refer to the authorities cited. See footnote for other sources of information.

Genus, subgenus or species group	Body Size & Weight H & B Length ♂♂ (mm)	H & B Length ♀♀ (mm)	Tail Length ♂♂ (mm)	Tail Length ♀♀ (mm)	Weight ♂♂ (g)	Weight ♀♀ (g)	Sexual Dimorphism H & B Length (♀ in % of ♂)
Colobus (Colobus)	617 (38) [543-699]	574 (38) [483-673]	755 (38) [521-1000]	735 (38) [528-959]	9072 (15) [6600-11 790]	7553 (23) [5448-10 230]	93
Colobus (Piliocolobus)	565 (31) [450-690]	523 (23) [465-584]	663 (31) [540-800]	666 (23) [515-750]	8905 (8) [5800-11 340]	6378 (10) [5100-9072]	93
Procolobus	480 (6) [455-510]	466 (14) [380-546]	559 (6) [500-610]	573 (14) [430-710]	4000 (1)	3500 (2) [2900, 4100]	97
Presbytis melalophos group	502 (33) [423-600]	506 (33) [420-570]	700 (33) [560-824]	706 (33) [590-820]	7031 (1)	6464 (1)	101
Presbytis cristata group	554 (81) [420-710]	520 (71) [420-660]	780 (81) [570-1025]	740 (71) [560-955]	7320 (25) [4310-9080]	6560 (29) [4752-8626]	94
Presbytis vetulus group	570 (10) [495-645]	514 (8) [484-560]	807 (10) [618-965]	717 (8) [640-820]	8172 (3) [5675-9761]	5902 (3) [4994-7491]	90
Presbytis entellus group	641 (28) [501-780]	584 (33) [406-680]	912 (28) [714-1080]	862 (32) [693-1016]	12 748 (19) [7945-20 430]	9421 (20) [5448-17 252]	91
Pygathrix (Pygathrix)	586 (4) [550-630]	597 (1)	680 (4) [600-735]	597 (1)	—	—	102
Pygathrix (Rhinopithecus)	590 (2) [560, 620]	587(3) [510-730]	871 (2) [822, 920]	788 (3) [660-970]	—	—	99
Nasalis	745 (2) [730, 760]	620 (3) [610-640]	665 (2) [660, 670]	573 (3) [550-620]	20 344 (10)[1] [14 062-23 587]	9873 (15)[1] [8165-11 794]	83
Simias	515 (1)	500 (1)	155 (1)	140 (1)	—	7144 (1)	97

Weight (♀ in % of ♂)	Skull and Teeth Cranial Capacity	No. of Cusps on lower M3*	Skeleton Intermembral Index†	Brachial Index†	Crural Index†	Other Chromosome Diploid No.[4]	Reproduction Gestation Period (days)*
83	77.6cc (81)[2] [60-96]	5 (rarely 6)	79 (40) [76-83]	92 (40) [88-98]	87 (40) [85-93]	44	165 (5) [147-178]
72	76.6cc (95)[2] *(C. badius)* 56.9cc (13) *(C. kirkii)*	5 (rarely 6)	86 (21) [79-89]	94 (21) [91-99]	86 (21) [79-87]	44	—
86	54.5cc (13)[2] [49-62]	6 (rarely 5)	—	—	—	—	—
92	70.2cc (36)[3] [60-83]	4	76 (41) [73-79]	109 (41) [99-115]	89 (41) [86-94]	—	—
90	61.4cc (34)[3] [54-71]	5	82 (61) [78-87]	94 (61) [88-99]	86 (61) [82-89]	44	—
72	—	—	76 (6) [75-79]	101 (6) [98-104]	89 (6) [88-90]	44	198-220 (est.)
74	—	—	79 (6) [73-83]	100 (6) [97-104]	86 (6) [83-89]	44	196 (est.)
—	—	5 or 6	89 (5) [85-91]	100 (5) [98-103]	89 (5) [86-92]	44	165 (est.); 180-190 (est.)
—	—	5 or 6	95 (9) [94-97]	99 (9) [96-103]	85 (9) [83-90]	—	—
49[1]	92cc (25)[1]	5 (rarely 6)	93 (22) [89-97]	100 (22) [96-104]	86 (22) [83-91]	48	166 (est.)
—	—	—	98 (1)	104 (1)	83 (1)	—	—

[1] Schultz, A. H. (1942)

[2] Verheyen, W. N. (1962)

[3] Washburn, S. L. (1942)

[4] Chiarelli, A. B. (1973)

* See text for references

† Limb Proportions from Napier (1971) and unpublished data

† Intermembral Index: $\frac{\text{Humerus + Radius length} \times 100}{\text{Femur + Tibia length}}$

† Brachial Index: $\frac{\text{Radius length} \times 100}{\text{Humerus length}}$

† Crural Index: $\frac{\text{Tibia length} \times 100}{\text{Femur length}}$

Bibliography

Agrawal,V.C. 1974 Taxonomic status of Barbe's Leaf-monkey, *Presbytis barbei* Blyth. *Primates* **15**: 235-239

Aiello,L. 1981 Locomotion in the Miocene Hominoidea. *in* Stringer,C.B. (Ed.). *Aspects of Human Evolution*. London: Taylor and Francis.

Alexander,B. 1907 *From the Niger to the Nile*. London: Edward Arnold.

Allen,G.M. 1938 The mammals of China and Mongolia. *in* Granger,W. (Ed). *Natural History of Central Asia*. New York: American Museum of Natural History **11** (1).

——— 1939 A checklist of African mammals. *Bull. Mus. comp. Zool. Harv.* **83**: 1-763.

Allen,G.M. and Coolidge,H.J. 1940 Mammal and bird collections of the Asiatic Primate Expedition. Mammals. *Bull. Mus. comp. Zool. Harv.* **87**: 131-166.

Allen,J.A. 1925 Primates collected by the American Museum Congo Expedition. *Bull. Am. Mus. nat. Hist.* **47**: 283-499, 89 pls.

Anderson,J. 1879 *Anatomical and zoological researches: comprising an account of the zoological results of the two expeditions to western Yunnan in 1868 and 1875*. London: Bernard Quaritch.

——— 1881 *Catalogue of Mammalia in the Indian Museum, Calcutta. Part I. Primates, Prosimiae, Chiroptera, and Insectivora*. Calcutta: Trustees of the Indian Museum.

Andrews,P. 1981 Ecological polarity in primate evolution. *in* Joysey,K.A. and Friday,A.E. (Eds). *Problems of phylogenetic reconstruction*. London: Academic Press.

Anon (unpubl.) Inventory of Mammalia and parts of Mammalia in the possession of the Zoological Society of London.

——— (unpubl.)a Zoological Society's Osteological Catalogue.

——— (unpubl.)b Primates and Chiroptera. *Catalogue of Mammalia*. **1**.

——— 1964 Photograph of adult-infant relationship in *Colobus guereza kikuyuensis*. *Zoonooz* **43**: 12.

Ansell,W.F.H. 1959 The type localities of *Colobus angolensis sharpei* and *Colobus angolensis sandbergi* Lönnberg. *Rev. Zool. Bot. Afr.* **60**: 168-171.

Baldwin,L.A., Kavanagh,M. and Teleki,G. 1975 Field research on langur and proboscis monkeys: an historical, geographical and bibliographical listing. *Primates* **16**: 351-363.

Baldwin,L.A., Teleki,G. and Kavanagh,M. 1976 Field research on colobus, guenon, mangabey and patas monkeys: an historical, geographical and bibliographical listing. *Primates* **17**: 233-252.

Baptista,N.A. (unpubl.) Eastern Ghat Survey: Mammal Report.

Bauchop,T. 1978 Digestion of leaves in vertebrate arboreal folivores. *in* Montgomery,G.G. (Ed.). 1978.

Bauchop,T. and Martucci,R. 1968 Ruminant-like digestion of the langur monkey. *Science, N.Y.* **161**: 698-700.

Bennett,E.T. 1832 On several skins of Mammalia from Algoa Bay. *Proc. zool. Soc. Lond.* (1832): 122-123.

——— 1833 Characters of two new species of monkeys in the Society's collection. *Proc. zool. Soc. Lond.* (1833): 67-68.

Berenberg-Gossler,H.von 1911 Untersuchungen über Bau und Entwicklung des zusammengesetzten Magens der Schlankaffen. (Reference not seen, cited in Hill,W.C. Osman, 1958).

Biswas,B. 1967 Authorship of the name *Presbytis geei* (Mammalia: Primates). *J. Bomb. nat. Hist. Soc.* **63**: 429-431.

Blackburn-Maze,C.J. (unpubl.) Letters to G.Dollman, 22 May, 18 June and 27 June, 1914.

Blanford,W.T. 1888 *The fauna of British India, including Ceylon and Burma. Mammalia*. London: Taylor and Francis.

Blyth,E. 1841 Report for the month of September. Animal Kingdom. *J. asiat. Soc. Beng.* **10**: 836-842.

——— 1843 Revision of previous report to the Society, vol. 10, page 839. *J. asiat. Soc. Beng.* **12**: 169-176.

——— 1844 List of birds from Calcutta. *Ann. Mag. nat. Hist.* **13**: 312-313.

——— 1844a Notices of various Mammalia, with descriptions of many new species. Part 1. The primates. *J. asiat. Soc. Beng.* **13**: 470-494.

——— 1847 Supplementary report of the Curator of the zoological department. *J. asiat. Soc. Beng.* **16**: 728-737, 1 pl.

——— 1847a Report of the Curator, Zoological Department. *J. asiat. Soc. Beng.* **16**: 1271-1272, 1 pl.

——— 1851 Report on the Mammalia and more remarkable species of birds inhabiting Ceylon. *J. asiat. Soc. Beng.* **20** 153-185.

——— 1859 Report of the Curator, Zoological Department, for February to May meetings, 1859. *J. asiat. Soc. Beng.* **28**: 271-298.

——— 1863 *Catalogue of the Mammalia in the Museum of the Asiatic Society of Bengal*. Calcutta: Asiatic Society of Bengal.

——— 1875 Catalogue of the mammals and birds of Burma. *J. asiat. Soc. Beng.* **44** (2, extra number): 1-167.

Boddaert,P. 1784[1785] *Elenchus animalium*. Rotterdam: C.R.Hake **1**.

Boggess,J. 1979 Troop male membership changes and infant killing in langurs (*Presbytis entellus*). *Folia primatol.* **32**: 65-107.

Bonaparte,C. 1856 Excursions dans les divers musées d'Allemagne, de Hollande et de Belgique, et Tableaux parallèliques de l'ordre des Echassiers. *C. r. hebd. Séanc. Acad. Sci., Paris* **43**: 410-421.

Bonhote,J.L. 1901 On the mammals collected during the "Skeat" Expedition to the Malay Peninsula, 1899-1900. *Proc. zool. Soc. Lond.* (1900): 869-883.

——— 1903 Report on the mammals. *in* Annandale,N and Robinson,H.C. (Eds.). *Fasciculi Malayenses. Zoology*. London: Longmans, Green and Co. for University of Liverpool Press (1).

Booth,A.H. 1957 Observations on the natural history of the Olive Colobus Monkey, *Procolobus versus* (van Beneden). *Proc. zool. Soc. Lond.* **129**: 421-430.

——— 1958 The zoogeography of West African primates. *Bull. Inst. fr. Afr. noire* (A) **20**: 587-622.

Brandon-Jones,D. (unpubl.) Primates collected by A.H.Booth in Ivory Coast. Field List reconstructed October, 1980.

——— 1978 Comment on the proposal to conserve Colobidae Blyth, 1875, as the family-group name for the leaf-eating monkeys (Mammalia, Primates). Z.N.(S.) 2094. *Bull. zool. Nomencl.* **35**: 69-70.

——— 1984 Colobus and Leaf monkeys. *in* MacDonald,D. (Ed.). *The encyclopaedia of mammals*. London and Sydney: George Allen and Unwin. **1**.

Broom,R. and Robinson,J.T. 1950 A new sub-fossil baboon from Kromdraai, Transvaal. *Ann. Transv. Mus.* **21**: 242-245.

Buffon,G.L.L.de and Daubenton,J.L.M. 1766 *Histoire naturelle, génerale et particulière, avec la description du Cabinet du Roi*. Paris: Imprimerie Royale **14**.

Cantor,T. (unpubl.) Packing lists of Malayan collections, case no.4. India Museum, List of collections (1853). (BMNH Archives, Department of Zoology).

——— 1845 Observations on two Malayan species of *Semnopithecus*. *Proc. linn. Soc. Lond.* **1** (1 Apr 1845): 235-236.

——— 1846 Catalogue of Mammalia inhabiting the Malayan Peninsula and Islands. *J. asiat. Soc. Beng.* **15**: 171-203.

——— 1846a On a species of *Semnopithecus* from the Peninsula of Malacca. *Ann. Mag. nat. Hist.* **17**: 335-340.

Cautley,P.T. and Falconer,H. 1840 Notice on the remains of a fossil monkey from the Tertiary strata of the Sewalik Hills in the north of Hindoostan. *Trans. geol. Soc. Lond.* (2)**5**: 449-504.

——— 1868 On the fossil Quadrumana of the Sewalik Hills. 1. Notice on the remains of a monkey from the Tertiary strata of the Sewalik Hills in the north of Hindostan. *in* Murchison,C. (Ed.), 1868.**1**: 292-297.

Chasen,F.N. 1935 On mammals from Siam. *J. nat. Hist. Soc. Siam* **10**: 31-57.

——— 1940 A handlist of Malaysian mammals. *Bull. Raffles Mus.* **15**: i-xx, 1-209, 1 folding map.

Chasen,F.N. and Kloss,C.Boden 1927 Spolia Mentawiensia. Mammals. *Proc. zool. Soc. Lond.* (1927): 797-840, 5 pls, 1 map.

——— 1931 On a collection of mammals from the lowlands and islands of North Borneo. *Bull. Raffles Mus.* **6**: 1-82.

Chen Ho-yi 1974 A visit to the Wanglang Nature Reserve. *Wildlife* **16**: 7-8.

Chen,Y., Luo,L., Shan,X. and Cao,X. 1978 The karyotype of the golden monkey, (*Rhinopithecus r. roxellanae*). *Acta Zool. Sin.* **24**: 127-135.

Chiarelli,B. 1973 Numerical data on the chromosomes of the Old World Primates. *J. hum. Evol.* **2**: 297-300.

Clutton-Brock,T.H. 1975 Feeding behaviour of red colobus and black and white colobus in East Africa. *Folia primatol.* **23**: 165-207.

Collett,R. 1893 On a new monkey from north-east Sumatra. *Proc. zool. Soc. Lond.* (1892): 613-617, 1 pl.

Colyer,J.F. (unpubl.) Catalogue of private collection, Royal College of Surgeons (Odontological Museum).

——— (unpubl.)a Notes of the dental conditions of a collection of monkeys from Uganda.

——— 1936 *Variations and diseases of the teeth of animals*. London: Bale, Sons and Danielsson,Ltd.

——— 1940 Variations of the teeth of Preuss's colobus. *Proc. R. Soc. Med.* **33**: 756-768.

——— 1943 Variations of the teeth of the colobus monkey. *Dent.Rec.* **63**: 183-188.

Cooke,H.B.S. and Maglio,V.J. 1972 Plio-Pleistocene stratigraphy in East Africa in relation to proboscidean and suid evolution. *in* Bishop,W.W. and Miller,J.A. (Eds.). *Calibration of hominoid evolution*. Scottish Academic Press.

Coppens,Y., Howell,F.C., Isaac,G.L. and Leakey,R.E.F. (Eds.) 1976 *Earliest man and environments in the Lake Rudolf Basin.* Chicago: University of Chicago Press.

Curtin,H.C. 1977 Niche separation in sympatric Malaysian leaf-monkeys (*Presbytis obscura* and *Presbytis melalophos). Yb. phys. Anthrop.* (1976): 421-439.

Curtin,H.C. and Chivers,D.J. 1978 Leaf-eating primates of peninsular Malaysia: the siamang and the dusky leaf-monkey. *in* Montgomery,G.G. (Ed.), 1978.

Cuvier,F. 1824 *Des dents des mammiferes considerees comme caractères zoologiques.* Paris: Le Normant.

Dandelot,P. 1971 *The mammals of Africa. An identification manual.* Washington,D.C.: Smithsonian Institution Press (Part 3, Order Primates).

Dandelot,P. and Groves,C.P. (in prep.)

Dào Van Tien 1970 Sur les formes de semnopithè que noir, *Presbytis francoisi* (Colobidae, Primates) au Vietnam et déscription d'une forme nouvelle. *Mitt. zool. Mus. Berl.* **46**: 61-65.

——— 1977 Sur quelques rares mammifères au nord du Vietnam. *Mitt. zool. Mus. Berl.* **53**: 325-330.

David,G.F.X. and Ramaswami,L.S. 1969 Studies on menstrual cycles and other related phenomena in the langur *(Presbytis entellus entellus). Folia primatol.* **11**: 300-316.

Davis,D.Dwight 1958 Mammals of the Kelabit Plateau, northern Sarawak. *Fieldiana Zool.* **39**: 119-147.

——— 1962 Mammals of the lowland rainforest of north Borneo. *Bull. natn. Mus. St. Singapore* **31**: 1-129.

Delson,E. (unpubl.). Cast Exchange List, 1973.

——— (unpubl.)a Letter to P.H. Napier. (20 Jun 1974).

——— (unpubl.)b Letter to P.H. Napier. (15 Oct 1974).

——— (unpubl.)c Cast Exchange List, 1977.

——— 1975 Evolutionary history of the Cercopithecidae. *Contrib. Primatol.* **5**: 167-217.

——— 1975a Paleoecology and zoogeography of the Old World Monkeys. *in* Tuttle,R. (Ed.). *Primate functional morphology and evolution.* The Hague: Mouton.

——— 1976 The family-group name of the leaf-eating monkeys (Mammalia, Primates): a proposal to give Colobidae Blyth, 1875, precedence over Semnopithecidae Owen, 1843 and Presbytina Gray, 1825. Z.N.(S.) 2094. *Bull. zool. Nomencl.* **33**: 85-89.

——— 1977 Catarrhine phylogeny and classification: principles, methods and comments. *J. hum. Evol.* **6**: 433-459.

——— 1979 *Prohylobates* (Primates) from the Early Miocene of Libya: a new species and its implications for cercopithecid origins. *Geobios* **12** (5): 725-733, 1 pl.

Delson,E. and Andrews,P. 1975 Evolution and interrelationships of the catarrhine primates. *in* Luckett,W.P. and Szalay,F.S. (Eds.). *Phylogeny of the primates. A multidisciplinary approach.* New York and London: Plenum Press.

Depéret,C. 1889 Sur le *Dolichopithecus ruscinensis* nouveau singe fossile du Pliocène du Roussillon. *C. r. hebd. Séanc. Acad. Sci., Paris* **109**: 982-983.

——— 1890 Les animaux pliocènes du Roussillon. *Mem. Soc. Géol. France, Paléontol.* **3**: 1-126, 18 pls.

——— 1929 *Dolicopithecus arvensis* Depéret, nouveau singe du pliocène supérieur du Sénèze (Haute Loire). *Trav. Lab. Géol. Univ. Lyon* **15** (12): 5-12, 1 pl.

Deraniyagala,P.E.P 1954 A new race of long-tailed bear monkey. *Ceylon Adm. Rep.* (1954) (4. Education, Science and Art (E)): 1-26.

——— 1955 A new race of leaf monkey from Ceylon. *Spolia zeylan.* **27**: 293-294.

Desmarest,A.G. 1822 *Mammalogie ou description des espè ces de mammifè res. pt.* 2. Paris: Agasse.

Dietrich,W.O. 1942 Ältestquartiäre Säugetiere aus der Südlichen Serengeti, Deutsch-ostafrika. *Palaeontographica* (A) **94**: 43-133, 11 pls.

Dollman,G. 1909 A new species of *Presbytis*, allied to *P. rubicundus. Ann. Mag. nat. Hist.* (8) **4**: 204-205.

——— 1909a Two new species of *Colobus* from Central Africa, collected by Mr.R.Grauer. *Ann. Mag. nat. Hist.* (8)**4**: 474-476.

——— 1912 *Rhinopithecus avunculus, sp.n.* (Abstract No.106). *Proc. zool. Soc. Lond.* (1912): 18.

——— 1912a A new snub-nosed monkey. *Proc. zool. Soc. Lond.* (1912): 503-504.

——— 1932 *in* Ward,F. Kingdon, 1932.

Dufresne,C. 1797 Sur une nouvelle espèce de singe. *Bull. Soc. philom. Paris* **1** (7): 49.

Eck,G. 1976 Cercopithecoidea from Omo Group deposits, *in* Coppens,Y., Howell,F.C., Isaac,G.L. and Leakey,R.E.F. (Eds.), 1976.

——— 1977 Diversity and frequency distribution of Omo Group Cercopithecoidea. *J. hum. Evol.* **6**: 55-63, 6 pls

Edwards,A.Milne 1870 Note sur quelques mammifères du Thibet oriental. *C. r. hebd. Séanc. Acad. Sci., Paris* **70**: 341-342.

——— 1871 Note sur une nouvelle espèce de semnopithèque provenant de la Cochinchine. *Nouv. Archs Mus. Hist. nat. Paris* **6 (Bulletin)**: 7-9, pl.1.

——— 1872 Mémoire sur la faune mammalogique du Tibet oriental et principalement de la principauté de Moupin. *in* Edwards,M.H.Milne and Edwards,A.Milne. *Recherches pour servir à l'histoire naturelle des mammifères.* Paris: G.Masson **Teste et Atlas.**

——— 1876 Sur quelques mammifères et crustacés nouveaux. *Bull. Soc. philomath. Paris* (6) **13**: 8-10.

——— 1886 *in* Rivière,E., 1886.

——— 1897 Note sur une nouvelle espè ce du genre Rhinopithèque provenant de la haute vallée du Mékong. *Bull. Mus. Hist. nat. Paris* **3**: 156-160.

Edwards,A.Milne and Pousargues,E.de 1898 Le rhinopithè que de la vallée du haut Mékong (*Rhinopithecus bieti*, A.M.-E.). *Nouv. Archs Mus. Hist. nat. Paris* (3) **10**: 121-142. 4 pls.

Eisenberg,J.F. and McKay,G.M. 1970 An annotated checklist of the recent mammals of Ceylon with keys to the species. *Ceylon J.Sci.* (Bio. Sci) **8**: 69-99.

Eisentraut,M. 1973 Die Wirbeltierfauna von Fernando Poo und Westkamerun. *Bonn zool. Monogr.* **3**: 1-428, 5 pls.

Ellerman,J. and Morrison-Scott,T.C.S. 1951 *Checklist of palaearctic and Indian mammals*, 1758-1946. London: Trustees of the British Museum.

Elliot,D.G. 1906 Descriptions of an apparently new species of monkey of the genus *Presbytis* from Sumatra, and of a bat of the genus *Dermanura* from Mexico. *Proc. biol. Soc. Wash.* **19**: 49-50.

——— 1907 Descriptions of apparently new species and subspecies of mammals belonging to the families Lemuridae, Cebidae, Callitrichidae and Cercopithecidae in the collection of the Natural History Museum. *Ann. Mag. nat. Hist* (7)**20**: 185-196.

——— 1909 Descriptions of apparently new species and subspecies of monkeys of the genera *Callicebus, Lagothrix, Papio, Pithecus, Cercopithecus, Erythrocebus and Presbytis. Ann. Mag. nat. Hist.* (8)**4**: 244-274.

——— 1910 Descriptions of some new species of monkeys of the genera *Pithecus* and *Pygathrix* collected by Dr. W.L.Abbott and presented to the United States National Museum. *Proc. U.S. natn. Mus.* **38** (1751): 343-352.

——— 1913 *A review of the Primates.* New York: Amer. Mus. nat. Hist. Monogr.**3**.

Elliot,W. 1839 A catalogue of the species of Mammalia found in the southern Mahratta country with their synonyms in the native languages in use there. *Madras J. Lit. Sci.* **10** (July-December): 92-108.

Erxleben,J.C.P. 1777 *Systema regni animalis, Classis I. Mammalia.* Lipsiae.

Eschscholtz,J.F.von 1821 Beschreibung einer neuen Affengattung, *Presbytis mitrata. in* Kotzebue,O.von, 1821.

Everett,A.H. 1879 2nd quarterly report on the Bornean Cave Exploration. *in* Report of the Committee appointed for exploring certain caves in Borneo. *Rep. Br. Ass. Advmt Sci.* (1879): 149-155.

Falconer,H. and Cautley,P.T. 1837 On additional fossil species of the Order Quadrumana from the Sewalik Hills. *J. asiat. Soc. Beng.* **6**: 354-360, 1 pl.

——— 1868 On the fossil Quadrumana of the Sewalik Hills. 3. An additional fossil species of the Order Quadrumana of the Sewalik Hills. *in* Murchison,C. (Ed.), 1868.

Feiler,A. and Eck,S. 1979 Ein weiteres Stück von *Simias concolor* im staatlichen Museum für Tierkunde in Dresden. *Säugetierk. Mitt* **26**: 317-318.

Fischer,J.B. 1829 *Synopsis mammalium.* Stuttgart: J.G.Cotta.

Fleagle,J.G. 1977 Locomotor behaviour and skeletal anatomy of sympatric Malaysian leaf-monkeys (*Presbytis obscura* and *Presbytis melalophos*). *Yb. phys. Anthrop.* (1976): 440-453.

——— 1977a Locomotor behaviour and muscular anatomy of sympatric Malaysian leaf-monkeys (*Presbytis obscura* and *Presbytis melalophos*). *Am. J. phys. Anthrop.* **46**: 297-308.

——— 1978 Locomotion, posture and habitat utilization in two sympatric Malaysian leaf-monkeys (*Presbytis obscura* and *Presbytis melalophos*). *in* Montgomery,G.G. (Ed.), 1978.

Flower,S.S. 1900 On the Mammalia of Siam and Malay Peninsula. *Proc. zool. Soc. Lond.* (1900): 306-379.

Fooden,J. 1964 Rhesus and crab-eating macaques: intergradation in Thailand. *Science, N.Y.* **143**: 363-365.

——— 1971 Report on primates collected in Western Thailand January - April, 1967. *Fieldiana Zool.* **59**: 1-62.

——— 1976 Primates obtained in peninsular Thailand June - July, 1973, with notes on the distribution of continental southeast Asian leaf-monkeys (*Presbytis*). *Primates* **17**: 95-118.

Fourtau,R. 1918 *Contribution à l'étude des vertébrés miocènes de l'Egypte.* Cairo: Survey Department, Ministry of Finance.

Freedman,L. 1957 The fossil Cercopithecoidea of South Africa. *Ann. Transv. Mus.* **23**: 121-262, 53 pls.

——— 1961 Some new fossil cercopithecoid specimens from Makapansgat, South Africa. *Palaeont. afr.* **7** (1960): 7-58.

——— 1965 Fossil and subfossil primates from the limestone deposits at Taung, Bolt's Farm and Witkrans, South Africa. *Palaeont. afr.* **9**: 19-48.

——— 1970 A new checklist of fossil Cercopithecoidea of South Africa. *Palaeont. afr.* **13**: 109-110.

Freedman,L. and Brain,C.K. 1972 Fossil cercopithecoid remains from the Kromdraai Australopithecine site (Mammalia: Primates). *Ann. Transv. Mus.* **28**: 1-16.

Fry,T.B. 1929 Bombay Natural History Society's Mammal Survey of India, Burma and Ceylon. Report No 46 (Supplementary). On the second, third and fourth collections from Toungoo, Burma, made by Mr J.M.D. Mackenzie, I.F.S., between the dates February 9 1927 to March 2, 1928. *J. Bomb. nat. Hist. Soc.* **33**: 636-652.

Furuya,Y. 1963 The social life of the silvered leaf monkeys (*Trachypithecus cristatus*). *Primates* **3** (1961-62): 41-60.

Gairdner,K.G. (unpubl.) Letter to Oldfield Thomas, June 1916, with list of mammals sent to British Museum, 1915.

——— (unpubl.)a Field notes on mammals collected during 1911-13 in Rat Buri and Phet Buri areas of Thailand.: 1-11.

——— 1914 Notes on the fauna and flora of Ratburi and Petchaburi Districts. *J. nat. Hist. Soc. Siam* **1**: 27-40.

——— 1915 Notes on the fauna and flora of Ratburi and Petchaburi Districts (continued). *J. nat. Hist. Soc. Siam* **1**: 131-156, 1 map.

——— 1915a Additions to the mammalian fauna of Ratburi. *J. nat. Hist. Soc. Siam* **1**: 252-255.

Gaudry,A. 1862 *Animaux fossiles et géologie de l'Attique.* Paris: F.Savy.

Gee,E.P. 1956 A new species of langur in Assam. *J. Bomb. nat. Hist. Soc.* **53**: 252-254, 1 map.

——— 1961 The distribution and feeding habits of the golden langur, *Presbytis geei* Gee (Khajuria, 1956). *J. Bomb. nat. Hist. Soc.* **58**: 1-12, 3 pls, 2 maps.

Geoffroy Saint-Hilaire,E. 1812 Tableau des quadrumanes, ou des animaux composant le premier Ordre de la Classe des Mammifères. *Annis Mus. Hist. nat. Paris* **19**: 85-122.

Geoffroy Saint-Hilaire,I. 1830 Descriptions de plusieurs espèces nouvelles de singes, appartenant aux genres semnopithèque et macaque. *Bull. Sci. nat. Géol.* **22**: 317-318.

——— 1831 Mammifères. *in* Bélanger,C. *Voyage aux Indes-Orientales par le nord de l'Europe ... pendant les années* 1825-1829. *Zoologie.* Paris: A.Bertrand.

——— 1842 Sur les singes de l'ancien monde, spécialement sur les genres gibbon et semnopithèque. *C. r. hebd. Séanc. Acad. Sci., Paris* **15**: 716-720.

——— 1843 Description des mammifè res nouveaux ou imparfaitement connus de la collection du Muséum d'Histoire naturelle, et remarques sur la classification et les caractères des mammifères. *Archs Mus. Hist. nat. Paris* **2**: 485-592, 6 pls.

——— 1851 *Catalogue méthodique de la collection des mammifères.* Paris: Gide et Baudry.

Gerrard,E. 1862 *Catalogue of the bones of Mammalia in the collections of the British Museum.* London: Trustees of the British Museum.

Gervais,P. 1849 *Zoologie et Paléontologie françaises, Ed. 1.* Paris: Bertrand.

——— 1859 *Zoologie et Paléontologie françaises, Ed. 2. 2 vols, Texte et Atlas.* Paris: A.Bertrand.

Gittins,S.P. and Akonda,A.W. 1982 What survives in Bangladesh?. *Oryx* **16**: 275-281.

Gloger,C.W.L. 1841 *Gemeinnütziges Hand- und Hilfsbuch der Naturgeschichte.* Breslau: A.Schulz and Co. **1** (1).

Gochfeld,M. 1974 Douc langurs. *Nature, Lond.* **247**: 167.

Granvik,H. 1925 On mammals from the eastern slopes of Mount Elgon, Kenya Colony. *Acta Univ. lund.* (2)**21** (3): 1-32, 2 pls.

Gray,J.E. (unpubl.) Early registers of Mammalia.

——— 1842 Descriptions of some new genera and fifty unrecorded species of Mammalia. *Ann. Mag. nat. Hist.* ([1])**10**: 255-267.

——— 1843 *List of the specimens in the collection of the British Museum. Additional species.* London: Trustees of the British Museum.

——— 1866 Notice of two new species of *Colobus* from western Africa. *Ann. Mag. nat. Hist.* (3)**17**: 77.

——— 1868 Notice of a new species of *Colobus* and of another monkey from Zanzibar. *Proc. zool. Soc. Lond.* (1868): 180-182, 1 pl.

——— 1871 *Catalogue of monkeys, lemurs and fruit-eating bats in the collection of the British Museum.* London: Trustees of the British Museum (1870).

Griffith,W. 1840 Journal of the mission which visited Bootan in 1837-38, under Captain R. Boileau Pemberton. Part 2. *J. asiat. Soc. Beng.* **8**: 251-291, 1 map.

Groves,C.P. 1970 The forgotten leaf-eaters, and the phylogeny of the Colobinae. *in* Napier,J.R. and Napier,P.H. (Eds.), 1970.

——— 1973 Notes on the ecology and behaviour of the Angola colobus (*Colobus angolensis* P.L.Sclater, 1860) in N.E. Tanzania. *Folia primatol.* **20**: 12-26.

Grubb,P. and Booth,A.H. (unpubl.) List of monkey skulls in Zoology Department, University of Ghana, collected by A.H.Booth.

H.M.Customs and Excise (unpubl.) Letters to Department of Zoology, August 1980.

Haddow,A.J. (unpubl.) Letters in Odontological Museum, Royal College of Surgeons.

——— (unpubl.)a Uganda colobus records and field notes (BMNH Archives, Department of Zoology).

——— 1952 Field and laboratory studies on an African monkey, *Cercopithecus ascanius schmidti* Matschie. *Proc. zool. Soc. Lond.* **122**: 297-394.

Haddow,A.J., Dick,G.W.A., Lumsden,W.H.R. and Smithburn,K.C. 1951 Monkeys in relation to the epidemiology of yellow fever in Uganda. *Trans. R. Soc. trop. Med. Hyg.* **45**: 189-224.

Haddow,A.J., Smithburn,K.C., Mahaffy,A.F. and Bugher,J.C. 1947 Monkeys in relation to yellow fever in Bwamba County, Uganda. *Trans. R. Soc. trop. Med. Hyg.* **40**: 677-700.

Harrisson,T.H. 1933 The Oxford University Expedition to Sarawak. *Geogrl. J.* **82**: 385-410, 1 map.

Hayman,R.W. 1936 On a collection of mammals from the Gold Coast. *Proc. zool. Soc. Lond.* (1935): 915-937, 1 pl.

Heller,E. 1913 New races of ungulates and primates from Equatorial Africa. *Smithson. misc. Collns* **61** (17): 1-12.

Hick,U. 1975 The breeding and maintenance of Douc langurs at Cologne Zoo. *in* Martin,R.D. (Ed.). *Breeding endangered species in captivity.* London, New York and San Francisco: Academic Press.

Hill,C.A. (unpubl.) Letter to Curator of mammals, 24 March 1975.

——— 1975 The longevity record for *Colobus*. *Primates* **16**: 235.

Hill,J.E. 1960 The Robinson Collection of Malaysian mammals. *Bull. Raffles Mus.* **29**: 1-112.

Hill,W.C.Osman 1934 A monograph on the purple-faced leaf-monkeys (*Pithecus vetulus*). *Ceylon J. Sci.* (B)**19**: 23-88.

——— 1936 Supplementary observations on purple-faced leaf-monkeys (genus *Kasi*). *Ceylon. J. Sci.* (B)**20** (1): 115-133, 1 pl.

——— 1936a On a hybrid leaf-monkey; with remarks on the breeding of leaf-monkeys in general. *Ceylon J. Sci.* (B)**20** (1): 135-148, 1 pl.

——— 1936b Notes on Barbe's leaf-monkey (*Trachypithecus barbei* Blyth). *Proc. zool. Soc. Lond.* (1936): 105-109.

——— 1937 The primates of Travancore. *Proc. zool. Soc. Lond.* (B) (1937): 205-216, 1 pl.

——— 1937a The type of *Semnopithecus thersites* Blyth. *Ceylon J. Sci.* (B)**20** (2): 207-209, 1 pl.

——— 1937b On the breeding and rearing of certain species of primates in captivity. *Ceylon J. Sci.* (B)**20** (3): 369-389, 3 pls.

——— 1952 The external and visceral anatomy of the olive colobus monkey. *Proc. zool. Soc. Lond.* **122**: 127-186.

——— 1958 Verdauungstrakt. IIA. Pharynx, oesophagus, stomach, small and large intestine. *in* Hofer.H., Schultz,A.H. and Starck,D. (Eds.). *Primatologia.* Basel: S. Karger **3** (1).

——— 1958a B. External genitalia. *in* Hofer,H., Schultz,A.H. and Starck,D. (Eds.). *Primatologia.* Basel: S. Karger **3** (1).

——— 1966 Catarrhini: Cercopithecoidea, Cercopithecinae. *Primates. Comparative Anatomy and Taxonomy.* Edinburgh: University Press **6**.

——— 1970 Cynopithecinae: *Papio, Mandrillus, Theropithecus. Primates. Comparative Anatomy and Taxonomy.* Edinburgh: University Press **8**.

Hill,W.C.Osman and Booth,A.H. 1957 Voice and larynx in African and Asian Colobidae. *J. Bomb. nat. Hist. Soc.* **54**: 309-321.

Hinton,M.A.C. 1923 The nomenclature and subspecies of the purple-faced langur, with notes on other langurs inhabiting Ceylon. *Ann. Mag. nat. Hist.* (9)**11**: 506-515.

——— 1923a Scientific results from the Mammal Survey. No. 36. On the capped langurs (*Pithecus pileatus* Blyth and its allies). *J. Bomb. nat. Hist. Soc.* **29**: 77-83.

Hinton,M.A.C. and Fry,T.B. 1923 Bombay Natural History Society's Mammal Survey of India, Burma and Ceylon. Report No. 37. Nepal. *J. Bomb. nat. Hist. Soc.* **29**: 399-428.

Hinton,M.A.C. and Lindsay,H.M. 1926 Bombay Natural History Society's Mammal Survey of India, Burma and Ceylon. Report No. 41. Assam and Mishmi Hills. *J. Bomb. nat. Hist. Soc.* **31**: 383-402.

Hladik,C.M. 1977 A comparative study of the feeding strategies of two sympatric species of leaf monkeys: *Presbytis senex* and *Presbytis entellus*. *in* Clutton-Brock,T.H. (Ed.). *Primate Ecology.* London, New York and San Francisco: Academic Press.

——— 1978 Adaptive strategies of primates in relation to leaf-eating. *in* Montgomery,G.G. (Ed.), 1978.

Hladik,C.M. and Hladik,A. 1972 Disponibilités alimentaires et domaines vitaux des primates à Ceylan. *Terre Vie* **26**: 149-215.

Hodgson,B.H. 1832 On the Mammalia of Nepal. *J. asiat. Soc. Beng.* **1**: 335-349.

——— 1841 Three new species of monkey: with remarks on the genera *Semnopithecus et Macacus*. *J. asiat. Soc. Beng.* **9** (1840): 1211-1213.

——— 1842 Classified catalogue of mammals of Nepal, (corrected to end of 1841, first printed in 1832). *J. asiat. Soc. Beng.* **10** (1841): 907-916.

——— 1846 *Catalogue of the specimens of Mammalia and birds of Nepal and Thibet presented by B.H. Hodgson, Esq. to the British Museum*. London: Trustees of the British Museum.

Hollihn,U. 1973 Remarks on the breeding and maintenance of colobus monkeys (*Colobus guereza*), proboscis monkeys (*Nasalis larvatus*) and Douc langurs (*Pygathrix nemaeus*) in zoos. *Int. Zoo. Yb.* **13**: 185-188.

Hooijer,D.A. 1948 A new race of the leaf-monkey *Presbytis aygula* (L.) from Deli, northeastern Sumatra. *Proc. K. ned. Akad. Wet.* **51**: 234-237.

——— 1960 Race and sex in the silvered leaf monkey (*Trachypithecus cristatus*). *Archs neerl. Zool.* **13**: 580-581.

——— 1962 Quaternary langurs and macaques from the Malay Archipelago. *Zool. Verh. Leiden* (No. 55): 1-64, 2 pls.

Horsfield,T. 1823 *Zoological researches in Java, and the neighbouring islands*. London: Kingsbury, Parbury and Allen (7): no pagination.

——— 1830 Catalogue of zoological specimens. *in* Raffles,S., 1830: 633-697.

——— 1851 *A catalogue of the Mammalia in the Museum of the Hon. East-India Company*. London: The Honourable East-India Company.

Horwich,R.H. and Manski,D. 1975 Maternal care and infant transfer in two species of *Colobus* monkeys. *Primates* **16**: 49-73.

Hose,C. 1893 A journey up the Baram River to Mt. Dulit, and the highlands of Borneo. *Geog. J.* **1**: 198-208, 1 map.

Hose,C. 1893a *A descriptive account of the mammals of Borneo*, London: Edward Abbott.

Hrdy,S.B. 1977 *The langurs of Abu*. Cambridge, Mass.: Harvard University Press.

Hull,D.B. 1978 Aberrations in the coat colour patterns of black and white *Colobus* monkeys. *E. Afr. Wildl. J.* **16**: 21-27.

——— 1979 A craniometric study of the black and white *Colobus* Illiger, 1811 (Primates: Cercopithecoidea). *Am. J. phys. Anthrop.* **51**: 163-181.

Husson,A.M. and Holthuis,L.B. 1955 The dates of publication of "Verhandelingen over de Natuurlijke Geschiedenis der Nederlandsche overzeesche Bezittingen" edited by C.J.Temminck. *Zool. Meded. Leiden,* **34**: 17-24.

Hutton,T. 1844 *in* Blyth,E., 1844a.

Illiger,C. 1811 *Prodromus systematis mammalium et avium*. Berlin.

International Commission on Zoological Nomenclature 1956 Opinion 417. Rejection for nomenclatorial purposes of Volume 3 (Zoologie) of the work by Lorenz Oken entitled "Okens Lehrbuch der Naturgeschichte" published in 1815-1816. *Opin. Decl. int. Commn zool. Nom.* **14**: 3-42.

——— 1964 *International Code of Zoological Nomenclature*. London: International Trust for Zoological Nomenclature.

James,W.Warwick 1960 *The jaws and teeth of primates*. London: Pitman Medical Publishing Co., Ltd.

Jay,P.C. 1965 The common langur of North India. *in* DeVore,I. (Ed.). *Primate behavior*. New York: Holt, Rinehart and Winston.

Jolly,C.J. 1967 The evolution of the baboons. *in* Vagtborg,H. (Ed.). *The baboon in medical research*. Austin and London: University of Texas Press.

Jouffroy,F.K. and Lessertisseur,J. 1977 Processus de reduction des doigts (main et pied) chez les primates. Modalités, implications génétiques. *Colloques int. Cent. natn. Rech. scient.* (266): 381-392.

Kano,T. 1971 Distribution of the primates on the eastern shore of Lake Tanganyika. *Primates* **12**: 281-304.

Kawabe,M. and Mano,T. 1972 Ecology and behaviour of the wild proboscis monkey, *Nasalis larvatus* (Wurmb), in Sabah, Malaysia, *Primates* **13**: 213-228.

Kay,R.F. 1977 The evolution of molar occlusion in the Cercopithecidae and early Catarrhines. *Am J. phys. Anthrop.* **46**: 327-352.

Kay,R.F. and Hylander,W.L. 1978 The dental structure of mammalian folivores with special reference to Primates and Phalangeroidea. *in* Montgomery,G.G. (Ed.), 1978.

Kay,R.F. and Simons,E.L. 1980 The ecology of Oligocene African Anthropoidea. *Int. J. Primatol.* **1**: 21-37.

Keenan,J. (unpubl.) Letter to J.E.Hill, 12 March 1963, including report and map.

Kelaart,E.F. 1850 Description of new species and varieties of mammals found in Ceylon. *J. Ceylon Brch R. asiat. Soc.* **2**: 321-328.

——— 1851 *in* Blyth,E., 1851.

——— 1852 *Prodromus faunae zeylanicae; being contributions to the zoology of Ceylon*. Colombo, Ceylon.

Kern,J.A. 1964 Observations on the habits of the proboscis monkey, *Nasalis larvatus* (Wurmb), made in the Brunei Bay area, Borneo. *Zoologica, N.Y.* **49**: 183-192, 4 pls.

Kerr,R. 1792 *The animal kingdom or zoological system of the celebrated Sir Charles Linnaeus, Class I. Mammalia*. London: J. Murray and R. Faulder.

Khajuria,H. 1956 A new langur (Primates: Colobidae) from Goalpara District, Assam. *Ann. Mag. nat. Hist.* (12)**9**: 86-88.

Khajuria,H. and Agrawal,V.C. 1979 On the types of *Presbytis barbei* Blyth. *Primates* **20**: 317-319.

Kloss,C. Boden 1911 Diagnoses of new mammals from the Trengganu Archipelago, east coast of the Malay Peninsula. *Ann. Mag. nat. Hist.* (8)**7**: 115-119.

——— 1916 On a collection of mammals from Siam. *J. nat. Hist. Soc. Siam* **2**: 1-32.

——— 1916a On a collection of mammals from the coast and islands of south-east Siam. *Proc. zool. Soc. Lond.* (1916): 27-75, 1 map.

——— 1919 On mammals collected in Siam. *J. nat. Hist. Soc. Siam* **3**: 333-407.

——— 1926 A new race of monkey from Annam. *Ann. Mag. nat. Hist.* (9)**18**: 214.

——— 1930 A note on Javanese monkeys. *Ann. Mag. nat. Hist.* (10)**6**: 697.

Knottnerus-Meyer,T. 1933 Ein neuer Schlankaffe aus Siam. *Zool. Gart. Lpz.* **6**: 259.

Knox.R. 1681 *An historical relation of the island Ceylon, in the East Indies: together with an account of the detaining in captivity the author and divers other Englishmen now living there, and of the author's miraculous escape*. London: Richard Chiswell for the Royal Society.

Kotzebue,O.von 1821 *Entdeckungs-Reise in die Süd-See und nach den Behrings-Strasse zur Erforschung einer nordöstlichen Durchfahrt unternommen in der Jahren 1815-1818*. Weimar: Gebrüdern Hoffman.

Krumbiegel,I. 1942 Zur Kenntnis der Säugetierfauna von Fernando Poo. *Arch. Naturgesch.* (NF) **11** (4): 305-349.

——— 1978 Die kurzschwanz-stumpfnase, *Simias concolor* Miller, 1903 und die übrigen Nasenaffen. *Säugetierk. Mitt.* **26**: 59-75.

Kuhl,H. 1820 *Beiträge zur Zoologie und vergleichenden Anatomie*. Frankfurt am Main: Hermannschen Buchhandlung.

Kuhn,H.-J. 1964 Zur Kenntnis von Bau und Funktion des Magens der Schlankaffen (Colobinae). *Folia primatol.* **2**: 193-221, 1 pl.

——— 1967 Zur Systematic der Cercopithecidae. *in* Starck,D., Schneider,R. and Kuhn,H.-J. (Eds.). *Progress in Primatology*. Stuttgart: Gustav Fischer.

——— 1972 On the perineal organ of male *Procolobus*. *J. hum. Evol.* **1**: 371-378, 4 pls.

Lacépède,B.G.E.de la V. 1799 *Tableau des divisions, sous-divisions, ordres et genres des mammifères*. Paris: Plassan.

Lapouge,G.de 1894 Note sur un nouveau singe pliocène (*Anthropodus rouvillei*). *Bull. Soc. scient. méd. Ouest* **3**: 202-208.

Laurie,E.M.O. and Hill,J.E. 1954 *List of land mammals of New Guinea, Celebes and adjacent islands, 1758-1952*. London: Trustees of the British Museum.

Leakey,M.G. 1976 Cercopithecoidea of the East Rudolf succession. *in* Coppens,Y., Howell,F.C.,Isaac,G.L. and Leakey,R.E.F. (Eds.), 1976.

Leakey,M.G. and Leakey,R.E.F. 1973 New large Pleistocene Colobinae (Mammalia, Primates). *Fossil Vertebr. Afr.* **3**: 121-138, 5 pls.

Leakey,R.E.F. 1969 New Cercopithecidae from the Chemeron Beds of Lake Baringo, Kenya. *Fossil Vertebr. Afr.* **1**: 53-69, 5 pls.

Leconte,J. 1858 Descriptions of several new mammals from Western Africa. *Proc. Acad. nat. Sci. Philad.* **9** (1857): 10-11.

Leigh,C. 1926 Weights and measurements of the Nilgiri langur (*Pithecus johnii*). *J. Bomb. nat. Hist. Soc.* **31**: 223.

Lereboullet,M. 1838 Tableaux des orders, des familles et des genres des mammifères adoptés pour le cours de zoologie de la Faculté des Sciences, par M. Duvernoy. *Mém. soc. Mus. Hist. nat. Strasbourg* **2**: 1-10, 5 tables.

Li,Z. and Ma.S. 1978 A revision of the white-headed langur. *Acta Zootaxonomica Sinica* **5**: 440-442.

Li,Z. Ma,S., Hua,C. and Wang,Y 1980 Distribution and habit of Yunnan snub-nosed monkey. *Zoological Research* **2**: 9-16.

Lindsay,H.M. 1926 Bombay Natural History Society's Mammal Survey of India, Burma and Ceylon. Collection No. 43. Nelliampathy Plateau and Palni Hills. *J. Bomb. nat. Hist. Soc.* **31**: 591-597.

——— 1926a Bombay Natural History Society's Mammal Survey of India, Burma and Ceylon. Collection No. 44. Kangra and Chamba. *J. Bomb. nat. Hist. Soc.* **31**: 597-606.

——— 1926b Bombay Natural History Society's Mammal Survey of India, Burma and Ceylon. Report No 39. Mergui Archipelago. *J. Bomb. nat. Hist. Soc.* **31**: 42-48.

Link,H.F. 1795 *Beyträge zur Naturgeschichte.* Rostock and Leipzig: K.C.Stiller.

Linnaeus,C. 1758 Regnum Animale. *Systema naturae (10th Ed.).* Holmiae: Laurentii Salvii **1**.

——— 1771 *Mantissa plantarum altera.* Holmiae: L.Salvii.

Lippold,L.K. 1977 The douc langur: a time for conservation. *in* H.S.H. Prince Rainier III of Monaco and Bourne,G.H. (Eds.). *Primate Conservation.* New York, San Francisco, London: Academic Press.

Lippold,L.K. and Brockman,D.K. 1974 San Diego's douc langurs. *Zoonooz* **47**: 4-11.

Lönnberg,E. 1908 On a new guereza (*Colobus angolensis sandbergi*), and remarks on other black and white guerezas. *Ark. Zool.* **4**(15): 1-13.

——— 1912 Some new mammals from British East Africa. *Ann. Mag. nat. Hist.* (8)**9**: 63-67.

——— 1919 Contribution to the knowledge about the monkeys of the Belgian Congo. *Revue Zool. Bot. afr.* **7** (2): 107-154.

Lorenz von Libernau,L.R. 1914 Noch zwei neue Formen von Stummel-affen aus der Sammlung R. Grauer's *Anz. Akad. Wiss. Wien* **51**: 507-509.

——— 1914a Einige neue Stummelaffen von Innerafrika aus der Sammlung R.Grauer's. *Anz. Akad. Wiss. Wien* **51**: 383-386.

Lowe,W.P. (unpubl.) Delacour and Lowe Mammal Collections in French Indo-China, 1925-26, 1926-27 and 1927. Zoology Library, BM(NH), Mammalia MS. Notebooks Nos. 8 and 9.

——— (unpubl.)a Indo China Mammal Catalogue, 1929-30. Mammal Section Archives, BM(NH).

——— (unpubl.)b Field notes, Ivory Coast Expedition, 19 Nov 1922. Zoology Library, BM(NH), Mammalia MS Notebooks, No. 11.

——— (unpubl.)c Field notes, Ashanti, Gold Coast Expedition, 19 Dec 1933. Zoology Library, BM(NH), Mammalia MS Notebooks, No. 12.

Lydekker,R. 1884 Indian Tertiary and post-Tertiary Vertebrata. Rodents, ruminants and synopsis of Mammalia. *Mem. geol. Surv. India Palaeont. indica* (10)**3** (3): 105-134.

——— 1885 *Catalogue of the fossil Mammalia in the British Museum (Natural History).* London: Trustees of the British Museum.

——— 1886 Indian Tertiary and post-Tertiary Vertebrata. Siwalik Mammalia. *Mem. geol. Surv. India Palaeont. indica* (10)**4** (1): 1-22, 6 pls.

——— 1886a Indian Tertiary and post-Tertiary Vertebrata. The fauna of the Karnul Caves. *Mem. geol. Surv. India Palaeont. indica* (10)**4** (2): 23-58, 5 pls.

——— 1905 An undescribed guereza. *Ann. Mag. nat. Hist.* (7)**16**: 432.

——— 1906 Colour evolution in guereza monkeys. *Proc. zool. Soc. Lond.* (1905): 325-329.

Lyon,M.W. 1907 Mammals collected in western Borneo by Dr.W.L. Abbott. *Proc. U.S. natn. Mus.* **33**: 547-571, 1 map.

——— 1908 Mammals collected in eastern Sumatra by Dr.W.L.Abbott during 1903, 1906 and 1907, with descriptions of new species and subspecies. *Proc. U.S. natn. Mus.* **34**: 619-679, 4 pls.

——— 1911 Mammals collected by Dr.W.L.Abbott on Borneo and some of the small adjacent islands. *Proc. U.S. natn. Mus.* **40**: 53-146, 7 pls.

Machado,A.de Barros 1969 Mamiféros de Angola ainda não citados ou pouca conhecidos. *Publçoes cult. Co. Diam. Angola* (No. 46): 93-232.

Maglio.V.J. and Cooke,H.B.S. 1978 *Evolution of African mammals.* Cambridge, Mass. and London: Harvard University Press.

Maier,W. 1970 Neue Ergebnisse der Systematik und der Stammesgeschichte der Cercopithecoidea. *Z. Säugethierk.* **35**: 193-214.

——— 1971 New fossil Cercopithecoidea from the lower Pleistocene cave deposits of the Makapansgat Limeworks, South Africa. *Palaeont. afr.* **13** (1970): 69-107.

Mallinson,J.J.M. 1973 Establishing mammal gestation periods. *Rep. Jersey Wildl. Preserv. Trust* (1972) (9): 62-65.

Manley,G. 1976 "Wanderers" in *Presbytis senex. in* Chiver,D.J. and Herbert,J. (Eds.). *Recent Advances in Primatology.* London, New York, San Francisco: Academic Press **1**.

Marler,P. 1968 *Colobus guereza.* Territoriality and group compositon. *Science,N.Y.* **163**: 93-95.

Martin,W.C.L. 1838 Monograph of the genus *Semnopithecus. Mag. nat. Hist.* (NS)**2**: 320-326, 434-441.

——— 1838a Notes on the anatomy of the proboscis monkey (*Simia nasalis*). *Proc. zool. Soc. Lond.* (1837): 70-73.

——— 1841 *A natural history of quadrupeds and other mammiferous animals comprising a description of the class Mammalia including the principal varieties of the human race.* London: Whitehead and Co.

Matschie,P. 1900 Einige Formen der Gattung *Colobus. Sber. Ges. naturf. Freunde Berl.* (1900) (8): 181-189.

——— 1913 Neue Affen aus Afrika nebst einigen Bemerkungen über bekannte Formen. *Annls Soc. r. zool. malacol. Belg.* **47**: 45-81.

——— 1913a Über Affen aus dem belgischen Congo. *Revue zool. afr.* **2**: 203-212.

——— 1914 Ein kleiner Beitrag zur Kenntnis der *Colobus* des unteren Sanga-Gebietes in Neukamerun. *Sber. Ges. naturf. Freunde Berl.* (1914) (7): 342-349.

——— 1914a Neue Affen aus Mittelafrika. *Sber. Ges. naturf. Freunde Berl.* (1914) (7): 323-342.

——— 1917 Die Untergattung: *Stachycolobus* Rochebr. *Sber. Ges. naturf. Freunde Berl.* (1917) (2): 152-163.

Matthew,W.D. and Granger,W. 1923 New fossil mammals from the Pliocene of Sze-Chuan, China. *Bull. Am. Mus. nat. Hist.* **48**: 563-598.

McCann,C. 1934 Observations on some of the Indian langurs. *J. Bomb. nat. Hist. Soc.* **36**: 618-628.

Mearns,C.S. and Pidgeon,M.S. 1978 Breeding and pelage development of black-and-white colobus monkey *Colobus polykomos polykomos. Dodo* (15): 61-69.

Medway,Lord (unpubl.) Gunong Benom (Malaya) Expedition, 1967. Field notes, lists, etc. Mammals.

——— (unpubl.)a Letter to J.E.Hill, 28 Dec 1970.

——— (unpubl.)b Field notes, Perak, 1967.

——— (unpubl.)c Photographs of *Presbytis hosei hosei.* Primate archives.

——— 1965 *Mammals of Borneo. Field Keys and an annotated checklist.* Singapore: Malaysian Branch of the Royal Asiatic Society.

——— 1969 *The wild mammals of Malaya and offshore islands including Singapore.* Kuala Lumpur and Singapore: Oxford University Press.

——— 1970 The monkeys of Sundaland. *in* Napier,J.R. and Napier,P.H. (Eds.), 1970.

Menzies,J.I. (unpubl.) Letters to R.W.Hayman, 17 Feb and 17 Jun 1967.

——— 1970 An eastward extension to the known range of the olive colobus monkey (*Colobus verus* Van Beneden). *J.W. Afr. Sci. Ass.* **15**: 83-84.

Merfield,F.G. (unpubl.) Lists and field notes, 1932-1939.

Mertens,R. 1925 Verseichnis der Säugetier-Typen des Senckenbergischen Museums. *Senckenbergiana* **7**: 18-37.

Meyer,A.B. 1893 Letter containing remarks upon a specimen of *Semnopithecus nemaeus* from Hainan. *Proc. zool Soc. Lond.* (1892): 665.

Miller,G.S. 1903 Seventy new Malayan mammals. *Smithson. misc. Collns* **45** (1420): 1-73, 19 pls.

——— 1903a Mammals collected by Dr.W.L.Abbott on the coast and islands of north-west Sumatra. *Proc. U.S. natn. Mus.* **26**: 437-484, 2 pls.

——— 1906 The mammals collected by Dr.W.L.Abbott in the Rhio-Linga Archipelago. *Proc. U.S. natn. Mus.* **31**: 247-286, 1 map.

——— 1907 Mammals collected by Dr.W.L.Abbott in the Karimata Islands, Dutch East Indies. *Proc. U.S. natn. Mus.* **31**: 55-66, 1 map.

——— 1913 Fifty-one new Malayan mammals. *Smithson. misc. Collns* **61** (21): 1-30.

——— 1934 The langurs of the *Presbytis femoralis* group. *J. Mammal.* **15**: 124-137.

——— 1934a A new monkey from Dutch Northeast Borneo. *Proc. biol. Soc. Wash.* **47**: 15-17.

——— 1942 Zoological results of the George Vanderbilt Sumatran Expedition 1936-1939. Part V. Mammals collected by Frederick A. Ulmer, Jr. on Sumatra and Nias. *Proc. Acad. nat Sci. Philad.* **94**: 107-165.

Mills,J.P. 1923 Bombay Natural History Society's Mammal Survey of India, Burma and Ceylon. Report No. 36 Naga Hills. *J. Bomb. nat. Hist. Soc.* **29**: 221-229.

Mitchell,A. and Weitzel,V. Monkeys and men in the land of mud. *Hemisphere* **27**: 308-314.

Mollett,O.D.v.d.S. 1947 Fossil mammals from the Makapan Valley, Potgietersrust. I. Primates. *S. Afr. J. Sci.* **43**: 295-303.

Montgomery,G.G. (Ed.) 1978 *The ecology of arboreal folivores.* Washington, D.C.: Smithsonian Institution Press.

Moreau,R.E. (unpubl.) Letter to R.W.Hayman, 7 Jan 1944.

Moreau,R.E., Hopkins,G.H.E. and Hayman,R.W. 1946 The type-localities of some African mammals. *Proc. zool. Soc. Lond.* **115**: 387-447.

Mukherjee,R.P. 1978 Further observations on the golden langur (*Presbytis geei* Khajuria, 1956) with a note to capped langur (*Presbytis pileatus* Blyth, 1843) of Assam. *Primates* **19**: 737-747.

Mukherjee,R.P. and Saha,S.S. 1974 The golden langurs (*Presbytis geei* Khajuria, 1956) of Assam. *Primates* **15**: 327-340.

Müller,S. 1838 Over eenige nieuwe zoogdieren van Borneo. *Tijdschr. Natuurl. Gesch. Physiol* **5**: 134-150.

——— 1839-40 Over de zoogdieren van den Indischen Archipel. *in* Temminck,C.J. (Ed.), 1839-44.

Müller,S. and Schlegel,H. 1840-1843 Monographisch overzigt van het Geslacht *Semnopithecus*. *in* Temminck,C.J. (Ed.), 1839-44.: 57-84, 7 pls.

Murchison,C. (Ed.) 1868 *Palaeontological memoirs and notes of the late Hugh Falconer*. London: Robert Hardwicke **1**.

Napier,J.R. 1971 *The roots of mankind*. London: George Allen and Unwin, Ltd.

Napier,J.R. and Napier,P.H. 1967 *A handbook of living primates*. London: Academic Press.

——— (Eds.) 1970 *Old World Monkeys*. New York and London: Academic Press.

Napier,P.H. and Groves,C.P. *Simia fascicularis* Raffles, 1821 (Mammalia, Primates): request for the suppression under the plenary powers of *Simia aygula* Linnaeus, 1758, a senior synonym. Z.N. (S.) 2399. *Bull. zool Nomencli.* **40**: 117-118.

Necrasov,O., Samson,P. and Radulesco,C. 1961 Sur un nouveau singe catarhinien fossile, découvert dans un nid fossilifère d'Olténie (R.P.R.). *Anal. Stiint. Univ. Al. I. Cuza* **7** (2): 401-416, 5 pls.

Neumann,O. 1899 Drei neue afrikanische Säugethiere. *Sber. Ges. naturf. Freunde Berl.* (1899) (1): 15-22.

——— 1902 Ueber neuer nordost- und ostafrikanische Säugethiere. *Sber. Ges. naturf. Freunde Berl.* (1902) (3): 49-59, 2 pls.

Nishida,T., Itani,J., Hiraiwa,M. and Hasegawa,T. 1981 A newly discovered population of *Colobus angolensis* in East Africa. *Primates* **22**: 557-563.

Oates,J.F. & Trocco,T.F. 1983 Taxonomy and phylogeny of black-and-white colobus monkeys. *Folia primatol.* **40**: 83-113.

Ogilby,W. 1835 Descriptions of Mammalia and birds from the Gambia. *Proc. zool. Soc. Lond.* (1835): 97-105

——— 1838 On a new gibbon (*Hylobates choromandus*), and a new species of *Colobus* (*C.leucomeros*). *Proc. zool. Soc. Lond.* (1837): 68-69.

——— 1838a The menageries. The natural history of monkeys, opossums and lemurs. *Library of entertaining knowledge*. London: Knight.

Oken,L. 1816 *Lehrbuch der Naturgeschichte*. Leipzig and Jena.

Osbeck,P. 1757 *Daghof ofwer en Ostindisk Resa aren 1750, 1751, 1752*. Stockholm: L.L.Grefing.

Osgood,W.H. 1932 Mammals of the Kelley-Roosevelts and Delacour Asiatic expeditions. *Publs. Field Mus. nat. Hist.* (Zoology)**18**: 191-339, 1 map.

Otto,A.W. 1825 Über eine neue Affenart, den *Cercopithecus* (?) *leucoprymnus*. *Nova Acta physico-med.* **12** (2): 503-518, 2 pls.

P'eng,H., Li,C. and Yang,T. 1968 A survey of the habits and habitats of *Rhinopithecus brelichi* (Abstract). Abstracts of papers from China Zoological Society's 30th Anniversary Scientific Conference. *Communist Chinese Scientific Abstracts*. Washington,D.C.:U.S.A. Department of Commerce-Joint Publications Research Service **2**.

Peters,W.C.H. 1868 Über neue Säugethiere und Amphibien. *Mber. K. preuss. Akad Wiss.* (1868): 637-642.

——— 1879 Über die von Herrn Dr. G.A.Fischer auf einer in J. 1878 in Ost-Afrika, von Mombas bis in das Pokômoland und das südliche Galla-land, unternommenen Reise eingesammelten Säugethiere. *Mber. K. preuss. Akad. Wiss.* (1879): 829-832, 7 pls.

Pfeffer,P. 1969 Considérations sur l'écologie des forêts claires du Camboge oriental. *Terre Vie* **23**: 3-24.

Phillips,W.W.A. (unpubl.) Letter to J.E.Hill, 22 Aug 1975.

——— 1926 Guide to the mammals of Ceylon. IV. Monkeys. *Spolia zeylan.* (B)**13**: 269-280, 4 pls., 1 map.

——— 1926a The white langur or wanderoo in Ceylon. *Spolia zeylan.* (B)**13**: 291-294.

——— 1927 A new *Pithecus* monkey from Ceylon *Pithecus philbricki* sp. n. *Ceylon J. Sci.* (B)**14**: 57-59.

——— 1927a A further note upon the white (albino) langur in Ceylon. *Spolia zeylan.* (B) **15**: 335-337, 1 pl.

——— 1935 *Manual of the mammals of Ceylon*. Ceylon: The Director, Colombo Museum.

Pilbeam,D.R. and Walker,A.C. 1968 Fossil monkeys from the Miocene of Napak, North-East Uganda. *Nature, Lond.* **220**: 657-660.

Pocock,R.I. 1924 A new genus of monkeys. *Proc. zool. Soc. Lond.* (1924): 330-331.

——— 1928 The langurs, or leaf monkeys, of British India. *J. Bomb. nat. Hist. Soc.* **32**: 472-504, 2 pls., 660-677, 2 pls.

——— 1935 The monkeys of the genus *Pithecus* (*or Presbytis*) and *Pygathrix* found to the east of the Bay of Bengal. *Proc. zool. Soc. Lond.* (1934): 895-961, 2 pls.

——— 1936 The external characters of a female red colobus monkey (*Procolobus badius waldroni*). *Proc. zool. Soc. Lond.* (1935): 939-944.

——— 1939 Primates and Chiroptera (in part). *The fauna of British India including Ceylon and Burma. Mammalia*. London: Taylor and Francis, Ltd **I**.

——— 1941 Notes on some British Indian mammals received too late for entry in Vol I and in the main text of Vol. II. Carnivora. *The fauna of British India including Ceylon and Burma*. London: Taylor and Francis, Ltd. **II**.

Poirier,F.E. 1969 The Nilgiri langur (*Presbytis johnii*) troop: its composition, structure, function and change. *Folia primatol* **10**: 20-47.

Pousargues,E.de 1895 Sur une collection de mammifères provenant du voyage de M. Max Moskowitz au pays de Kong. *Bull. Mus. natn. Hist. nat., Paris* **1**: 98-101.

——— 1896 Étude sur les mammifères du Congo français. *Annls Sci. nat.* (8)**3**: 129-416.

——— 1898 Note préliminaire sur un nouveau semnopithèque des frontières du Tonkin et de la Chine. *Bull. Mus. natn. Hist. nat., Paris* **4**: 319-321.

——— 1899 Sur un nouveau colobe découvert par M. Ed. Foa sur la rive occidentale du Lac Tanganyika. *Bull. Mus. natn. Hist. nat., Paris* **5**: 278-280.

Raffles,S. 1830 *Memoir of the life and public services of Sir Thomas Stamford Raffles, F.R.S.* London: John Murray.

Raffles,T.S. 1821 Descriptive catalogue of a zoological collection, made on account of the Honourable East India Company, in the island of Sumatra and its vicinity, under the direction of Sir Thomas Stamford Raffles, Lieutenant-Governor of Fort Marlborough: with additional notices of the natural history of those countries. *Trans. Linn. Soc. Lond.* **13**: (1): 239-276.

Rahm,U. 1959 Un *Colobus polycomos* nouveau au Congo. *Folia scient. Afr. cent.* **5** (2): 34.

——— 1970 Ecology, zoogeography and systematics of some African forest monkeys. *in* Napier,J.R. and Napier,P.H. (Eds.), 1970.

Rahm,U. and Christiaensen,A.R. 1960 Notes sur *Colobus polycomos cordieri* (Rahm) du Congo Belge. *Revue Zool. Bot. Afr.* **61**: 215-220.

Ray,J. 1693 *Synopsis methodica animalium quadrupedum et serpenti generis*. London: S.Smith and B.Walford.

Reichenbach,H.G.L. 1862 *Central-Atlas für zoologische Gärten* (*Fortsetzung der vollständigsten Naturgeschichte der Affen*). Dresden und Leipzig **Lief 14-16. Affen der alten Welt.**

Reid,J. 1837 On a new species of monkey (*Semnopithecus obscurus*). *Proc. zool. Soc. Lond.* (1837): 14.

Remane,A. 1960 Zähne und Gebiss. *in* Hofer,H., Schultz,A.H. and Starck,D. (Eds.). *Primatologia*. Basel and New York: S.Karger **3** (2).

Ripley,S. 1967 The leaping of langurs: a problem in the study of locomotor adaptation. *Am. J. phys. Anthrop.* **26**: 149-170.

——— 1980 Infanticide in langurs and man: adaptive advantage or social pathology?. *in* Cohen,M.N., Malpass,R.S. and Klein,H.G. (Eds.). *Biosocial mechanisms of population regulation*. New Haven and London: Yale University Press.

Ristori,G. 1890 Le Scimmie fossili italiani. *Bull. R. Com. Geol. Italiana* **21**: 178-196.

Rivière,E. 1886 Exposition de la Mission Brazzae au Muséum. *Revue scient. Paris* (3)**12** (1): 13-23.

Robinson,H.C. and Kloss,C.Boden 1911 On six new mammals from the Malay Peninsula and adjacent islands. *J. fed. Malay St. Mus.* **4**: 169-174.

——— 1915 On a collection of mammals from the Siamese province of Bandon, N.E. Malay Peninsula. *J. fed. Malay St. Mus.* **5**: 111-127.

——— 1918 Results of an expedition to Korinchi Peak, Sumatra. Vertebrates. I. Mammals. *J. fed. Malay St. Mus.* **8**: 1-80, 1 pl.

——— 1919 On five new mammals from Java. *Ann. Mag. nat. Hist.* (9)**4**: 374-378.

Rochebrune,A.-T.de 1886-87 *Faune de la Sénégambie. Supplément*. Paris: O.Doin **1, Mammifères.**

Rodgers,W.A. The distribution and conservation status of colobus monkeys in Tanzania. *Primates* **22**: 33-45.

Roonwal,M.L. 1979 Field study of geographical, subspecific and clinal variations in tail carriage in the Hanuman langur, *Presbytis entellus* (Primates), in South Asia. *Zool. Anz.* **202**: 235-255.

Roosevelt,T.and Roosevelt,K. 1929 *Trailing the Giant Panda*. New York and London: Charles Scribner's Sons.

Roth,J. and Wagner,A. 1854 Die fossilen Knochenüberreste von Pikermi in Greichenland. *Abh. bayer. Akad. Wiss.* **7**: 371-388.

Royle,J.F. 1832 Account of the Honourable Company's Botanic Garden at Seharánpúr. *J. asiat. Soc. Beng.* **1**: 41-58.

Rudran,R. 1973 Adult male replacement in one-male troops of purple-faced langurs (*Presbytis senex senex*) and its effect on population structure. *Folia primatol.* **19**: 166-192.

——— 1973a The reproductive cycles of two subspecies of purple-faced langurs (*Prebytis senex*) with relation to environmental factors. *Folia primatol.* **19**: 41-60.

Rüppell,W.P.E.S. 1835 *Neue Wirbelthiere zu der Fauna von Abyssinien gehörig*. Frankfurt-am-Main: Sigmund Schmerber.

Ryley,K.V. 1913 Bombay Natural History Society's Mammal Survey of India, Burma and Ceylon. Report No.9. Mysore. With field notes by the collector,G.C.Shortridge. *J. Bomb. nat. Hist. Soc.* **22**: 283-295.

——— 1913a Bombay Natural History Society's Mammal Survey of India, Burma and Ceylon. Report No. 10. Kathiawar. With field notes by the collector, C.A.Crump. *J. Bomb. nat. Hist. Soc.* **22**: 464-486.

——— 1913b Bombay Natural History Society's Mammal Survey of India, Burma and Ceylon. Report No. 11. Coorg. With field notes by the collector, G.C.Shortridge. *J. Bomb. nat. Hist. Soc.* **22**: 486-513.

——— 1914 Bombay Natural History Society's Mammal Survey of India, Burma and Ceylon. Report No. 12. Palanpur and Mount Abu. With field notes by the collector, C.A.Crump. *J. Bomb. nat. Hist. Soc.* **22**: 684-699.

——— 1914a Bombay Natural History Society's Mammal Survey of India, Burma and Ceylon. Report No.13. S.Ceylon. *J. Bomb. nat. Hist. Soc.* **22**: 700-710.

——— 1914b Bombay Natural History Society's Mammal Survey of India, Burma and Ceylon. Report No. 14. N. Shan States, Burma. With field notes by the collector, G.C.Shortridge. *J. Bomb. nat. Hist. Soc.* **22**: 710-725.

Sabater Pi,J. 1973 Contribution to the ecology of *Colobus polykomos satanas* (Waterhouse, 1838) of Rio Muni, Republic of Equatorial Guinea. *Folia primatol.* **19**: 193-207.

Sanderson,I.T. (unpubl.) Primate lists, 1932-33.

——— 1940 The mammals of the North Cameroons forest area. Being results of the Percy Sladen Expedition to the Mamfe Division of the British Cameroons. *Trans. zool. Soc. Lond.* **24** (7): 623-725, 22 pls.

Schlegel,H. 1876 Les singes: Simiae. *Mus. Hist. nat. Pays-bas* **7** (Monogr. 40): 1-365.

Schreber,J.C.D.von 1774-1811 *Die Säugthiere in Abbildungen nach der Natur mit Beschreibungen.* Leipzig: T.D.Weigel.

Schultz,A.H. 1942 Growth and development of the proboscis monkey. *Bull. Mus. comp. Zool. Harv.* **89**: 279-314.

——— 1970 The comparative uniformity of the Cercopithecoidea. *in* Napier,J.R. and Napier,P.H. (Eds.), 1970.

Schwarz,E. 1927 A new black-and-white guereza from the Ivory Coast. *Ann. Mag. nat. Hist.* (9) **19**: 155.

——— 1928 Bemerkungen über die roten Stummelaffen. *Z. Säugetierk.* **3**: 92-97.

——— 1928a Die Sammlung afrikanischer Affen im Congo Museum. *Revue Zool. Bot. Afr.* **16**: 105-152.

——— 1929 On the local races and distribution of the black and white colobus monkeys. *Proc. zool. Soc. Lond.* (1929): 585-598.

Schweinfurth,G., Ratzel,F., Felkin,R.W. and Hartlaub,G. 1888 *Emin Pasha in Central Africa.* London: G.Philip and Son.

Sclater,P.L. 1860 List of Mammalia collected by Mr.J.Monteiro in Angola. *Proc. zool. Soc. Lond.* (1860): 245-247.

Setzer,H.W. 1965 Mammals of the Anglo-Egyptian Sudan. *Proc. U.S. natn. Mus.* **106**: 447-587.

Shaw,G. 1800 Mammalia. *General zoology or systematic natural history.* London: G.Kearsley **1** (1).

Sherborn,C.D. 1934 Dates of publication of catalogues of Natural History (post 1850) issued by the British Museum. *Ann. Mag. nat. Hist.* (10)**13**: 308-312.

Simons,E.L. 1967 A fossil *Colobus* skull from the Sudan (Primates,Cercopithecidae). *Postilla* **111**: 1-8, 2 pls.

——— 1969 Miocene monkey (*Prohylobates*) from northern Egypt, *Nature, Lond.* **223**: 687-689.

——— 1972 *Primate evolution. An introduction to man's place in nature.* New York: The Macmillan Company.

——— 1974 *Parapithecus grangeri* (Parapithecidae, Old World higher primates): new species from the Oligocene of Egypt and the initial differentiation of Cercopithecoidea. *Postilla* **166**: 1-12.

Simons,E.L. and Delson,E. 1978 Cercopithecidae and Parapithecidae. *in* Maglio,V.J. and Cooke,H.B.S. (Eds.), 1978.

Simpson,G.G. 1945 The principles of classification and a classification of mammals. *Bull. Am. Mus. nat. Hist.* **85**: 1-350.

Sody,H.J.V. 1930 *Pithecus aygula fredericae* n. subsp. *Trop. Natuur.* **19**: 68.

——— 1930a On some new or insufficiently known mammals from Java, Borneo and Celebes. *Natuurk. Tijdschr. Ned. -Indië* **90**: 258-273, 1 pl.

——— 1931 Six new mammals from Sumatra, Java, Bali and Borneo. *Natuurk. Tijdschr. Ned. -Indië* **91**: 349-360.

——— 1937 Some notes on *Trachypithecus pyrrhus, Presbytis aygula*, and *Cuon javanicus* from Bali, Java and Sumatra. *Proc. zool. Soc. Lond.* (1937): 257-261.

Stromer,E. 1913 Mitteilungen über Wirbeltierreste aus dem Mittelpliocän des Natrontales (Ägypten). 1. Affen. *Z. dt. geol. Ges.* **65**: 350-372, 3 pls.

——— 1920 Mitteilungen über Wirbeltierreste aus dem Mittelpliocän des Natrontales (Ägypten). 5. Nachtrag zu Affen. *Sber. bayer. Akad. Wiss.* (1920) (2): 345-360, 1 pl.

Struhsaker,T.T. (unpubl.) Letter to G.B.Corbet, 4 April 1971.

——— 1974 Correlates of ranging behaviour in a group of red colobus monkeys (*Colobus badius tephrosceles*). *Amer. Zool.* **14**: 177-184.

——— 1975 *The red colobus monkey.* Chicago and London: University of Chicago Press.

Struhsaker,T.T. and Oates,J.F. 1975 Comparison of the behavior and ecology of red colobus and black-and-white colobus monkeys in Uganda: a summary. *in* Tuttle,R.H. (Ed.). *Socioecology and psychology of primates.* The Hague, Paris: Mouton.

Sugiyama,Y. 1964 Group composition, population density and some sociological observations of Hanuman langurs (*Presbytis entellus*). *Primates* **5** (3-4): 7-37.

Sugiyama,Y., Yoshiba,K. and Parthasarathy,M.D. 1965 Home range, mating season, male group and intertroop relations in Hanuman langurs (*Presbytis entellus*). *Primates* **6**: 73-106.

Swindler,D.R. and Sirianni,J.E. 1969 Variability of maxillary incisors among primates. *Int. Congr. anthrop. ethnol. Sci.* **I**: 308-311.

Szalay,F.S. and Delson,E. 1979 *Evolutionary history of the Primates.* New York: Academic Press.

T'an Pang-Chieh 1957 Rare catches by Chinese animal collectors. *Zoo Life* **12** (2): 61-63.

Temminck,C.J. 1807 *Catalogue systêmatique du cabinet d'ornithologie et de la collection de quadrumanes.* Amsterdam: C.Sepp Janz. (Ed.).

——— 1839-44 *Verhandelingen over de Natuurlijke Geschiedenis der Nederlandsche overzeesche Bezittingen. Zoologie.* Leiden: Luchtmans and Van der Hoek.

Tenaza,R.R. 1975 Territory and monogamy among Kloss' gibbons (*Hylobates klossii*) in Siberut Island, Indonesia. *Folia primatol.* **24**: 60-80.

Thomas,O. (unpubl.) Letter from Charles Hose, 17 Feb 1895.

——— 1885 Report on the mammals obtained and observed by Mr.H.H.Johnston on Mount Kilimanjaro. *Proc. zool. Soc. Lond.* (1885): 219-222, 1 pl.

——— 1886 On the mammals presented by Allan O. Hume, Esq., C.B., to the Natural History Museum. *Proc. zool. Soc. Lond.* (1886): 54-79, 2 pls.

——— 1888 On a collection of mammals obtained by Emin Pasha and presented by him to the Natural History Museum. *Proc. zool. Soc. Lond.* (1888): 3-17, 2 pls.

——— 1889 Description of a new Bornean monkey belonging to the genus *Semnopithecus. Proc. zool. Soc. Lond.* (1889): 159-160, 1 pl.

——— 1889a On the mammals of Mount Kina Balu, North Borneo. *Proc. zool. Soc. Lond.* (1889): 228-236, 1 pl.

——— 1892 Description of a remarkable new *Semnopithecus* from Sarawak. *Ann. Mag. nat. Hist.* (6)**10**: 475-477.

——— 1893 Description of a new monkey of the genus *Semnopithecus* from northern Borneo. *Proc. zool. Soc. Lond.* (1892): 582-583, 1 pl.

——— 1893a Descriptions of two North-Bornean mammals. *Ann. Mag. nat. Hist.* (6)**12**: 230-232, 1 pl.

——— 1895 On some mammals collected by Dr. E.Modigliani in Sipora, Mentawei Islands. *Annali Mus. civ. Stor. nat. Genova* (2) **14**: 660-672.

——— 1898 On the mammals obtained by Mr. A.Whyte in Nyasaland, and presented to the British Museum by Sir H.H.Johnston,K.C.B.; being a fifth contribution to the mammal fauna of Nyasaland. *Proc. zool. Soc. Lond.* (1897): 925-939, 1 pl.

——— 1900 List of mammals obtained by Mr. H.J.Mackinder during his recent expedition to Mount Kenya, British East Africa. *Proc. zool. Soc. Lond.* (1900): 173-180.

——— 1901 List of mammals obtained by Dr. Donaldson Smith during his recent journey from Lake Rudolf to the Upper Nile. *Proc. zool. Soc. Lond* (1900): 800-807.

——— 1901a On the more notable mammals obtained by Sir Harry Johnston in the Uganda Protectorate. *Proc. zool. Soc. Lond.* (1901): 85-90, 1 pl.

——— 1902 On some new mammals from northern Nyasaland. *Proc. zool. Soc. Lond.* (1902): 118-121.

——— 1903 *Rhinopithecus brelichi*, sp. n. *Proc. zool. Soc. Lond.* (1903): 224-225, 1 pl.

——— 1903a On a collection of mammals from Abyssinia, including some from Lake Tsana, collected by Mr. Edward Degen. *Proc. zool. Soc. Lond.* (1902): 308-316.

——— 1905 On mammals from the island of Fernando Po, collected by Mr. E.Seimund. *Proc. zool. Soc. Lond.* (1904): 183-193, 1 pl.

——— 1910 A new monkey from the Malay Peninsula (Abstract No. 83). *Proc. zool. Soc. Lond.* (1910): 25.

——— 1911 The mammals of the 10th edition of Linnaeus: an attempt to fix the types of the genera and the exact bases and localities of the species. *Proc. zool. Soc. Lond.* (1911): 120-158.

——— 1918 A selection of lectotypes of Indian mammals, from the co-types described by Hodgson, Gray, Elliot and others. *J. Bomb. nat. Hist. Soc.* **25**: 368-372.

——— 1921 A new monkey and a new squirrel from the middle Mekong, on the eastern frontier of Siam. *Ann. Mag. nat. Hist.* (9)**7**: 181-183.

——— 1927 The Delacour exploration of French Indo-China - Mammals. *Proc. zool. Soc. Lond.* (1927): 41-58.

——— 1928 The Delacour exploration of French Indo-China - Mammals. II. On mammals collected during the winter of 1926-27. *Proc. zool. Soc. Lond.* (1928): 139-150.

——— 1929 The Delacour exploration of French Indo-China - Mammals. III. Mammals collected during the winter of 1927-28. *Proc. zool. Soc. Lond.* (1928): 831-841.

Thomas,O. and Hartert,E. 1894 First glimpses of the zoology of the Natuna Islands. III. List of the first collection of mammals from the Natuna Islands. *Novit. zool.* **1**: 652-660.

Thomas,O. and Wroughton,R.C. 1909 On a collection of mammals from western Java presented to the National Museum by Mr. W.E.Balston. *Proc. zool. Soc. Lond.* (1909): 371-392.

——— 1909a Diagnoses of new mammals collected by Mr. H.C.Robinson in the islands of the Straits of Malacca. *Ann. Mag. nat. Hist.* (8)**4**: 534-536.

——— 1909b Diagnoses of new mammals collected by Mr. H.C.Robinson in the Malay Peninsula and Rhio Archipelago. *Ann. Mag. nat. Hist.* (8)**3**: 439-441.

——— 1909c On mammals from the Rhio Archipelago and Malay Peninsula collected by Messrs. H.C.Robinson, C.Boden Kloss and E.Seimund and presented to the National Museum by the Government of the Federated Malay States. *J. fed. Malay St. Mus.* **4**: 99-129.

Thorington,R.W. and Groves,C.P. 1970 An annotated classification of the Cercopithecoidea. *in* Napier,J.R. and Napier,P.H. (Eds.), 1970.

Tilson,R.L. 1976 Infant coloration and taxonomic affinity of the Mentawei Islands leaf monkey, *Presbytis potenziani*. *J. Mammal.* **57**: 766-769.

——— 1977 Social organisation of Simakobu monkeys (*Nasalis concolor*) in Siberut Island, Indonesia. *J. Mammal.* **58**: 202-212.

Tilson,R.L. and Tenaza,R.R. 1976 Monogamy and duetting in an Old World monkey. *Nature, Lond.* **263**: 320-321.

Trouessart,E.-L. 1879 *Catalogus mammalium tam viventium quam fossilium*. Berlin: R.Friedländer und Sohn **1**.

——— 1906 Description des mammifères nouveaux d'Afrique et de Madagascar. *Bull. Mus. Hist. nat. Paris***12**: 443-447.

——— 1911 On a new species of *Semnopithecus (Semnopithecus poliocephalus)* from Tonkin. *Ann. Mag. nat. Hist.* (8)**8**: 271-272, 1 pl.

Udvardy,M.D.F. 1975 A classification of the biogeographical provinces of the world. *I.U.C.N. occ. Pap.* (18).

Van Beneden,P.-J. 1838 Notice sur une nouvelle espèce de singe d'Afrique. *Bull. Acad. r. Belg.* **5**: 344-348, 1pl.

Van Peenen,P.F.D., Light,R.H. and Duncan,J.F. 1971 Observations on mammals of Mt. Sontra, South Vietnam. *Mammalia* **35**: 126-143.

Verheyen,W.N. 1959 Une nouvelle sous-espèce de *Colobus polykomos* (Zimmermann, 1780). *Revue Zool. Bot. afr.* **60**: 120-121.

——— 1962 Contribution à la craniologie comparée des primates. *Ann. Mus. r. Afr. cent.* (8vo., Sci. zool.)**105**: 1-255, 32 pls.

Vigors,N.A. and Horsfield,T. 1828 Observations on some of the Mammalia contained in the museum of the Zoological Society. *Zool. J. Lond.* **4**: 105-113.

Vogel,C. 1968 The phylogenetic evaluation of some characters, and some morphological trends in the evolution of the skull in catarrhine primates. *in* Chiarelli,B. (Ed.). *Taxonomy and phylogeny of Old World Primates with reference to the origin of man.* Torino:Rosenberg and Sellier.

——— 1971 Behavioral differences of *Presbytis entellus* in two different habitats. *Proc. 3rd int. Congr. Primat., Zurich* **3** (1970): 41-47.

Von Koenigswald,G.H.R. 1969 Miocene Cercopithecoidea and Oreopithecoidea from the Miocene of East Africa. *Fossil Vertbr. Afr.* **1**: 39-52, 1 pl.

Vrba,E.S. 1975 Some evidence of chronology and palaeoecology of Sterkfontein, Swartkrans and Kromdraai from the fossil Bovidae. *Nature, Lond.* **254**: 301-304.

Wagner,A. 1839 Fossile Ueberreste von einem Affenschädel und anderen Säugthieren aus Griechenland. *Gelehrte Anziegen Bayerisches Akad. Wiss.* **38**: 306-311.

——— 1840 Fossile Ueberreste von einem Affen und einigen andern Säugthieren aus Griechenland. *Abh. bayer Akad. Wiss.* **3**: 153-190.

——— 1857 Neue Beitrage zur Kenntnis der fossilen Säugthier-Ueberreste von Pikermi. *Abh. bayer Akad. Wiss.* **8**: 111-158, 7 pls.

Wagner,J.A. 1840 *Schreber's Die Säugthiere in Abbildungen nach der Natur, mit Beschreibungen.* Leipzig: T.D.Weigel **Supplementband 1**.

Walters,L.S. (unpubl.) Letters to P.H.Napier, 20 Nov 1978 and 3 Sep 1979.

——— (unpubl.) a Jersey Wildlife Preservation Trust Records.

Ward,F.Kingdon 1932 Explorations on the Burma-Tibet frontier. *Geogrl. J.* **80**: 465-483, 8 pls, 2 maps.

Washburn,S.L. 1942 Skeletal proportions of adult langurs and macaques. *Hum. Biol.* **14**: 444-472.

——— 1944 The genera of Malaysian langurs. *J. Mammal.* **25**: 289-294.

Watanabe,K. 1981 Variations in group composition and population density of the two sympatric Mentawaian leaf monkeys. *Primates* **22**: 145-160.

Waterhouse,G.R. 1838 On some new species of Mammalia from Fernando Po. *Proc. zool. Soc. Lond.* (1838); 57-61.

Weitzel,V. 1978 Dental morphology of two broadly sympatric subgenera of *Presbytis*. *Am. J. phys. Anthrop.* **48**: 447.

Wesmael,C. 1835 Description d'une nouvelle espèce de quadrumane. *Bull. Acad. r. Belg.* **2**: 236-238.

Wilson,C.C. and Wilson,W.L. 1977 Behavioral and morphological variation among primate populations in Sumatra. *Ybk. phys. Anthrop.* (1976)**20**: 207-233.

Wroughton,R.C. 1912 Bombay Natural History Society's Mammal Survey of India. Report on Collection No. 1 East Kandesh. *J. Bomb. nat. Hist. Soc.* **21**: 392-410.

——— 1912a Bombay Natural History Society's Mammal Survey of India. Report on Collection No. 2. Berars. *J. Bomb. nat. Hist. Soc.* **21**: 820-825.

——— 1912b Bombay Natural History Society's Mammal Survey of India. Report on Collection No. 4. Nimar. *J. Bomb. nat. Hist. Soc.* **21**: 844-851.

——— 1912c Bombay Natural History Society's Mammal Survey of India. Report on Collection No. 5. Dharwar. *J. Bomb. nat. Hist. Soc.* **21**: 1170-1195, 1 pl.

——— 1913 Bombay Natural History Society's Mammal Survey of India. Report on Collection No. 6. Kanara. *J. Bomb. nat. Hist. Soc.* **22**: 29-44.

——— 1914 Bombay Natural History Society's Mammal Survey of India, Burma and Ceylon. Report No. 15. Kumaon. *J. Bomb. nat. Hist. Soc.* **23**: 282-301.

——— 1915 Bombay Natural History Society's Mammal Survey of India, Burma and Ceylon. Report No. 16. Dry Zone, Central Burma and Mt. Popa. *J. Bomb. nat. Hist. Soc.* **23**: 460-480.

——— 1915a Bombay Natural History Society's Mammal Survey of India, Burma and Ceylon. Report no. 17. S. Tenasserim. *J. Bomb. nat. Hist. Soc.* **23**: 695-720.

——— 1915b Bombay Natural History Society's Mammal Survey of India, Burma and Ceylon. Report No. 18. Ceylon. *J. Bomb. nat. Hist. Soc.* **24**: 79-96.

——— 1915c Bombay Natural History Society's Mammal Survey of India, Burma and Ceylon. Report No. 19. Bengal, Bihar and Orissa. *J. Bomb. nat. Hist. Soc.* **24**: 97-110.

——— 1915d Scientific results from the mammal survey. No. XXI. J. A new monkey from the Chindwin River. *J. Bomb. nat. Hist. Soc.* **24**: 55-57.

——— 1916 Bombay Natural History Society's Mammal Survey of India, Burma and Ceylon. Report No. 20. Chindwin River. *J. Bomb. nat. Hist. Soc.* **24**: 291-309.

——— 1916a Bombay Natural History Society's Mammal Survey of India, Burma and Ceylon. Report No. 22. Koyna Valley. *J. Bomb. nat. Hist. Soc.* **24**: 311-316.

——— 1916b Bombay Natural History Society's Mammal Survey of India, Burma and Ceylon. Report No. 23. Sikkim and Bengal Terai. *J. Bomb. nat. Hist. Soc.* **24**: 468-493, 1 map.

——— 1916c Scientific results from the mammal survey. No. XIV. F. The langurs of Assam. *J. Bomb. nat. Hist. Soc.* **24**: 653-655.

——— 1916d Bombay Natural History Society's Mammal Survey of India, Burma and Ceylon. Report No. 25. Chin Hills. *J. Bomb. nat. Hist. Soc.* **24**: 758-773.

——— 1917 Scientific results from the mammal survey. No. XV. C. A new "leaf monkey" from the Shan States. *J. Bomb. nat. Hist. Soc.* **25**: 46-48.

——— 1918 Summary of the results from the Indian Mammal Survey of the Bombay Natural History Society. Part 1. *J. Bomb. nat. Hist. Soc.* **25**: 547-598.

——— 1921 Bombay Natural History Society's Mammal Survey of India, Burma and Ceylon. Report No. 34. Travancore. *J. Bomb. nat. Hist. Soc.* **27**: 549-553.

——— 1921a Bombay Natural History Society's Mammal Survey of India, Burma and Ceylon. Report No. 35. Prome. *J. Bomb. nat. Hist. Soc.* **27**: 553-554.

Wroughton,R.C. and Davidson,W.M. 1918 Bombay Natural History Society's Mammal Survey of India, Burma and Ceylon. Report No. 29. Pegu. *J. Bomb. nat. Hist. Soc.* **25**: 472-481.

Wroughton,R.C. and Ryley,K.V. 1913 Bombay Natural History Society's Mammal Survey of India. Report on Collection No. 7. Central Provinces. *J. Bomb. nat. Hist. Soc.* **22**: 45-58.

——— 1913a Bombay Natural History Society's Mammal Survey of India. Report on Collection No. 8. Vijayanagar, Bellary. *J. Bomb. nat. Hist. Soc.* **22**: 58-66.

Wurmb,F.B.van 1784 Der Kahau. *Magazin Neueste Phys. Naturgesch.* (2): 3-10.

Yalden,D.W., Largen,M.J. and Kock,D. 1977 Catalogue of the mammals of Ethiopia. 3. Primates. *Monitore zool. ital.* (N.S.)**9** (Suppl.): 1-52.

Zimmermann,E.A.W. 1780 *Geographische Geschichte des Menschen, und der vierfussigen Thiere.* Leipzig: Weygand **2**.

Zingeser,M.R. 1970 The morphological basis of the underbite trait in langurs (*P. melalophos and T. cristatus*) with an analysis of adaptive and evolutionary implications. *Am. J. phys. Anthrop.* **32**: 179-186.

Index

Page numbers in **bold** indicate a major reference in the text